Recent Developments in Atomic Force Microscopy and Raman Spectroscopy for Materials Characterization

Edited by Chandra Shakher Pathak and Samir Kumar

Published in London, United Kingdom

IntechOpen

Supporting open minds since 2005

Recent Developments in Atomic Force Microscopy and Raman Spectroscopy for Materials Characterization
http://dx.doi.org/10.5772/intechopen.94185
Edited by Chandra Shakher Pathak and Samir Kumar

Contributors
Yi Zhu, Leonid I. I. Goray, Marco Coisson, Paola Tiberto, Gabriele Barrera, Federica Celegato, Chandra Shakher Pathak, Luciano H. Honorato Chagas, Sandra Shirley Ximeno Chiaro, Alexandre Amaral Leitão, Renata Diniz, Nikiwe Mhlanga, Phumlani Tetyana, Sanele Nyembe, Lucky Sikhwivhilu, Mahfujur Rahaman, Dietrich R. T. Zahn, M.H. Wathsala N. Jinadasa, Maths Halstensen, Klaus-Joachim Jens, Amila C. Kahawalage, Nils-Olav Skeie, Shan Yang, Yahachi Saito, Koji Asaka, Lucia Kiyomi Noda, Karthikeyan Krishnamoorthy, Sang-Jae Kim, Dorota Puchowicz, Małgorzata Cieślak

First published in London, United Kingdom, 2022 by IntechOpen
IntechOpen is the global imprint of INTECHOPEN LIMITED, registered in England and Wales, registration number: 11086078, 5 Princes Gate Court, London, SW7 2QJ, United Kingdom
Printed in Croatia

British Library Cataloguing-in-Publication Data
A catalogue record for this book is available from the British Library

Additional hard and PDF copies can be obtained from orders@intechopen.com

Recent Developments in Atomic Force Microscopy and Raman Spectroscopy for Materials Characterization
Edited by Chandra Shakher Pathak and Samir Kumar
p. cm.
Print ISBN 978-1-83968-229-2
Online ISBN 978-1-83968-230-8
eBook (PDF) ISBN 978-1-83968-231-5

For EU product safety concerns:
IN TECH d.o.o., Prolaz Marije Krucifikse Kozulić 3, 51000 Rijeka, Croatia, info@intechopen.com or visit our website at intechopen.com.

Meet the editors

Dr. Chandra Shakher Pathak received his Ph.D. in Physics from the Indian Institute of Technology, Delhi, India. He obtained an MSc in Physics from Kumaun University Nainital, Uttarakhand, and an MTech (Gold medalist) in Advanced Materials Science and Technology from the National Institute of Technology, Durgapur, India. His research work focuses on nanomaterials, thin-film growth, characterization, and photovoltaics. He has worked on nanomaterials, organic-inorganic heterojunctions, Schottky diodes, 2D materials, and photovoltaics. Currently, Dr. Pathak is working on the fabrication and stability of perovskite solar cells.

Samir Kumar received his Ph.D. in Physics from the Indian Institute of Technology, Delhi, India, in 2017. After a three-and-a-half-year postdoctoral fellowship at Kyoto University, Japan, he is currently a research professor at the Nano Bio-Photonics Lab, Korea University. Dr. Kumar has been actively involved in research on novel sculptured thin films for plasmonics, surface-enhanced spectroscopy, photocatalysis, water-repellent surfaces, and biological/chemical sensing applications. Currently, his research interests include the design, fabrication, and characterization of nano- and microelectromechanical systems for sensor and actuator applications.

Contents

Preface

This book covers recent developments in atomic force microscopy (AFM) and Raman spectroscopy. AFM is a widely used technique in air, liquids, or vacuum to produce high-resolution topographic images of a surface with nanometer-scale resolution. AFM provides information on morphology and phase composition. Advances in AFM modes enable its application in various research areas such as photovoltaics, 2D materials, nanomaterials, and more. Raman spectroscopy is a widely used technique for the identification of organic and inorganic chemical substances. Over the last century, significant advances in lasers, spectrometers, detectors, and holographic optical components have made Raman spectroscopy a useful tool for various applications, including basic chemical and materials research, medical diagnostics, life sciences, and in situ process monitoring.

This book provides a comprehensive overview of advanced methods in AFM such as Kelvin probe force microscopy (KPFM), conducting atomic force microscopy (CAFM), magnetic force microscopy (MFM), AFM nanoindentation and its various applications, as well as advances in Raman spectroscopy, from advanced topics such as the deep learning approach to Raman spectroscopy to fundamental applications.

This book is divided into three sections. The first section presents an introduction to AFM and its various advanced characterization techniques, such as CAFM, KPFM, MFM, and other basic techniques and their applications.

The second section contains a detailed introduction to Raman spectroscopy and its various applications in different fields.

The third section discusses the application of Raman spectroscopy in low-dimensional nanostructures such as 2D materials and carbon nanotubes (CNTs).

We thank all the authors for their contributions and patience in editing the book. We also thank Author Service Manager Mia Vulovic and IntechOpen for allowing us to edit this fascinating book project.

Chandra Shakher Pathak
Ben-Gurion University of the Negev,
Israel

Samir Kumar
Korea University,
Sejong, Republic of Korea

Section 1

Atomic Force Microscopy and Its Advanced Applications

Chapter 1

Application of Atomic Force Microscopy in Organic and Perovskite Photovoltaics

Chandra Shakher Pathak

Abstract

Atomic force microscopy (AFM) has become widely used technique in air, liquids, or vacuum to generate high-resolution topographic images of a surface having a nanometer-scale resolution. AFM gives the information about the morphology, phase composition etc. Photovoltaic materials have been attracting intense interest due to their performance and the morphology as well as quality of these materials affects their performance. AFM is now a day widely used technique for morphology and other electronic properties measurements at nanoscale for photovoltaic materials to understand their relation with device performance. This chapter describe the brief introduction of Kelvin probe force microscopy (KPFM) and conducting atomic force microscopy (CAFM) and their application in electrical characterization at nanoscale of organic and perovskite photovoltaic materials.

Keywords: AFM, KPFM, CAFM, Photoconductive AFM, Nano scale, Electrical properties, Surface potential

1. Introduction

Scanning tunneling microscopy (STM) was first introduced in 1982 and it is based on tunneling of electron between the metallic tip and sample surface and it is limited to the study of conducting surface [1]. After four years in 1986, the solution of this restriction was provided by atomic force microscopy (AFM) and it is based on the detection of attractive or repulsive forces [2]. AFM was used to analyze the surface morphology and used to measure the force of interaction between the AFM tip and the sample. AFM consists of a sharp tip having nanometer dimension which is attached to a cantilever is used to scan the sample surface. A laser beam focused on the cantilever which detects the bending of cantilever. The reflection of the laser beam is focused on photodiode detector. Deflection of the cantilever is monitored during the scanning and converted into surface image. AFM is generally operated in contact mode and tapping mode. AFM is the widely used tool for the characterization of materials surface at the nanoscale. AFM was developed to a very versatile technique by combination with other measurements methods. This chapter will cover the application of electrical mode of AFM specifically Kelvin probe force microscopy (KPFM) and conducting atomic force microscopy (CAFM).

1.1 Kelvin probe force microscopy

Lord Kelvin proposed macroscopic Kelvin probe method in 1898 to determine the contact potential difference (V_{CPD}) between a metallic plate and sample [Kelvin L., Contact electricity of metals *Phil. Mag.* 1898, 46, 82–120]. Later in 1991, KPFM was first introduced by Nonnenmacher *et al.* [3]. The KPFM mode is basically two-scan process. In the first scan topography of the surface is acquired in tapping mode along a single line profile. Following this, the mechanical excitation of the cantilever is turned off and in the second scan this topography is retraced at a certain lift height (LH) above the sample surface, recording local variations in contact potential difference (CPD). During the second scan, the tip-sample distance is constant and it is equal to d_{AFM} + LH, where d_{AFM} represent the tip-sample distance during the topographic scan. When the AFM tip is brought close to the sample surface, the electrostatic force is generated and it is proportional to the difference between their Fermi levels. If the work functions of the sample and the tip are different, electrons flow from the lower work function to the higher work function material to align their Fermi level and the system reach in charge equilibrium condition. The vacuum levels of the tip and sample are not the same and the surfaces of the sample and the tip are charged and an apparent V_{CPD} will form. Due to the V_{CPD} an electrical force acts on the contact area and this force can be nullified by applying a DC voltage and it is equal to the work function difference between tip and sample. The work function of the sample can be calculated when the tip work function is known from scanning a reference sample.

1.2 Conductive atomic force microscopy

Conductive atomic force microscopy (CAFM) is usually used to analyze the local variations in current of the sample. The AFM controller is used for applying dc bias through the substrate during measurements. The CAFM tip is connected through a low noise current amplifier to the AFM controllers to generate the current image. We can collect the current–voltage (*I-V*) characteristics by CAFM tip which can be contacted randomly at various positions of sample surface and *I-V* characteristics are collected at each position.

KPFM generates 3D mapping of surface electric potential and measure the local work function while CAFM generates the current map. CAFM is one of the simplest ways of characterizing electrical properties at high resolution is by applying voltage between the sample and a CAFM probe. Photoconductive AFM; current mapping under illumination is useful for correlating high resolution current mapping with the photovoltaic device performance. Now a day these two modes of AFM in electrical mode are widely used in all area of research; nanoelectronics field, solar cell, 2D materials and semiconductor industries, biology etc. This chapter consists the collection of some published work as well as some new results.

2. Organic photovoltaics

Conducting polymers have been attracting attention after their discovery by Shirakawa, MacDiarmid, and Heeger in 1977, who were awarded the Nobel Prize in Chemistry in 2000 for the discovery of conductive polymers [4, 5]. They used organic polymer polyacetylenes, which is a conjugate polymer and insulator and increased the conductivity of polyacetylene films by several orders of magnitude by chemical doping [4]. In recent years, there has been lot of research activity in the field of polymer electronics and attracted a lot of attention because of its high

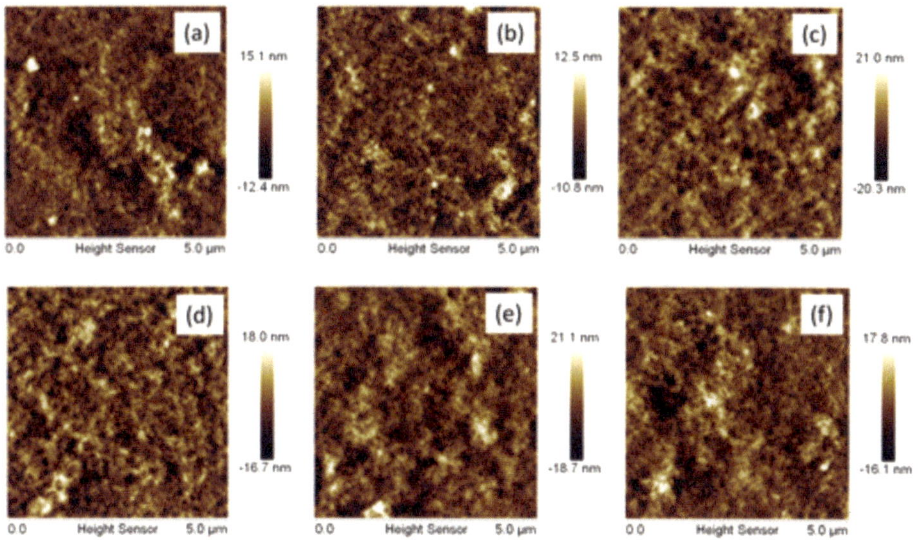

Figure 1.
AFM images of PEDOT:PSS films modified with (a) 0% NMP 1% MeOH, (b) 0.1% NMP 1% MeOH, (c) 0.3% NMP 1%MeOH, (d) 0.5% NMP 1% MeOH, (e) 0.8% NMP 1%MeOH, and (f) 0% MeOH 0.5% NMP.Reproduced with permission from [16].

flexibility, light weight and solution process ability [6]. Applications of conducting polymers include organic light emitting diodes, organic thin film transistors, organic solar cells, actuators and sensors etc. [7–13]. AFM is usually used to measure the roughness, morphology and phase analyses of conductive polymer [13]. KPFM is used to measure the work function [14–20] and CAFM [18, 21, 22] used to measure the current of the organic solar cell materials. These two methods gives valuable insights in the structure and working mechanism of organic photovoltaic devices.

Poly(3,4-ethylenedioxythiophene)-poly (styrenesulfonate) (PEDOT:PSS) is a well-known conducting polymer because of its high conductivity, excellent thermal stability, transparency, structural stability and processability [23–25]. PEDOT:PSS polymer is a promising candidate as a transparent electrode for optoelectronic devices. PEDOT is made from ethylenedioxythiophene (EDOT) monomers. PEDOT is insoluble in many common solvents, and it is unstable in its neutral state. To improve its processability, water-soluble polystyrene sulfonate (PSS), can be added and the addition of PSS causes it to become soluble. During the polymerization, PSS acts as charge balancing dopant to yield PEDOT:PSS.

Figure 1 shows the height images and **Figure 2** shows the surface potential images of PEDOT:PSS films with co-solvents (N-Methyl-2-pyrrolidone (NMP) and methanol (MeOH)). The average root-mean- square roughness (RMS) values were found in the range 3.2–5.5 nm. Pristine PEDOT:PSS film was quite smooth with a RMS roughness of 2.6 nm. RMS roughness values of PEDOT:PSS films increases after co-solvents addition. Variation in the RMS roughness after addition of co-solvents indicates the morphological change that arises from the conformation of the polymer chain [16]. The estimated average work function of co-solvents doped PEDOT:PSS thin films range from 4.63 to 4.82 eV as compared to 4.9 eV for the pristine PEDOT:PSS film [16]. Work function can be calculated with the following Equation [16, 27].

$$V_{CPD} = \frac{\phi_{tip} - \phi_{sample}}{-e} \tag{1}$$

where V_{CPD} is the contact potential different measured by KPFM in volts, ϕ_{tip} and ϕ_{sample} are the work functions of tip and sample in eV, respectively, while e is the electronic charge.

Figure 3 shows the current map of bare and NMP doped PEDOT:PSS films. Bare PEDOT:PSS has less current as compared to NMP doped PEDOT:PSS film as shown in **Figure 3** [28]. This shows the reduction of more PSS from the surface for NMP doped PEDOT:PSS film. Hosseini *et al.* [22] also reported higher current for ethylene glycol (EG) doped PEDOT:PSS film.

Figure 4 shows topography, surface potential images and corresponding line profile of bare and NMP doped PEDOT:PSS films. Bare PEDOT:PSS has work function of 4.90 eV and 4.77 eV for NMP doped PEDOT:PSS film [16, 28]. Line profile shows the homogenous distribution of surface potential. We also showed that the work function was reduced with the addition of dimethyl sulfoxide (DMSO) solvent in PEDOT:PSS [15].

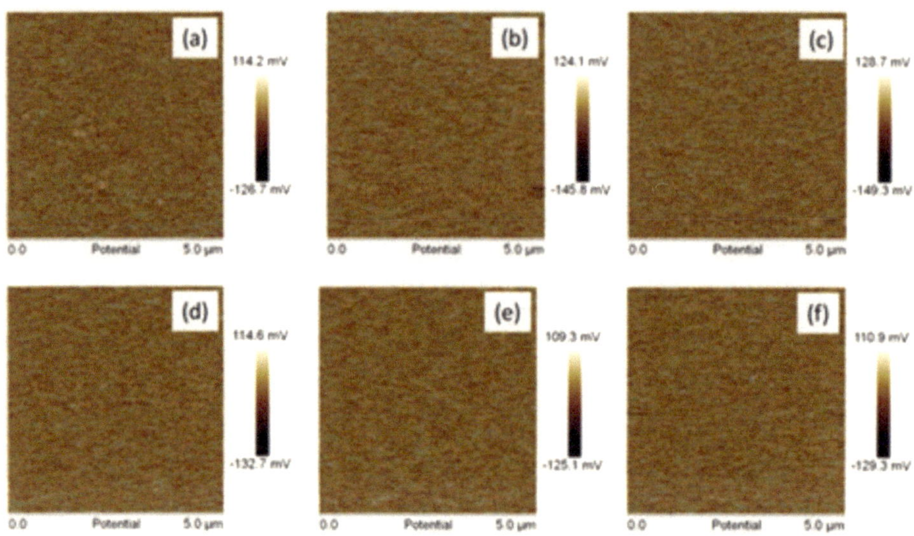

Figure 2.
Potential images of PEDOT:PSS films modified with (a) 0% NMP 1% MeOH, (b) 0.1% NMP 1% MeOH, (c) 0.3% NMP 1% MeOH, (d) 0.5% NMP 1% MeOH, (e) 0.8% NMP 1% MeOH, and (f) 0% MeOH 0.5% NMP. Reproduced with permission from [26].

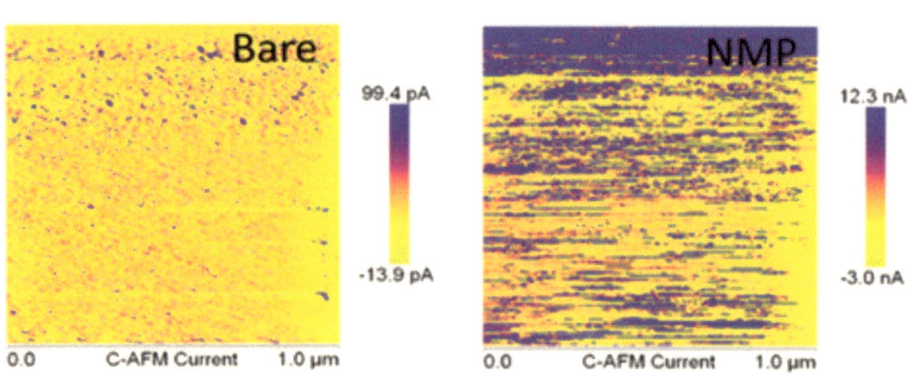

Figure 3.
CAFM images of bare and NMP doped PEDOT:PSS films. Reproduced with permission from [28].

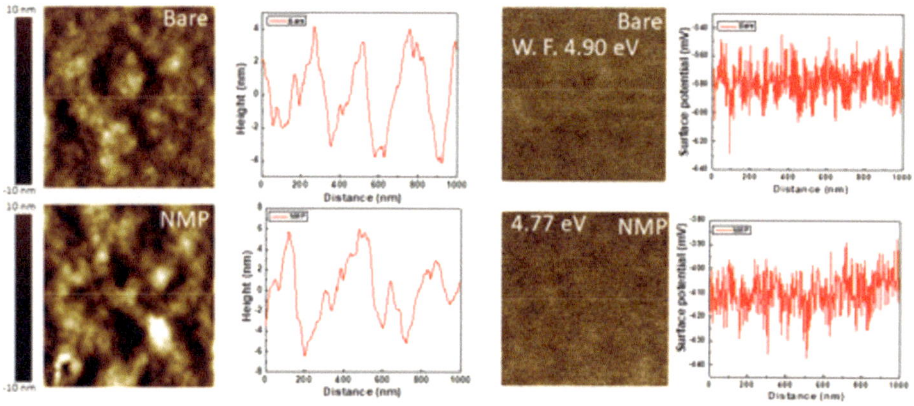

Figure 4.
Topography, CPD images and corresponding line profile of bare and NMP doped PEDOT:PSS films. Reproduced with permission from [28].

3. Perovskite photovoltaic

The last 10 years has seen a new photovoltaic (PV) technology being discovered and developed at a rate greater than any previous energy harvesting technology with research fuelled by the facile, low cost large area solution processing routes available for device fabrication. These devices, known as organic–inorganic metal halide perovskites have certified power conversion efficiencies (PCEs) > 25% [29], comparable values to the incumbent technologies but available at a fraction of the materials and processing costs. The perovskite thin films are typically polycrystalline ones, comprising microstructures such as grains and grain boundaries (GBs). Recently some of research groups have performed the microscopic investigation and suggested that the grain boundaries (GBs) in planar perovskite solar cells have beneficial [30, 31]. Hence engineering of the perovskite films and the microscopic investigation is essential for the further improvement in the properties of perovskite photovoltaics. CAFM and KPFM have been widely used to characterize the local properties of perovskite thin films at nanoscale to see the changes in electrical properties specially to improve the performance and stability of perovskite photovoltaics. CAFM [30–37] and KPFM [32, 38–40] have been used to characterize the local properties of perovskite thin films. Such studies suggested that GBs have beneficial effects due to efficient photogenerated charge carrier separation and collection at GBs [30, 31]. Downward- as well as upward- band bending at GBs were reported from KPFM measurements, depending on the GB composition [41]. Zhao *et al.* [32] measured the photocurrent of FTO/compactTiO_2/mesoporousTiO_2 + CH_3 NH_3PbI_3/$CH_3NH_3PbI_3$ film and found the photo current intensities were higher on the center of the grain than on the defect position, whereas the larger grain size leads to the higher photocurrent on the center. They also reported that the dark current intensities increased dramatically in the defect position, suggesting a high conductive character for defect position. Li *et al.* [31] fabricated $CH_3NH_3PbI_3$/mTiO2/c-TiO2/FTO typical device and performed c-AFM measurements under steady illumination of power 14 mW/cm^2 on $CH_3NH_3PbI_3$ surface from the top and showed that the photocurrent flows through the GBs are negligible at 0 V bias, while the major photocurrents form on the grains. However, with the low bias the photocurrents at the GBs become much higher than those of the grains. Xu *et al.* [35] also observed higher current near GBs for $CH_3NH_3PbI_3$- PCBM structure under high vacuum and dark conditions.

Kutes *et al.* [34] performed photoconductive AFM for methylammonium lead iodide ($CH_3NH_3PbI_3$)/c-TiO_2/FTO/glass structure and the schematic diagram of the photoconductive (CAFM under illumination) AFMs configuration shown in **Figure 5**. They observed higher current in grains than grain boundary (GB) as shown in **Figure 6**.

Yun *et al.* [30] also used CAFM and KPFM technique to investigate the role of GBs and the schematics shown in **Figure 7**. They observed the higher CPD at grain boundary with illumination and current near GBs for $CH_3NH_3PbI_3$/TiO_2/FTO/glass heterojunction structure than in the grains as shown in **Figures 8** and **9**.

They found that the higher current collection near GBs is consistent with KPFM results, which indicates that photogenerated carriers are more efficiently separated and transported along the GBs. Lower CPD at the GBs under the dark condition implies that downward band bending is present at GBs. KPFM is widely used for photovoltaics to analyses the work function of perovskite materials.

We prepared the MAPI films as described by Liu *et al.* [42] and they dipped PbI_2 film on CH_3NH_3I solution. Here we spin casted CH_3NH_3I in 1 ml isopropanol on mesoporous PbI_2 film and annealed the prepared films at 70° to 110°C for 10 minutes. **Figure 10** shows the height image and current–voltage characteristics of $CH_3NH_3PbI_3$ (MAPI) films on FTO annealed at 70°, 90° and 110° C for 10 min respectively. Which shows the higher grain size for 70° C annealed perovskite film (**Figure 10a**) and the current was in pA for all three samples.

After the temperature variation CAFM studies, We prepared mesoporous PbI_2 scaffolds MAPI films on FTO and annealed at 100°C were analyzed in details. CAFM and KPFM with and without illumination measurements were done and the schematics of these measurements are shown in **Figure 11** and results are shown in **Figures 12** and **13**. At dark and with the illumination of 532 nm laser having intensity of 0.6 W/cm^2 shows homogeneous higher current near GBs. KPFM also shows

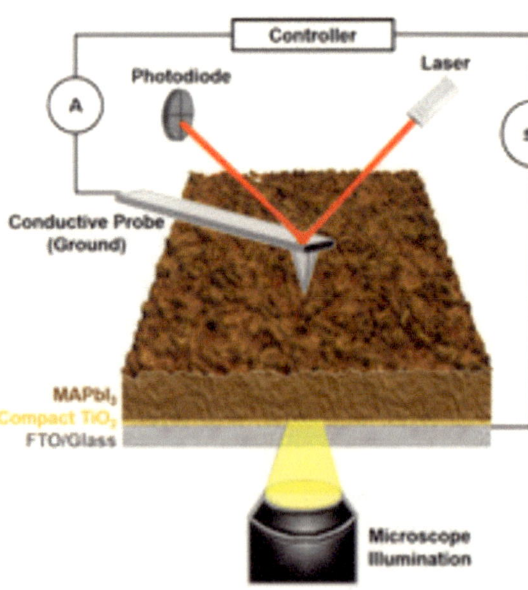

Figure 5.
Schematic diagram of the photoconductive AFMs configuration where the perovskite solar cell (PSC) is illuminated from below through a transparent-conducting cathode (FTO/glass) while measuring local current with a positionable conductive AFM probe anode from above. This diagram includes a 3D rendered, 3 μm × 3 μm, topographic AFM image of a methylammonium lead triiodide ($MAPbI_3$) thin film, along with a schematic cross section of the PSC containing a compact TiO_2 electron transport layer reproduced with permission from [34].

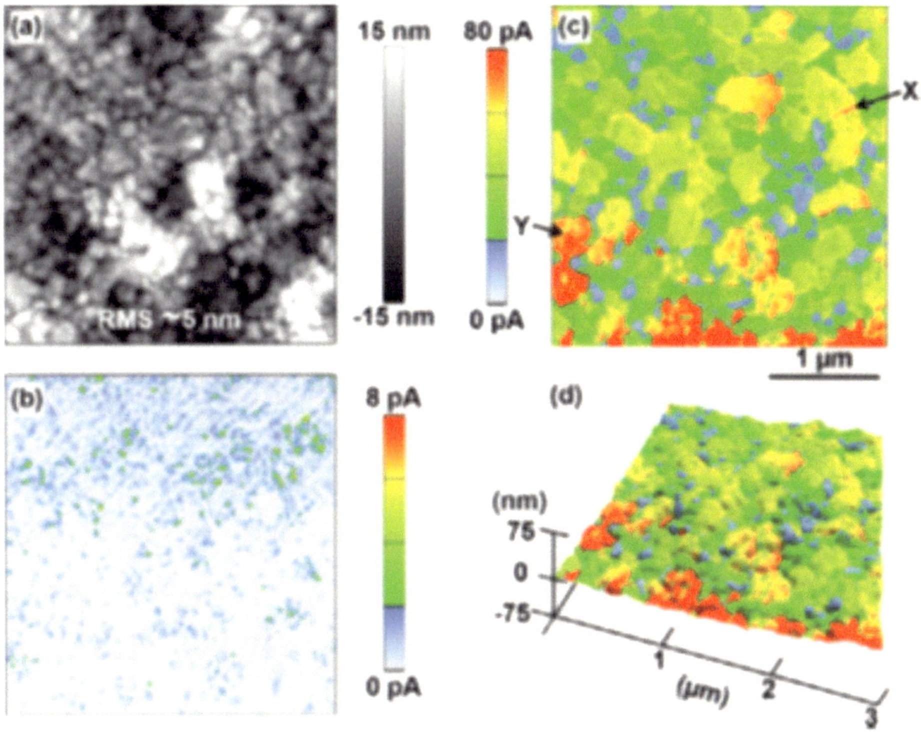

Figure 6.
Two-dimensional images of 3 μm × 3 μm region of a MAPbI3 thin film (same magnification): (a) topography, (b) dark Isc, and (c) Isc under 0.07 W.cm − 2 illumination. (d) Three-dimensional representation of the topography, overlaid by the illuminated Isc color contrast collected over the same area, revealing the microstructure-specific response. Same current scale for (c,d). Reproduced with permission from [34].

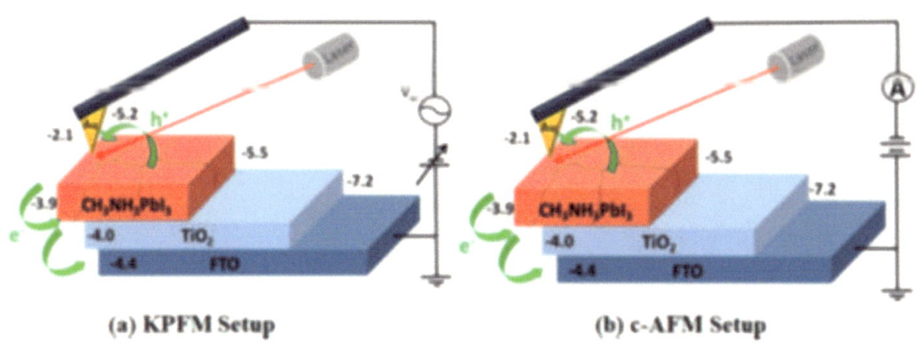

Figure 7.
Schematics of the (a) KPFM and (b) C-AFM set up. Relevant vacuum energy levels (in eV) for corresponding materials are indicated. Reproduced with permission from [30].

lower CPD value at GBs and mostly homogeneous mapping which are consistent with the results by Yun *et al.* [30]. Here we illuminated the sample from glass side not from the top (perovskite) side.

Figure 13 shows the topography and surface potential images and line profile of the perovskite films under illumination. The change in the CPD value reflects the change in the work function of the perovskite surface. Charge generation occurs significantly in GBs and the higher photocurrent near GBs suggests that GBs acts as channels for current flow than strong recombination centers. As we know that the higher number of defects are present at the grain boundaries which will increase

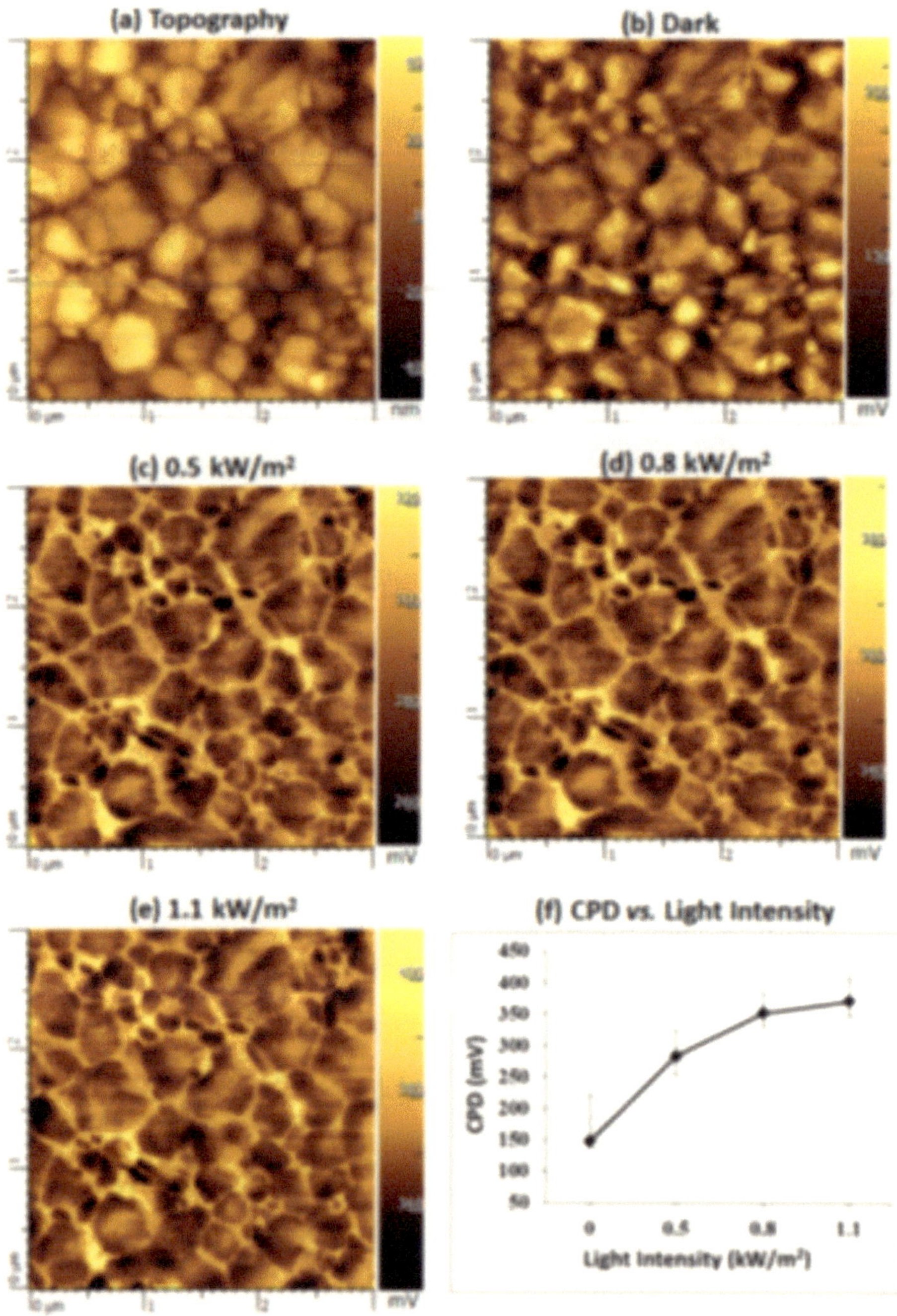

Figure 8.
KPFM measurements performed on a CH3NH3PbI3/TiO2/FTO/glass structure over an area of 3 μm². (a) Topography map and (b) CPD images taken in the dark. (c – e) CPD maps under various laser illumination intensities at a wavelength of 500 nm. (f) Intensity dependence of CPD of the sample at a wavelength of 500 nm as measured by KPFM. Reproduced with permission from [30].

the non-radiative recombination of electrons and holes. The charge accumulation or depletion between grains and GBs may cause band bending, which induces the charge carrier separation. Hence the investigation of electronic properties at grain boundaries is crucial. KPFM has been used to determine the surface potential at

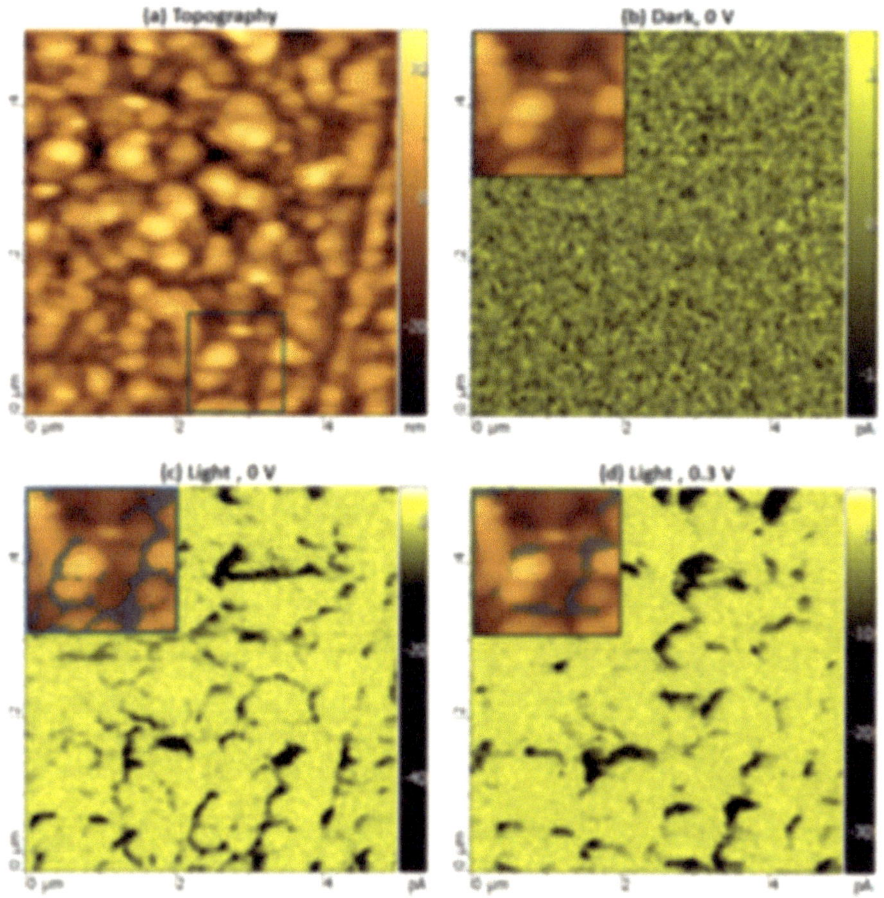

Figure 9.
CAFM measurements performed on a $CH_3NH_3PbI_3/TiO_2/FTO$/glass structure over an area of 5 μm^2. (a) Topographic image and (b) current image are taken in the dark at 0 V. current images (c) under illumination at 0 V and (d) under illumination at 0.3 V. insets in panels b – d are overlap of corresponding CAFM maps and a topography map of the region with a white outline in panel a. wavelength and intensity of the illumination were 500 nm and 1.1 kW/cm^2, . Reproduced with permission from [30].

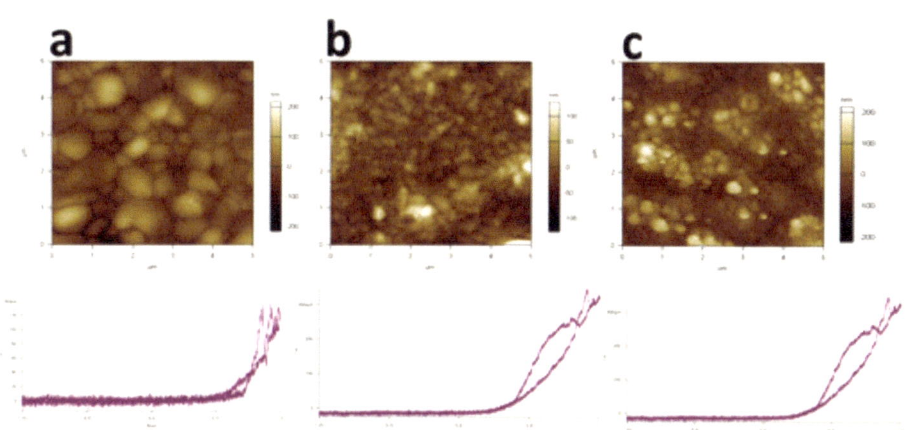

Figure 10.
Topography, corresponding current-bias characteristics of MAPI/film annealed at (a) 70°, (b) 90° and (c) 110° C.

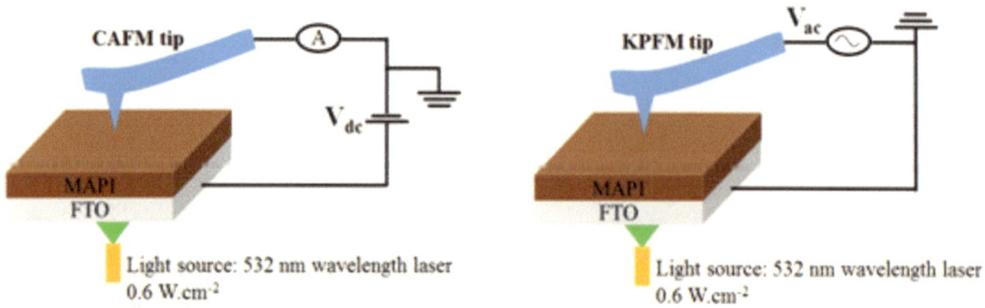

Figure 11.
Schematic representation of CAFM and KPFM measurements.

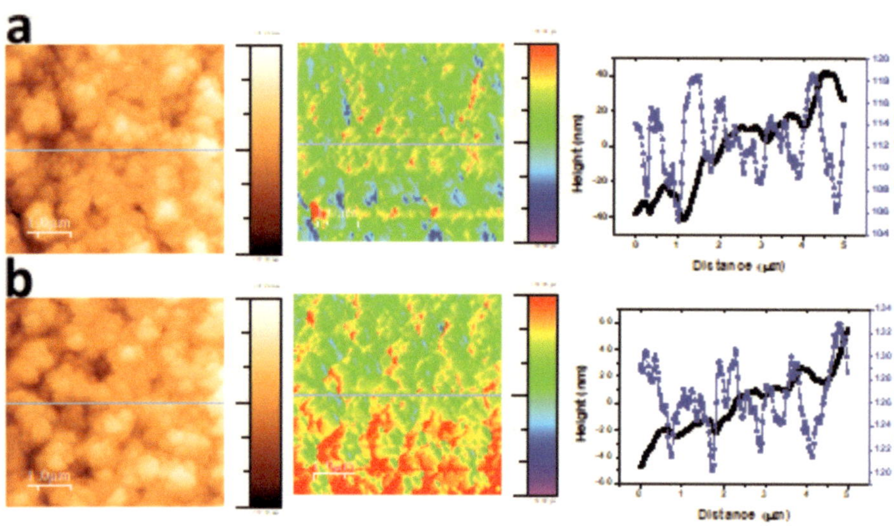

Figure 12.
Topography, current map and line profile of MAPI/FTO (a) dark and (b) under illumination of 532 nm laser having intensity of 0.6 W/cm^2.

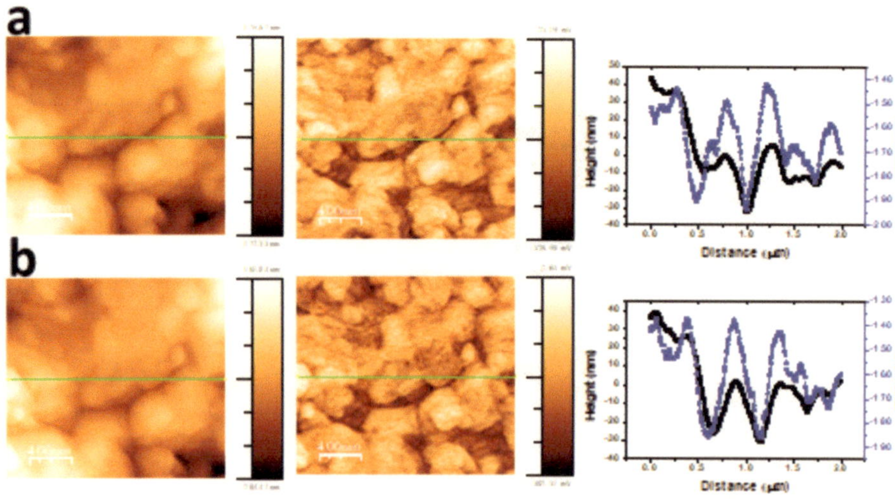

Figure 13.
Topography, CPD map and line profile of MAPI/FTO (a) dark and (b) under illumination of 532 nm laser having intensity of 0.6 W/cm^2.

grains and GBs. From the topography and CPD map we can see the individual grains and the CPD is higher in grains compared to GBs. The line profile plots are useful in quantitative analysis of CPD variations across topographical feature in perovskite films. CPD is higher at higher region (grains) and lower in lower regions (GBs) and it might be due to the presence of built in potential around the GBs.

4. Conclusion

AFM provides lots of required and interesting results with advanced modes and widely accepted technique to characterize all type of materials. CAFM and KPFM provides information about grains and GBs of photovoltaic materials which help us to understand the current transport and band bending to improve the performance and life time of photovoltaic materials. Combination of such characterization at nanoscale with macroscopic analysis can link the photovoltaic materials properties and optimization of device performance.

Acknowledgements

The author thanks to Prof. Iris Visoly-Fisher, Ben- Gurion University of the Negev, Israel for guidance and support.

Author details

Chandra Shakher Pathak[1,2]

1 Ben-Gurion National Solar Energy Center, Department of Solar Energy and Environmental Physics, Jacob Blaustein Institutes for Desert Research, Ben-Gurion University of the Negev, Israel

2 Department of Chemical Engineering, Ben-Gurion University of the Negev, Beer Sheva, Israel

*Address all correspondence to: cspathak12@gmail.com

References

[1] Binning G, Rohrer H, Gerber C, Weibel E. Surface Studies by Scanning Tunneling Microscopy. Phys Rev Lett 1982;49:57-61.

[2] G. Binning, C.F. Quate CG. Atomic Force Microscope. Phys Rev Lett 1986;56:930-3. https://doi.org/10.1201/9781420075250.

[3] Nonnenmacher M, O'Boyle MP, Wickramasinghe HK. Kelvin probe force microscopy. Appl Phys Lett 1991;58:2921-3. https://doi.org/10.1063/1.105227.

[4] Shirakawa H, Macdiarmid AG. Electrical conductivity in doped polyacetylene. Phys Rev Lett 1977;39:1098-101.

[5] Shirakawa H, Louis EJ, MacDiarmid AG, Chiang CK, Heeger AJ. Synthesis of electrically conducting organic polymers: Halogen derivatives of polyacetylene, (CH)x. J Chem Soc Chem Commun 1977:578-80. https://doi.org/10.1039/C39770000578.

[6] Groenendaal L, Jonas F, Freitag D, Pielartzik H, Reynolds JR. Poly(3,4-ethylenedioxythiophene) and Its Derivatives: Past, Present, and Future. Adv Mater 2000;12:481-94. https://doi.org/10.1002/(SICI)1521-4095(200004)12:7<481::AID-ADMA481>3.0.CO;2-C.

[7] Xia Y, Zhang H, Ouyang J, Yang R, Yang W, Huang F, et al. Highly conductive PEDOT:PSS films prepared through a treatment with zwitterions and their application in polymer photovoltaic cells. J Mater Chem 2010;20:9740-7. https://doi.org/10.1039/c0jm01593h.

[8] Pathak CS, Singh JP, Singh R. Effect of dimethyl sulfoxide on the electrical properties of PEDOT:PSS/n-Si heterojunction diodes. Curr Appl Phys 2015;15:528-34. https://doi.org/10.1016/j.cap.2015.01.020.

[9] Palumbiny CM, Heller C, Schaffer CJ, Körstgens V, Santoro G, Roth S V., et al. Molecular Reorientation and Structural Changes in Cosolvent-Treated Highly Conductive PEDOT:PSS Electrodes for Flexible Indium Tin Oxide-Free Organic Electronics. J Phys Chem C 2014;118:13598-606. https://doi.org/10.1021/jp501540y.

[10] Xia Y, Ouyang J. PEDOT:PSS films with significantly enhanced conductivities induced by preferential solvation with cosolvents and their application in polymer photovoltaic cells. J Mater Chem 2011;21:4927. https://doi.org/10.1039/c0jm04177g.

[11] Levermore PA, Jin R, Wang X, Chen L, Bradley DDC, de Mello JC, et al. High efficiency organic light-emitting diodes with PEDOT-based conducting polymer anodes. J Mater Chem 2008;18:4414-20. https://doi.org/10.1039/b805994b.

[12] Sun K, Zhang S, Li P, Xia Y, Zhang X, Du D, et al. Review on application of PEDOTs and PEDOT:PSS in energy conversion and storage devices. J Mater Sci Mater Electron 2015;26:4438-62. https://doi.org/10.1007/s10854-015-2895-5.

[13] Fan X, Nie W, Tsai H, Wang N, Huang H, Cheng Y, et al. PEDOT:PSS for Flexible and Stretchable Electronics: Modifications, Strategies, and Applications. Adv Sci 2019;6:1900813. https://doi.org/10.1002/advs.201900813.

[14] Pathak CS, Kapoor R, Singh JP, Singh R. Investigation of the effect of organic solvents on the electrical characteristics of PEDOT:PSS/p-Si heterojunction diodes. Thin Solid Films 2017;622:115-21. https://doi.org/10.1016/j.tsf.2016.12.030.

[15] Pathak CS, Singh JP, Singh R. Modification of electrical properties of PEDOT:PSS/ p -Si heterojunction diodes by doping with dimethyl sulfoxide. Chem Phys Lett 2016;652:162-6. https://doi.org/10.1016/j.cplett.2016.04.029.

[16] Pathak CS, Singh JP, Singh R. Optimizing the electrical properties of PEDOT:PSS films by co-solvents and their application in polymer photovoltaic cells. Appl Phys Lett 2017;111:102107-5. https://doi.org/10.1063/1.4994317.

[17] Nardes AM, Kemerink M, de Kok MM, Vinken E, Maturova K, Janssen RAJ. Conductivity, work function, and environmental stability of PEDOT:PSS thin films treated with sorbitol. Org Electron 2008;9:727-34. https://doi.org/10.1016/j.orgel.2008.05.006.

[18] Berger R, Domanski AL, Weber SAL. file:///J:/AFM/PEDOTPSS for Flexible and Stretchable Electronics Modifications, Strategies, and Applications.pdf. Eur Polym J 2013;49:1907-15. https://doi.org/10.1016/j.eurpolymj.2013.03.005.

[19] Pathak CS, Singh JP, Singh R. Preparation of novel graphene-PEDOT : PSS nanocomposite films and fabrication of heterojunction diodes with n -Si. Chem Phys Lett 2018;694:75-81. https://doi.org/10.1016/j.cplett.2018.01.049.

[20] Yang S, Yang C, Zhang X, Zheng Z, Bi S, Zhang Y, et al. A conjugated microporous polymer film fabricated by: In situ electro-chemical deposition as a hole transporting layer in organic photovoltaics. J Mater Chem C 2018;6:9044-8. https://doi.org/10.1039/c8tc02933d.

[21] Kalachyova Y, Guselnikova O, Postnikov P, Fitl P, Lapcak L, Svorcik V, et al. Reversible switching of PEDOT:PSS conductivity in the dielectric-conductive range through the redistribution of light-governing polymers. RSC Adv 2018;8:11198-206. https://doi.org/10.1039/c7ra12624g.

[22] Hosseini E, Ozhukil Kollath V, Karan K. The key mechanism of conductivity in PEDOT:PSS thin films exposed by anomalous conduction behaviour upon solvent-doping and sulfuric acid post-treatment. J Mater Chem C 2020;8:3982-90. https://doi.org/10.1039/c9tc06311k.

[23] Kim JY, Jung JH, Lee DE, Joo J. Enhancement of electrical conductivity of poly(3,4-ethylenedioxythiophene)/poly(4-styrenesulfonate) by a change of solvents. Synth Met 2002;126:311-6. https://doi.org/10.1016/S0379-6779(01)00576-8.

[24] Dimitriev OP, Grinko DA, Noskov Y V., Ogurtsov NA, Pud AA. PEDOT:PSS films-Effect of organic solvent additives and annealing on the film conductivity. Synth Met 2009;159:2237-9. https://doi.org/10.1016/j.synthmet.2009.08.022.

[25] Pathak CS, Singh JP, Singh R. Effect of dimethyl sulfoxide on the electrical properties of PEDOT:PSS/n-Si heterojunction diodes. Curr Appl Phys 2015;15:528-34. https://doi.org/10.1016/j.cap.2015.01.020.

[26] Pathak CS. Investigation of PEDOT:PSS/Si heterojunctions and graphene/Si schottky diodes for potential application in photovoltaics. Published by Indian Institute of Technology Delhi, 2018.

[27] Melitz W, Shen J, Kummel AC, Lee S. Kelvin probe force microscopy and its application. Surf Sci Rep 2011;66:1-27. https://doi.org/10.1016/j.surfrep.2010.10.001.

[28] Pathak CS, Singh JP, Singh R. Organic-inorganic heterojunctions for optoelectronic applications. Published by The Institution of Engineering and

Technology, 2020. https://doi.org/10.1049/pbcs049e.

[29] best-research-cell-efficiencies.20200925.pdf n.d.

[30] Yun JS, Ho-baillie A, Huang S, Woo SH, Heo Y, Seidel J, et al. Benefit of Grain Boundaries in Organic – Inorganic Halide Planar Perovskite Solar Cells. J Phys Chem Lett 2015;6:875-80. https://doi.org/10.1021/acs.jpclett.5b00182.

[31] Li JJ, Ma JY, Ge QQ, Hu JS, Wang D, Wan LJ. Microscopic Investigation of Grain Boundaries in Organolead Halide Perovskite Solar Cells. ACS Appl Mater Interfaces 2015;7:28518-23. https://doi.org/10.1021/acsami.5b09801.

[32] Zhao Z, Chen X, Wu H, Wu X, Cao G. Probing the Photovoltage and Photocurrent in Perovskite Solar Cells with Nanoscale Resolution. Adv Funct Mater 2016;26:3048-58. https://doi.org/10.1002/adfm.201504451.

[33] Leblebici SY, Leppert L, Li Y, Reyes-Lillo SE, Wickenburg S, Wong E, et al. Facet-dependent photovoltaic efficiency variations in single grains of hybrid halide perovskite. Nat Energy 2016;1:1-7. https://doi.org/10.1038/nenergy.2016.93.

[34] Kutes Y, Zhou Y, Bosse JL, Steffes J, Padture NP, Huey BD. Mapping the Photoresponse of CH3NH3PbI3Hybrid Perovskite Thin Films at the Nanoscale. Nano Lett 2016;16:3434-41. https://doi.org/10.1021/acs.nanolett.5b04157.

[35] Xu J, Buin A, Ip AH, Li W, Voznyy O, Comin R, et al. Perovskite-fullerene hybrid materials suppress hysteresis in planar diodes. Nat Commun 2015;6:1-8. https://doi.org/10.1038/ncomms8081.

[36] Gomez A, Sanchez S, Campoy-quiles M, Abate A. Topological distribution of reversible and non-reversible degradation in perovskite solar cells. Nano Energy 2018;45:94-100. https://doi.org/10.1016/j.nanoen.2017.12.040.

[37] Yang J, Liu X, Zhang Y, Zheng X, He X, Wang H, et al. Comprehensive understanding of heat-induced degradation of triple-cation mixed halide perovskite for a robust solar cell. Nano Energy 2018;54:218-26. https://doi.org/10.1016/j.nanoen.2018.10.011.

[38] Formamidinium- P, Lee DS, Yun JS, Kim J, Mahboubi A, Chen S, et al. Passivation of Grain Boundaries by Phenethylammonium in Formamidinium-Methylammonium Lead Halide Perovskite Solar Cells. ACS Appl Mater Interfaces 2018;3:647-54. https://doi.org/10.1021/acsenergylett.8b00121.

[39] Faraji N, Qin C, Matsushima T, Adachi C, Seidel J. Grain Boundary Engineering of Halide Perovskite CH3NH3PbI3 Solar Cells with Photochemically Active Additives. J Phys Chem C 2018;122:4817-21. https://doi.org/10.1021/acs.jpcc.8b00804.

[40] Yang C, Du P, Dai Z, Li H, Yang X, Chen Q. Effects of Illumination Direction on the Surface Potential of CH 3 NH 3 PbI 3 Perovskite Films Probed by Kelvin Probe Force Microscopy. ACS Appl Mater Interfaces 2019;11:14044-50. https://doi.org/10.1021/acsami.8b21774.

[41] Chen Q, Zhou H, Song T, Luo S, Hong Z, Duan H, et al. Controllable Self-Induced Passivation of Hybrid Lead Iodide Perovskites toward High Performance Solar Cells. Nano Lett 2014;14:4158-63. https://doi.org/10.1021/nl501838y.

[42] Liu T, Hu Q, Wu J, Chen K, Zhao L, Liu F, et al. Mesoporous PbI2 Scaffold for High-Performance Planar Heterojunction Perovskite Solar Cells. Adv Energy Mater 2016;6:1-7. https://doi.org/10.1002/aenm.201501890.

Chapter 2

Nanomaterials Characterisation through Magnetic Field Dependent AFM

Marco Coïsson, Gabriele Barrera, Federica Celegato and Paola Tiberto

Abstract

Atomic force microscopy is a versatile technique allowing to exploit many different physical effects for measuring a number of materials properties. The magnetic properties of surfaces and thin films are traditionally accessed through magnetic force microscopy, which produces magnetic field gradient maps generated by the magnetisation distribution at the surface of the sample. However, more advanced techniques can be derived from this fundamental setup, allowing for a richer characterisation of magnetic samples. In this chapter, we will describe how to extend a magnetic force microscope to allow magnetic field-dependent characterisations. Magnetisation reversal processes, as well as full hysteresis loops, can be investigated with such a technique, with field resolution adequate for identifying significant features such as domains reversal, nucleation or annihilation of domains, and other irreversible mechanisms. The same principle can also be exploited for the measurement of magnetostriction on thin films, and can be taken as guideline for other advanced applications of atomic force microscopy.

Keywords: MFM, magnetostriction, hysteresis loops, thin films

1. Introduction

Magnetic force microscopy [1] is a technique developed more than thirty years ago that is derived from atomic force microscopy [2]. It exploits an AFM tip that is coated with a magnetic material, and therefore acquires a magnetic moment. The tip becomes sensitive to the magnetic force gradient acting between the surface of a magnetic sample and the tip itself [3, 4], and allows the microscope to build local magnetisation maps at scales well below the micrometer [5].

In the course of the years, MFM has proven to be an extremely powerful technique to study a variety of magnetic systems, including patterned media [6, 7], cellular automata [8], individual magnetic nanostructures [9], or phase separations in magnetic materials at the nanoscale [10]. While its most straightforward application is the imaging of magnetic domains configurations at the sample magnetic remanence, several attempts have been made to adapt MFM to the study of the magnetisation reversal processes induced by an applied magnetic field [11–15]. These approaches opened new ways of exploiting the powerful MFM technique to the study of time-evolving magnetic domains configurations, whose changes are triggered by variations in an applied magnetic field, letting MFM characterisations expand toward the typical domain of magneto-optic microscopy, although with a much higher space resolution [16] but also with much slower time scales.

In this chapter, we will summarise a magnetic field-dependent MFM approach that is based on the single-point technique [5]. We will show how this advanced AFM application can be exploited to investigate magnetic field-induced magnetisation reversal processes at the nanoscale on different sets of magnetic materials, and we will show that the basic principles of this technique are sufficiently general to allow many different experiment designs for the study e.g. of hysteresis loops, local magnetic anisotropies, magnetostriction.

2. Magnetic field-dependent MFM

Magnetic force microscopy is an application derived from atomic force microscopy, that belongs to the family of the so-called "2nd pass" techniques. Their name derives from the fact that the microscope scans each line two times (usually back and forth each time): in the first scan, or pass 1, the AFM operates in intermittent contact mode and acquires the morphology of the sample along the scanned profile; in the second scan, or pass 2, the AFM operates in "lift mode", i.e. the cantilever is kept in oscillation but in non-contact mode, by lifting the scan height at a certain constant level above the sample surface (usually a few tens of nanometers). In this way, the short-range interactions typical of the Lennard-Jones-type potential become negligible, and long-range interactions such as the elctrostatic or magnetostatic ones can be investigated, provided that the microscope tip is coated by a suitable material. In the case of the MFM technique, the tip must be coated with a magnetic material, whose magnetisation must also lie along a known (and constant) direction (usually along the tip axis, perpendicular to the sample plane). When interacting with the sample surface, in pass 2, the MFM tip senses the second derivative of the z (vertical) component of the magnetic fields fringing from the sample surface, with respect to the z direction (i.e. $\partial^2 H_z/\partial z^2$), and as a result the cantilever oscillation amplitude, phase and frequency are affected. Typically, either the cantilever oscillation frequency or its phase are taken as signals for representing the magnetostatic interaction with the sample. In the following, all MFM images that will be shown will use the phase signal.

As this interaction happens through $\partial^2 H_z/\partial z^2$, it is necessary that fringing fields with a H_z component emerge from the surface of the sample in proximity of the tip position. This is typically possible in a few different cases, such as:

- a domain wall (either Bloch or Néel) is located underneath the tip, with the rest of the magnetisation lying in the sample plane;
- the sample is characterised by a perpendicular anisotropy;
- the sample is characterised by an in-plane magnetic anisotropy, but it is patterned in such a way that the magnetisation vector is not parallel to some of the patterns sides.

In general, therefore, a MFM detects a magnetic domains configuration when surface magnetic charges or poles appear under the tip, either because the magnetisation is tilted off the sample plane, or because geometrical constraints force the magnetisation to develop free poles along some directions (e.g. obtained by lithography or patterning).

Normally, such a MFM is limited to the investigation of the magnetic domains configuration of samples at their magnetic remanence. Whatever their remanent state, this must have been obtained by submitting the sample to a magnetic field history by exploiting some other technique, and cannot be modified during the MFM investigations.

These constraints can be at least partly lifted if the microscope is equipped with a means of generating a magnetic field (i.e. an electromagnet or a coil), either in the sample plane, or perpendicular to it. The microscope should be adequately designed for this application, as it must not contain magnetic elements in the head and in the scanner, which could be damaged by the application of an external magnetic field, and which would probably induce unacceptable drifts in the operation of the microscope or artefacts in the acquisition of the images. With the possibility to apply a magnetic field during image acquisitions, the MFM becomes a tool for characterising samples not only at their magnetic remanence, but also at specific field values which are of interest to the experimenter.

An example of such an application is given in **Figure 1**, where square dots of $Fe_{50}Pd_{50}$ with a lateral size of 800 nm and a thickness of 35 nm have been

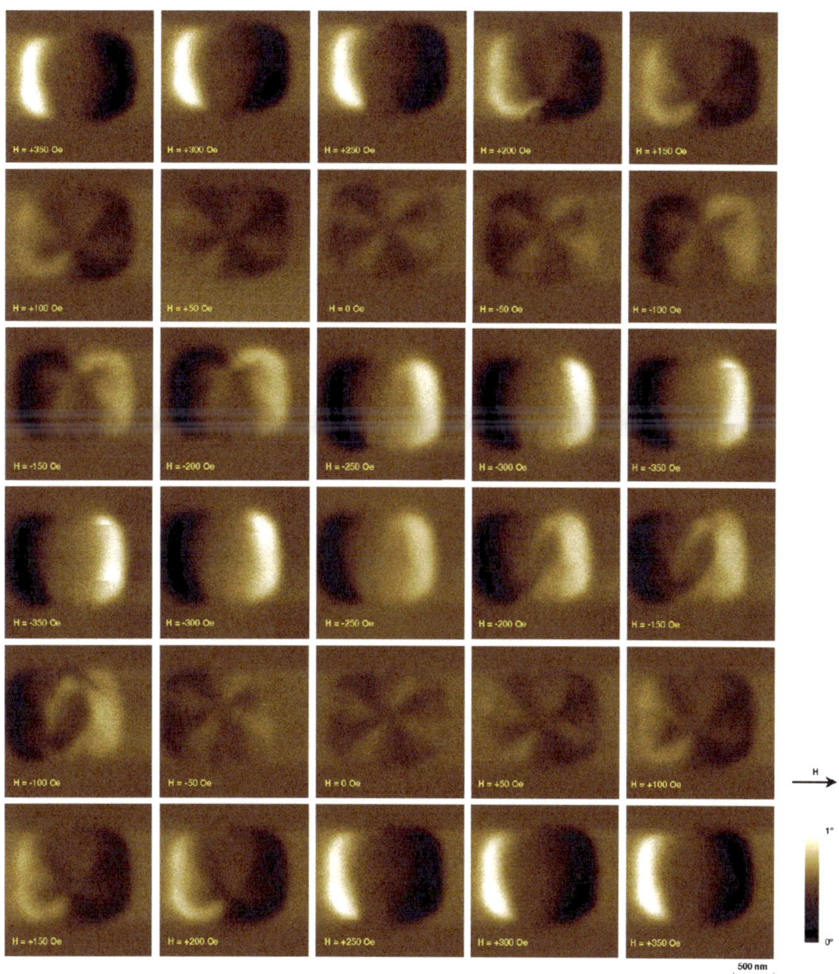

Figure 1.
MFM images (pass 2, phase channel) as a function of the applied magnetic field of $Fe_{50}Pd_{50}$ squares (lateral size 800 nm, thickness 35 nm). The images follow a whole hysteresis loop within the limits of ± 350 Oe, at steps of 50 Oe. The magnetic field is applied along the horizontal direction. The tip used was a Bruker MESP-HR.

investigated as a function of the applied magnetic field. The field has been cycled from +350 Oe to -350 Oe and back, at steps of 50, to complete a hysteresis loop. The MFM images of **Figure 1** show how the domain configuration of one of such squares evolves with the field, from a saturated state (bright and dark contrast at opposite sides of the square), to the nucleation of a vortex, the motion of its centre along the direction perpendicular to the applied field, its annihilation, and finally to the opposite saturated state (reversed bright and dark contrast), and then back, with the occasional development of a more complex domain structure (see the images at -200, -150 and -100 Oe when going from negative to positive saturation).

While powerful, this approach is limited to only a few images at different applied fields, because of time constraints or magnetic tip wear-out.

3. A "dynamic scanning mode" MFM

The magnetic field-dependent MFM technique discussed in Section 2 presents the undoubted advantage that the magnetic domain configuration of the studied sample can be obtained with as much detail as desired for any applied magnetic field value. In this way, even complex magnetic structures can be studied along their evolution with the magnetic field, in a way similar to what magneto-optics systems offer, but with a much improved space resolution, typical of AFM-based techniques. However, this advantage holds for a big part only in principle. In fact, it is difficult if not impossible to acquire the large number of MFM images required to study magnetisation processes with high magnetic field resolution. A first constraint comes from the time required to acquire each image, which is often of several minutes or even a few tens of minutes, that makes the capture of hundreds or even thousands of images impractical. A second constraint is due to the fact that even a well-tuned AFM in intermittent-contact mode (as is normally operated an AFM when performing magnetic characterisations) slightly wears out the tip in the course of time. Normally, only a few tens of images can be obtained with a single tip, even by optimally operating the microscope, before artefacts start to appear and the tip performance degrades (including the magnetic performance, due to the progressive damage or detachment of the magnetic coating). While replacing the tip is technically feasible, scanning exactly the same field of view after the tip replacement may not be easy for certain samples, and the obtained image quality would probably depend on the tip, as its shape, sharpness, and magnetic coating characteristics fluctuate among specimens even of the same batch. Therefore, while in principle a thorough field-dependent investigation of a magnetic sample is feasible with a conventional MFM, it is often impractical or even impossible due to the constraints discussed above.

To overcome these limitations, a different approach has been exploited [5, 17, 18], consisting in condensing the information on the magnetisation processes as a function of a sufficiently resolved magnetic field in just a single image. The difference between a MFM operating in "normal scanning mode" (a conventionally operating MFM) and one operating in "dynamic scanning mode" is schematically represented in **Figure 2**.

A MFM in "normal scanning mode" works by scanning a portion of the sample by moving the tip relative to the sample along a direction conventionally called x (the fast scan axis), that reproduces a scan line. Then, the scan line is moved by a certain step along the direction conventionally called y (the slow scan axis), so that at the end of a frame the whole field of view (the magenta area in **Figure 2**) is covered. Each scanned line (there can be several hundreds or thousands, in **Figure 2** they are 400) explores a different profile of the sample, and all scan lines together

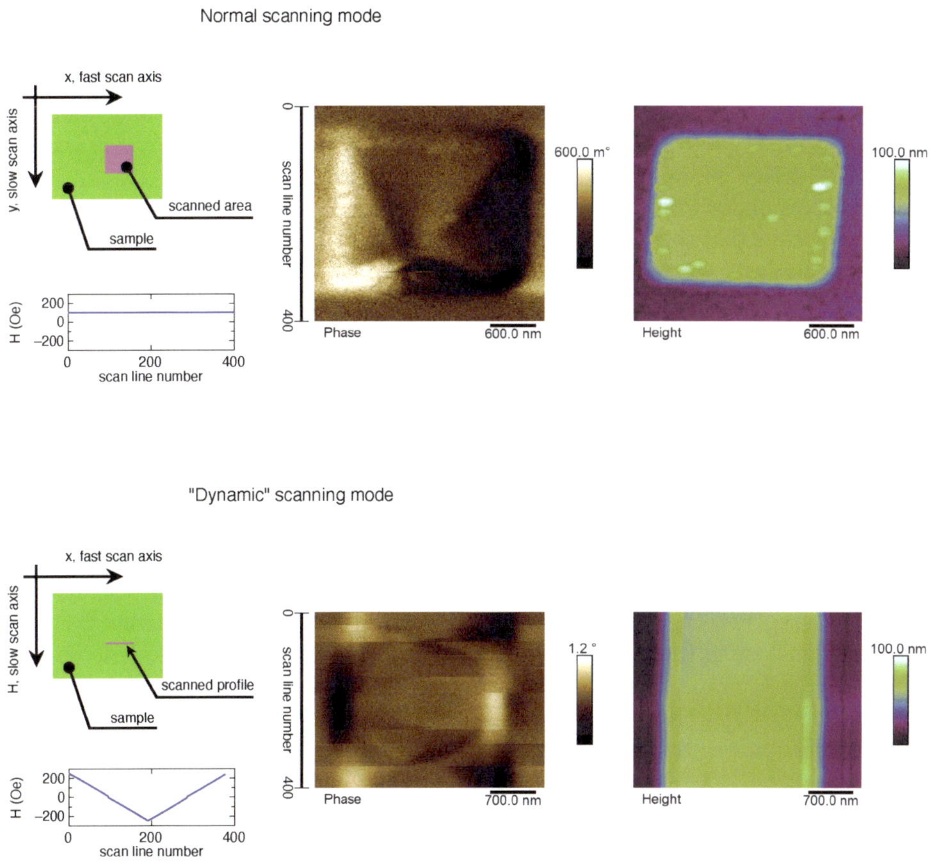

Figure 2.
Operation principle of an MFM in "normal" (upper panel) and "dynamic" (lower panel) scanning modes. Green rectangle: sample. Magenta square or line: area or line scanned by the MFM. The x position of the scan moves along the horizontal (fast) scan axis. In normal scanning mode: the y position of the scan moves along the vertical (slow) scan axis, while the applied magnetic field H is kept constant across each scanned line. In dynamic scanning mode: the y position of the scan does not move (slow scan axis disabled), whereas the applied magnetic field H changes at each scanned line. The right panels report typical MFM (phase channel, pass 2) and AFM (height channel, pass 1) images acquired in the two scanning modes for a $Fe_{50}Pd_{50}$ square dot (side 2 μm, thickness 35 nm) for a constant $H = 100$ Oe (normal mode) and for H cycling between ±250 Oe (dynamic mode). The tip was a Bruker MESP-HR.

make an image representing a certain surface area of the sample: for example, in the case of **Figure 2** in "normal scanning mode", a full AFM image of the investigated dot is obtained. In this operation mode, each scan line is acquired at the same applied magnetic field, e.g. 100 Oe in **Figure 2**. This is the way all MFM images of **Figure 1** have been captured.

A MFM in "dynamic scanning mode" works by disabling the slow scan axis. In this way, the field of view of the microscope is no longer a surface area, but just a single profile (**Figure 2**, bottom panel), since once the profile acquisition along the fast scan axis is complete the slow scan axis is not moved. However, as in the "normal scanning mode", the MFM image still consists of several hundreds or thousands of lines, although all along the same profile. If the geometrical *y* direction is removed from the image coordinates (as *y* remains constant), it is possible to assign the scan line number to a different input quantity that can be varied during the operation of the microscope. This input quantity could be e.g. the applied magnetic field. If the magnetic field is kept constant *during* the scan along the fast axis and is incremented *within* one profile scan and the next, then an image is acquired (see **Figure 2** bottom panel) that for each line reproduces the *same* profile but under different conditions, i.e. under different magnetic fields: for the same dot studied above, the corresponding AFM image is then just a repetition of the same morphology for each line, the whole dot image is no longer reproduced as the scanning is never moved along the *y* (slow scan) axis (see **Figure 2**). In this way, a *single* MFM image contains in itself the information on how the magnetic configuration evolves under hundreds or even thousands of different applied field values. The drawback is that this information is limited to a 1-D profile, and not to a 2-D surface area as in "normal scanning mode" MFM.

A "dynamic scanning mode" MFM requires a few additional hardware than a conventional MFM. Above all, its controller must give access to a way to know when the microscope has finished acquiring a line. Depending on how this information is obtained, a suitable hardware and/or software combination must be developed to ensure that the applied magnetic field variations are properly synchronised with the scan line change. For example [17, 18], a TTL level could be output by the AFM controller signalling each EOL, and used to trigger the magnetic field step through a suitably programmed current generator or ramp generator. If the AFM controller provides external input channels, the voltage reading e.g. of a gaussmeter could be fed to the controller, therefore directly linking each scanned line with the corresponding magnetic field value. In this way, the obtained 2-D image contains along the *x* (fast scan) direction the usual spatial dependence, and along the *y* (slow scan) direction a line-number dependence that is implicitly defined through the channel acquiring the external (gaussmeter) input. The image then visually represents the magnetic field dependence of the domains configuration along a single profile.

This operation mode is sometimes called "dynamic", as opposed to the "normal" or "static" operation mode of MFM systems. In "normal scanning mode", the MFM acquires a single image under a static (constant) applied field, whereas in "dynamic scanning mode" the applied field varies with time (although it is constant during the acquisition of each line). The term "dynamic", therefore, does not imply that a frequency-dependent characterisation is performed, and, in this context, should be used consciously.

A representative application of the "dynamic" operation mode is the measurement of local hysteresis loops of magnetic patterned micro- and nano-structures [17]. An example of a typical measurement is shown in **Figure 3** for a $Fe_{50}Pd_{50}$ patterned square, with a lateral size of 800 nm and a thickness of 35 nm (the same sample whose domains configurations were shown in **Figure 1** for selected applied field values). **Figure 3(a)** shows the height channel: the MFM tip scans along a

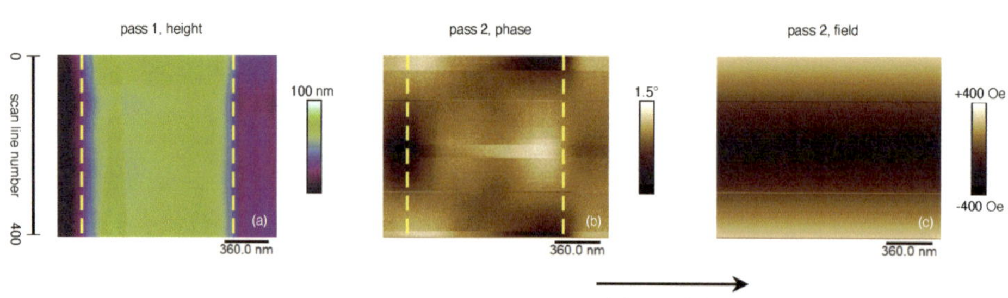

Figure 3.
Example of data acquisition in "dynamic scanning mode" for the measurement of a local hysteresis loop on a $Fe_{50}Pd_{50}$ square (lateral size 800 nm, thickness 35 nm). (a) Height channel (pass 1). The yellow dashed lines approximately mark the two lateral square edges. (b) Phase channel (pass 2), i.e. the magnetic response. The yellow dashed lines are placed in correspondence of those in panel (a). (c) AFM controller auxiliary input channel, set to receive the applied field value. The tip was a Bruker MESP-HR.

profile passing across the whole width of the FePd square. Since the microscope works in "dynamic" mode, the slow scan axis is disabled, and the same profile is repeatedly acquired. For this reason, the topography image in **Figure 3(a)** does not show significant variations for the different scan line numbers, as, ideally, exactly the same profile and therefore the same morphology are encountered during the scan of each line. Small fluctuations in the sample position, due e.g. to scanner drifts, are acceptable as long as the magnetic configuration under the tip is repeatable. The two dashed yellow lines in **Figure 3** approximately identify the position of the two side edges of the square, and are repeated in the same position in **Figure 3(b)**, where the phase channel in pass 2 is instead reported. This channel contains the magnetic response of the tip at the chosen lift scan height, and is therefore the channel to consider when looking at the magnetic domain configuration of the sample. **Figure 3(c)** reports the image acquired in pass 2 for the AFM controller auxiliary input, that, as described earlier, is fed with the gaussmeter reading. This channel, therefore, contains the applied field value for each scan line.

In order to build a local hysteresis loop from the acquired data, vertical sections must be extracted from the data reported in **Figure 3**, as described in **Figure 4**. From the magnetic response channel (**Figure 3(b)**), the phase values along the two dashed yellow lines are taken and their difference Δ_{phase} is calculated, as a function of the scan line number (**Figure 4(a)**). These data represent how the phase contrast evolves at the left and right sides of the $Fe_{50}Pd_{50}$ square as a function of the scan line number. Similarly, any vertical section of **Figure 3(c)** represents the evolution of the applied magnetic field with the scan line number, as shown in **Figure 4(b)**. Δ_{phase} and the applied field are both available as a function of a common parameter, i.e. the scan line number. If this parameter is taken out and the two quantities are plotted one against the other, the local hysteresis loop displayed in **Figure 4(c)** is obtained.

Local hysteresis loops obtained with this technique are not meant to be interpreted as the magnetic hysteresis loops measured e.g. with magnetometers, whose vertical axis is a magnetic moment or a magnetisation. Local hysteresis loops as those shown in **Figure 4(c)** do not plot a magnetisation as a function of the applied magnetic field, but a quantity that describes how the magnetisation is arranged in a specific portion of the sample, as a function of the applied magnetic field. In the case of the $Fe_{50}Pd_{50}$ square used in this example, the magnetic contrast at the left and right sides of the squares evolves with the applied field between "bright" and "dark" extrema, corresponding to the cases of uniform (saturated) magnetisation along the applied field direction (see **Figure 1**). All intermediate colour shades correspond to lower magnetic signals detected by the MFM tip in pass

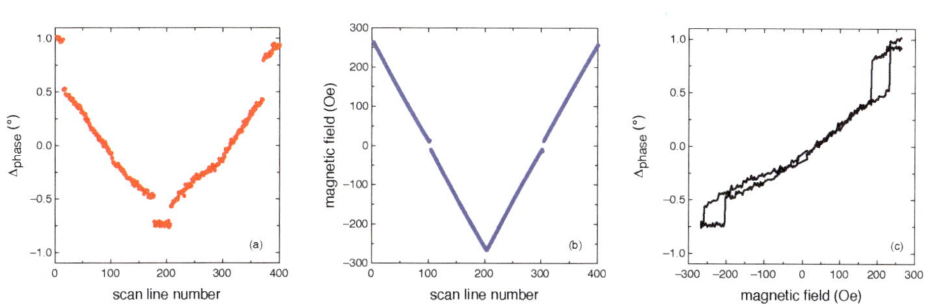

Figure 4.
*Construction of a local hysteresis loop from the data of **Figure 3**. (a) Difference of phase values as a function of the scan line number along the yellow lines of **Figure 3(b)**. (b) Applied field value as a function of scan line number, extracted from any vertical section of **Figure 3(c)**. (c) Merge of panels (a) and (b) by removing the common parameter (scan line number), resulting in the local hysteresis loop.*

2, phase channel, meaning that the magnetisation, at those field values, is less perpendicular to the square sides, i.e. less aligned to the applied magnetic field. Abrupt or peculiar features may appear, as in **Figure 4(c)**, in correspondence of specific features of the magnetisation reversal processes, for example vortex nucleation and expulsion, also putting in evidence their hysteretic behaviour.

The "dynamic scanning mode" operation of the MFM allows a great versatility, and is not limited to the field-dependent characterisations of patterned structures, such as the squares discussed so far. An example of an alternative application of the same method is shown in **Figure 5**, where a continuous thin film of $Fe_{78}Si_9B_{13}$ (thickness 230 nm) annealed at 250 °C for 1 h is investigated, through VSM and "normal scanning mode" MFM. Due to internal stresses quenched during the sputtering deposition [19], this sample is characterised by a non-negligible

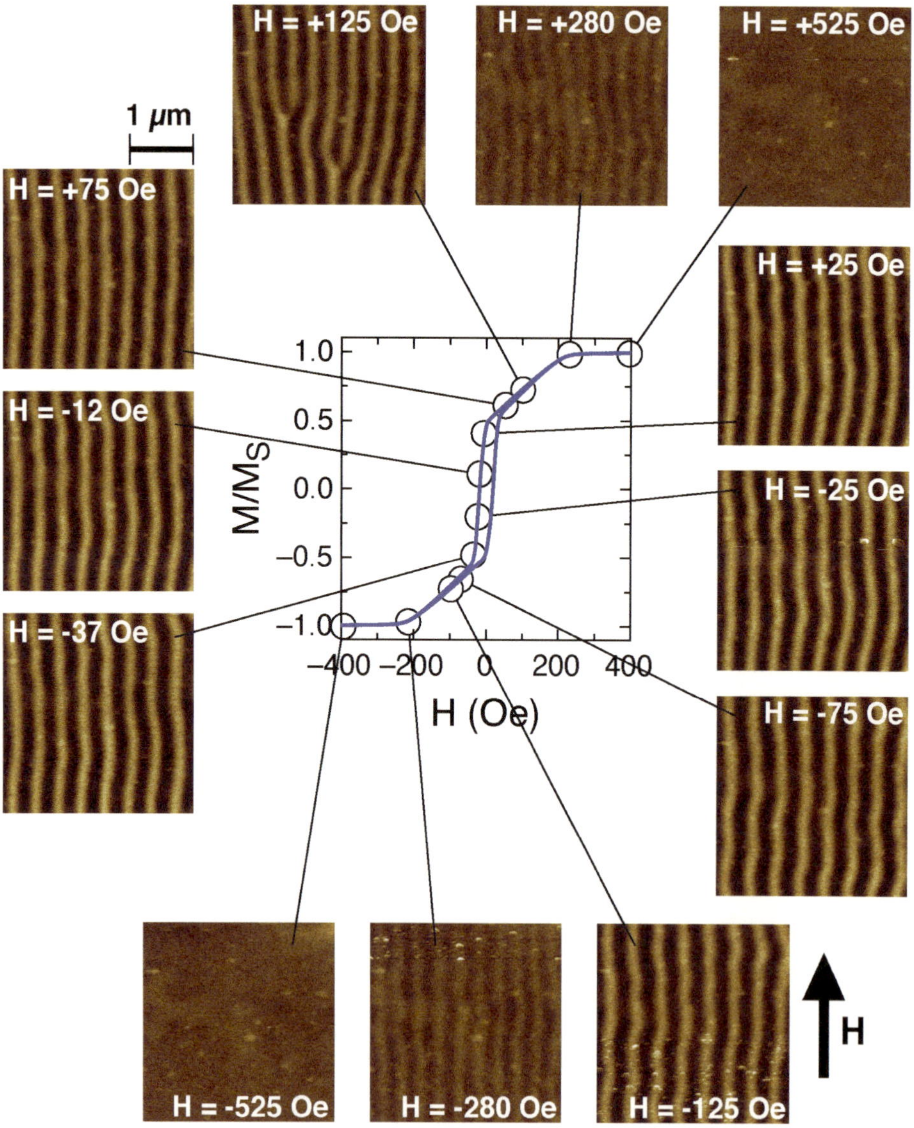

Figure 5.
Hysteresis loop and corresponding selected MFM images (acquired in "normal scanning mode") of the stripe domains of a $Fe_{78}Si_9B_{13}$ thin film (thickness 230 nm) annealed at 250 °C for 1 h. In the MFM images, the magnetic field is applied along the vertical direction. The tip is a Bruker MESP-HR.

out-of-plane magnetic anisotropy, that is responsible for the dense stripe domain configuration clearly visible in the MFM images, and for the characteristic hysteresis loop shape. In each stripe, the magnetisation is tilted with respect to the sample plane, having an in-plane component (parallel to the applied field) that gives no magnetic contrast, and an out-of-plane component, alternately up and own, that is the source of the magnetic contrast visible at the MFM. At low applied field values, where the colour contrast between the bright and dark stripes is particularly evident, the applied field (parallel to the stripes orientation) does marginally affect the magnetisation vector, by slightly rotating it toward the film plane. However, when the hysteresis loop changes slope, two effects take place at the same time, as visible at the MFM: the magnetic contrast gradually fades out, and the number of stripes visible in each MFM images changes. The gradual reduction of the magnetic contrast is due to the progressive tilt of the magnetisation vector toward the applied field direction: the out of plane component of the magnetisation reduces as the field is increased, eventually disappearing when the field is strong enough to overcome the out-of-plane anisotropy and force the magnetisation in the sample plane. The variation of the number of stripes visible in each image, instead, can be appreciated by directly counting them, and is due to bifurcation processes that can sometimes be identified in an MFM image, such as that acquired at an applied field of +125 Oe.

The "dynamic scanning mode" turns out to be particularly powerful to investigate this specific process [20]. Provided that the fast scan axis is set perpendicular to the stripes orientation, and therefore the applied field direction orthogonal to it, an image such as that shown in **Figure 6** can be obtained, detailing the evolution of the stripes domain configuration of the sample as a function of the applied field, for the same loop branch along which a few MFM images in "normal scanning mode" where discussed in **Figure 5**. The versatility of the "dynamic scanning mode" appears one more time evident: as the sample domain configuration is constituted by long, parallel stripes, the information obtained by acquiring full MFM images in "normal scanning mode" is somewhat redundant, a single profile being able to summarise a whole image. In fact, the single profile provides at the same time the magnetic contrast between the bright and dark stripes, and their number. But in "dynamic scanning mode" a single MFM image consists of the same profile acquired several hundreds of times (as compared to 12 images available in "normal scanning mode" in **Figure 5**), therefore giving with a significant magnetic field resolution the evolution of the dense stripe domain configuration. In particular, not only the magnetic contrast can be seen to gradually fade out, but the abrupt discontinuities in the stripes in **Figure 6** are a

Figure 6.
MFM image acquired in "dynamic scanning mode" of the same sample shown in ***Figure 5****. The tip is a Bruker MESP-HR.*

clear indication that at those applied field values bifurcations have appeared in the vicinity of the scanned profile, giving rise to an offset of the stripes position along the profile and possibly to a variation in their number.

As it is the case for the hysteresis loops measured in "dynamic scanning mode" on the $Fe_{50}Pd_{50}$ squares, also in the case of the dense stripe domain configuration the two MFM techniques should be exploited together, along with other measurement techniques, to gain a detailed understanding of the magnetisation processes [17], paving the way to more complex investigations such as chirality control in dots [21] and rotatable anisotropy experiments [20].

4. Variations

The possibility to apply a magnetic field to the AFM, combined with the "dynamic scanning mode", actually opens new possibilities, not limited to the investigation of magnetic domains configuration evolution or to the measurement of local hysteresis loops. As an example, **Figure 7** schematically illustrates how the "dynamic scanning mode" can be exploited to measure the magnetostrictive properties of thin films [22]. In this application, the studied material is not the sample scanned by the tip, as in conventional AFM characterisations. Instead, it is the coating applied to a soft AFM cantilever (blue layer in **Figure 7**), that is brought in contact with a flat surface (green rectangle), while keeping equal to zero the field of view (0 nm scan size along both fast and slow scan axes). In practice, the AFM tip touches the flat surface and does not move while "scanning". When a magnetic field is applied, the magnetostrictive coating on the bottom side of the cantilever extends, and the cantilever bends upward (**Figure 7(b)**). This configuration, however, is out-of-equilibrium, as the AFM feedback loop will quicly react by retracting the piezo scanner (the orange block in **Figure 7**) to compensate for the apparent height variation of the sample. The adhesion forces between the tip and the flat surface ensure that the cantilever will follow the piezo scanner retracting, and will restore its initial flat configuration. The vertical retraction of the piezo scanner gives a measurement of the vertical deflection of the cantilever due to the elongation of the magnetostrictive layer on its bottom surface. In this experiment, the "dynamic scanning mode" operates exactly as discussed before, by synchronising the magnetic field changes with the EOL signal of the AFM controller. In fact, even if the scan size of the tip is equal to zero, the AFM still operates by acquiring an image where, simply, all collected data points (along each line, and for all lines) belong to the same position on the flat surface. As each line is acquired under a different

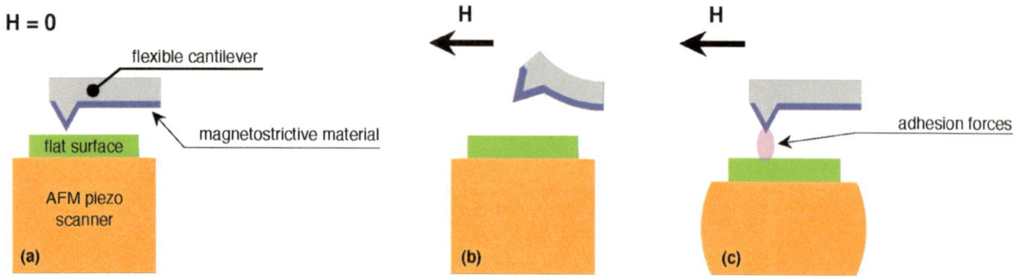

Figure 7.
Schematic representation of the operation principle of the magnetostriction measurement setup exploiting an AFM. (a) Flat cantilever in contact with a flat surface (green rectangle) with zero applied magnetic field. (b) Under an applied field H the cantilever bends upward because of the presence of a magnetostrictive material (blue layer) on its bottom. (c) The AFM feedback loop retracts the scanner (orange rectangle) to compensate for the apparent vertical displacement of the cantilever. Adhesion forces (region highlighted with a magenta shade) ensure that the cantilever follows the motion of the piezo scanner until its flat configuration is restored.

applied field, different cantilever deflection values are obtained, which can be combined to reproduce a deflection *vs.* field curve representative of the magnetostrictive response of the magnetic coating of the cantilever.

An example is provided in **Figure 8** for a $Fe_{81}Al_{19}$ coating with a thickness of 270 nm, which is compared to its hysteresis loop measured with a VSM. In **Figure 8(b)** a magnification of the region highlighted with a magenta shade shows the excellent correspondence between the peak of the deflection curve and the coercivity of the material.

The application briefly discussed in **Figures 7** and **8** provides a hint to the versatility of the AFM "dynamic scanning mode", that is in principle much more general than discussed in this chapter. In fact, the technique has only a few very general requirements:

- the material to be studied must have a property that can be investigated through an AFM (e.g. its magnetisation or its magnetostriction);
- this property must change through an externally controlled input quantity (e.g. a magnetic field);
- the externally controlled input quantity must be set fast enough to be compatible with the typical line trace times of the AFM being used; therefore, input quantities with a long inertia (e.g. a temperature) might be less convenient for these applications, whereas faster ones (e.g. electric or magnetic fields) are more appropriate;
- there must be a software or hardware arrangement enabling the control of the external input quantity triggered by the EOL signal of the AFM controller;
- there must be a way to feed the AFM controller with an analog signal proportional with the value of the externally controlled input quantity (e.g. a gaussmeter), so that its value for each line can be associated with the other channels acquired by the microscope during the "dynamic scanning".

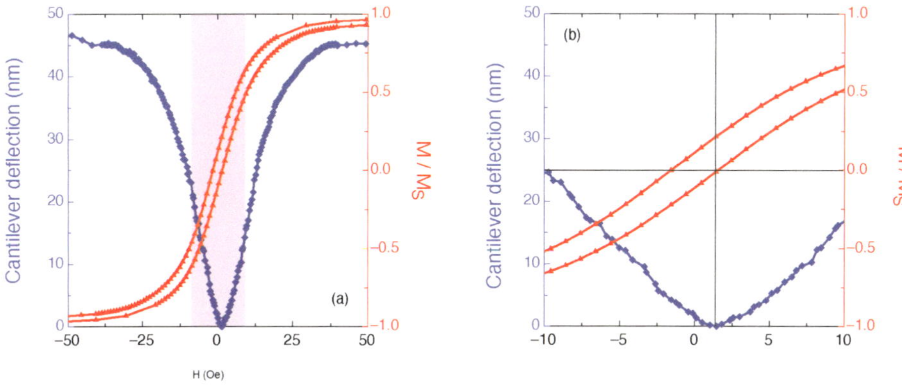

Figure 8.
(a) Cantilever deflection as a function of the applied magnetic field (blue symbols, axis and legend), compared with the magnetic hysteresis loop (red symbols, axis and legend) for a $Fe_{81}Al_{19}$ thin film with a thickness of 270 nm. (b) The same as in (a), but magnified in the magenta-shaded region. The horizontal and vertical lines mark the coincidence of coercivity with the peak of the deflection curve. The cantilevers were Nanosensors Pointprobes with a length of 445 μm, a width of 50 μm, a thickness of 2 μm, and a nominal value of elastic constant of 0.3 N/m.

These general requirements allow the design of complex and custom experiments, exploiting the versatility of an AFM, at the nanoscale, of which the study of local hysteresis loops, magnetic field evolution of domains in patterned media, and magnetostrictive response of thin films are just a few examples.

5. Conclusions

A magnetic field-dependent MFM, derived from the single-point technique, has been illustrated in its operating principles, showing a few examples of characterisations where the evolution of the magnetic domain configuration of a sample is studied as a function of the applied magnetic field. Using the same technique, local hysteresis loops have been investigated on patterned $Fe_{50}Pd_{50}$ dots, as well as dense stripe domains in $Fe_{78}Si_9B_{13}$ continuous thin films. The technique has been shown to be sufficiently general to allow significant variations, such as that for the measurement of magnetostriction on $Fe_{81}Al_{19}$ thin films. The requirements for the exploitation of the "dynamic scanning mode" of AFMs have been summarised, opening the way to the design of innovative experiments in laboratories equipped with suitable atomic force microscopes.

6. Materials and methods

$Fe_{50}Pd_{50}$ thin films with thickness of 35 nm have been prepared by rf sputtering on Si/Si-oxide substrates, and then patterned by electron beam lithography into squares with lateral sizes of 800 nm and 2 μm. The material, in its as-prepared state, has a polycrystalline fcc microstructure with a soft magnetic behaviour [23].

$Fe_{78}Si_9B_{13}$ thin films with thickness of 230 nm have been prepared by rf sputtering on Si_3N_4 substrates from targets made of amorphous ribbons. They are amorphous according to X-ray diffraction [19]. They have been annealed in furnace at 250 °C for 1 h in vacuum to partially relax the quenched-in stresses, in order to make their hysteresis loop softer [24].

$Fe_{81}Al_{19}$ thin films have been deposited on Si(100) substrates and on Si AFM cantilevers (length 445 μm, width 50 μm, thickness 2 μm, elastic constant ≈0.3 N/m) by dc magnetron sputtering at the Department of Materials Science and Metallurgy, Cambridge University, UK [22].

Acknowledgements

The Authors would like to thank Dr. Matteo Cialone and Dr. Wilhelm Hüttenes for helping with the magnetostriction measurements.

Conflict of interest

The authors declare no conflict of interest.

Abbreviations

AFM	Atomic force microscope /microscopy
EOL	End of line

MFM	Magnetic force microscope/microscopy
TTL	Transistor-to-transistor logic
VSM	Vibrating sample magnetometer/magnetometry

Author details

Marco Coïsson*, Gabriele Barrera, Federica Celegato and Paola Tiberto
INRIM, Advanced Materials and Life Sciences Division, Torino, Italy

*Address all correspondence to: m.coisson@inrim.it

References

[1] Hartmann U. High-resolution magnetic imaging based on scanning probe techniques. Journal of Magnetism and Magnetic Materials. 1996;157-158:545-549. https://doi.org/10.1016/0304-8853(95) 01264-8

[2] Binning G, Quate CF, Gerber Ch. Atomic force microscope. Physical Review Letters. 1986;56:930. https://doi.org/10.1103/PhysRevLett.56.930

[3] Vock S, Wolny F, Mühl T, Kaltofen R, Schultz L, Büchner B, Hassel C, Lindner J, Neu V. Monopolelike probes for quantitative magnetic force microscopy: calibration and application. Applied Physics Letters. 2010;97:252505. https://doi.org/10.1063/1.3528340

[4] Hug HJ, Stiefel B, van Schendel PJA, Moser A, Hofer R, Martin S, Güntherodt H-J. Quantitative magnetic force microscopy on perpendicularly magnetised samples. Journal of Applied Physics. 1998;83:5609. https://doi.org/10.1063/1.367412

[5] Zhu X, Grütter P, Metlushko V, Ilic B. Magnetization reversal and configurational anisotropy of dense permalloy dot arrays. Applied Physics Letters. 2002;80:4789. https://doi.org/10.1063/1.1489720

[6] Yang JKW, Chen Y, Huang T, Duan H, Thiyagarajah N, Hui HK, Leong SH, Ng V. Fabrication and characterisation of bit-patterned media beyond 1.5 Tbit/in^2. Nanotechnology. 2011;22:385301. https://10.1088/0957-4484/22/38/385301

[7] Piramanayagam SN, Ranjbar M, Sbiaa R, Tavakkoli AKG, Chong TC, Characterisation of high-density bit-patterned media using ultra-high resolution magnetic force microscopy. Physica Status Solidi Rapid Research Letters. 2012;6(3):141-143. https://doi.org/10.1002/pssr.201105537

[8] Imre A, Csaba G, Ji L, Orlov A, Bernstein GH, Porod W. Majority logic gate for magnetic quantum-dot cellular automata. Science. 2006;311(5768): 205-208. https://doi.org/10.1126/science.1120506

[9] Li H, Qi X, Wu J, Zeng Z, Wei J, Zhang H. Investigation of MoS_2 and graphene nanosheets by magnetic force microscopy. ACS Nano. 2013;7(3): 2842-2849. https://doi.org/10.1021/nn400443u

[10] Israel C, de Lozanne A. High-field magnetic force microscopy as susceptibility imaging. Applied Physics Letters. 2006;89:032502. https://doi.org/10.1063/1.2221916

[11] Sorop TG, Untiedt C, Luis F, Kröll M, Raşa M, de Jongh LJ. Magnetisation reversal of ferromagnetic nanowires studied by magnetic force microscopy. Physical Review B. 2003; 67:014402. https://doi.org/10.1103/PhysRevB.67.014402

[12] Zhu X, Grütter P, Metlushko V, Hao Y, Castaño FJ, Ross CA, Ilic B, Smith HI. Construction of hysteresis loops of single domain elements and coupled permalloy ring arrays by magnetic force microscopy. Journal of Applied Physics. 2003;93:8540. https://doi.org/10.1063/1.1540129

[13] Zhu X, Grütter P. Imaging, manipulation and spectroscopic measurements of nanomagnets by magnetic force microscopy. MRS Bulletin. 2004;29(7):457-462. https://doi.org/10.1557/mrs2004.139

[14] Rastei MV, Meckenstock R, Bucher JP. Nanoscale hysteresis loop of individual Co dots by field-dependent magnetic force microscopy. Applied

Physics Letters. 2005;87:222505. https://doi.org/10.1063/1.2138349

[15] Jaafar M, Serrano-Ramón L, Iglesias-Freire O, Fernández-Pacheco A, Ibarra MR, De Teresa JM, Asenjo A. Hysteresis loops of individual Co nanostripes measured by magnetic force microscopy. Nanoscale Research Letters. 2011;6:407. https://doi.org/10.1186/1556-276X-6-407

[16] Grütter P, Jung Th, Heinzelmann H, Wadas A, Meyer E, Hidber H-R, Güntherodt H-J. 10-nm resolution by magnetic force microscopy on FeNdB. Journal of Applied Physics. 1990;67:1437. https://doi.org/10.1063/1.345675

[17] Coïsson M, Barrera G, Celegato F, Enrico E, Manzin A, Olivetti ES, Tiberto P, Vinai F. Local field loop measurements by magnetic force microscopy. Journal of Physics D: Applied Physics. 2014;47:325003. https://doi.org/10.1088/0022-3727/47/32/325003

[18] Coïsson M, Barrera G, Celegato F, Tiberto P. Development and calibration of a MFM-based system for local hysteresis loops measurements. Journal of Physics: Conference Series. 2016;755:012002. https://doi.org/10.1088/1742-6596/755/1/012002

[19] Coïsson M, Celegato F, Olivetti E, Tiberto P, Vinai F, Baricco M. Stripe domains and spin reorientation transition in $Fe_{78}B_{13}Si_9$ thin films produced by rf sputtering. Journal of Applied Physics. 2008;104:033902. https://doi.org/10.1063/1.2960454

[20] Coïsson M, Barrera G, Celegato F, Tiberto P. Rotatable magnetic anisotropy in $Fe_{78}Si_9B_{13}$ thin films displaying stripe domains. Applied Surface Science. 2019;476:402-411. https://doi.org/10.1016/j.apsusc.2019.01.126

[21] Coïsson M, Barrera G, Celegato F, Manzin A, Vinai F, Tiberto P. Magnetic vortex chirality determination via local hysteresis loops measurements with magnetic force microscopy. Scientific Reports. 2016;6:29904. https://doi.org/10.1038/srep29904

[22] Coïsson M, Hüttenes W, Cialone M, Barrera G, Celegato F, Rizzi P, Barber ZH, Tiberto P. Measurement of thin films magnetostriction using field-dependent atomic force microscopy. Applied Surface Science. 2020;525:146514. https://doi.org/10.1016/j.apsusc.2020.146514

[23] Tiberto P, Celegato F, Barrera G, Coisson M, Vinai F, Rizzi P. Magnetization reversal and microstructure in polycrystalline $Fe_{50}Pd_{50}$ dot arrays by self assembling of polystyrene nanospheres. Science and Technology of Advanced Materials. 2016;17(1):462-472. https://doi.org/10.1080/14686996.2016.1201414

[24] Tiberto P, Celegato F, Coïsson M, Vinai F. Spin reorientation transition in amorphous FeBSi thin films submitted to thermal treatments. IEEE Transactions on Magnetics. 2008;44(11):3921-3924. https://doi.org/10.1109/TMAG.2008.2002255

Chapter 3

Diffraction Grating Groove Metrology Using AFM & STM

Leonid I. Goray

Abstract

AFM & STM metrology has been around for a long time, and especially intense since it has been awarded by the Nobel Prize in Physics in 1986. Since then, many AFM & STM groove profile measurements on surface relief diffraction gratings have been presented. However, a wide review of the results of the use of AFM & STM methods for groove metrology of various surface relief gratings has not really been undertaken. The following problems are discussed in this chapter: the cantilever tip deconvolution, geometry, and radius; groove shapes and abrupt groove slopes; roughness; PSD functions; etc. Also, the author demonstrates comparisons with other widely-used metrology techniques and examples of AFM & STM data of bulk, coated, and multilayer-coated ruled, or holographic, or lithographic gratings having realistic groove profiles. These gratings were chosen because high quality efficiency data exists, in particular, for space gratings or/and X-ray gratings characterized by synchrotron radiation sources; and their groove profiles, together with random nanoroughness, were measured by AFM or STM to be included in rigorous efficiency and scattered light intensity calculus. In the present chapter, both the earlier published results and the recent, non-published yet results are described and discussed.

Keywords: diffraction grating, AFM, STM, groove profile metrology, surface nanoroughness, PSD function, RMS roughness, diffraction efficiency, X-rays

1. Introduction

Scanning Tunneling Microscopy (STM) and Atomic Force Microscopy (AFM) metrology as parts of more general Scanning Probe Microscopy has been around for a long time, and especially intense since it has been awarded by the Nobel Prize in Physics in 1986. In 1988, our team was one the first who designed, manufactured, and used the specialized scanning tunneling microscope to measure newly developed big-size surface-relief holographic diffraction gratings obtained in non-organic photoresist and having in an order lower roughness (**Figure 1**) [1]. For today, STM & AFM profile measurements on surface-relief diffraction gratings are presented as a matter of routine, see, e.g., in Refs [2, 3]. On the other side, precise microscopic surface-relief patterns are used as grating standards to calibrate atomic force microscopes (see, i.e., in [4]).

However, a wide analysis of the use of the STM and AFM methods for surface relief grating groove metrology has not really been undertaken in details. The following problems are discussed here: the tip deconvolution, geometry, and radius; groove shapes and abrupt groove slopes; roughness; PSD functions; other. The

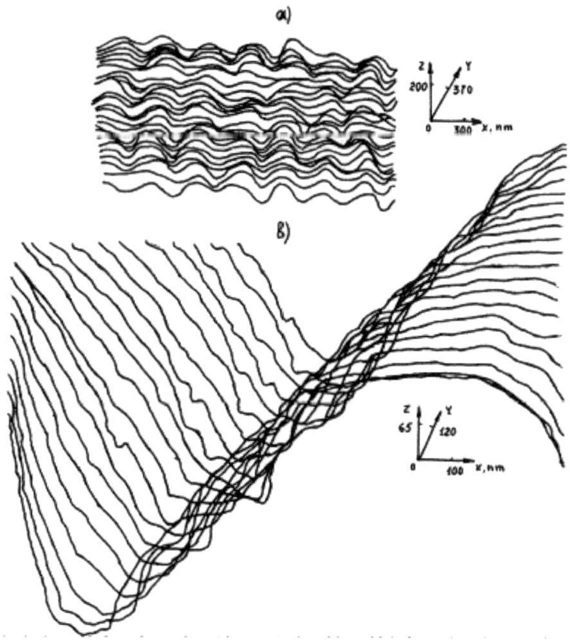

Figure 1.
STM images of holographic relief grating surfaces (Au-coated) obtained by (a) organic and (b) non-organic photoresists (after [1]).

author demonstrates examples of AFM & STM data and comparisons with other widely-used metrology techniques for bulk, coated, and multilayer-coated ruled, or holographic, or laser-lithographic, or electron-lithographic gratings having lamellar, or sinusoidal, or blazed, or other realistic groove profiles. These gratings were chosen because high quality efficiency data exists, in particular, for flight gratings or/and X-ray gratings characterized by synchrotron radiation sources; and their groove profiles, together with random nanoroughness, were measured by AFM or STM to be included in rigorous efficiency and scattered light intensity calculus.

Here the author briefly compares various standard techniques for exact determining the digital profile and 3D topography of a surface relief grating. Several widely used direct (or semi-direct – 'imaging') methods and respective instruments applicable for this purpose are compared. The advantage of using direct and exact groove metrology to predict efficiency and polarization characteristics of gratings now is well-known and widely used. The main purpose of such modern approach is rejecting unusable samples on earlier stages and decreasing expenses for their production and research. This is much more effective in compare to the earlier approaches, wherein: (1) a master grating is fabricated, whether by mechanical burnishing with a ruling engine, or holographic writing (interferometry), or direct laser recording (DWL), or various newer writing techniques, like as electron-beam lithography (EBL) and Si-etching, or their combinations; then, (2) it is replicated or/and coated, and, finally, (3) tested for the diffraction efficiency and scattering light intensity. For mechanically ruled gratings, a 'test' ruling can quickly be checked with this approach, whereas a complete ruling sometimes requires several days or even weeks of continuous use of expensive ruling machines [5]. Even for holographic or EBL gratings, considerable efforts of writing, etching and coating the grating with specialized coatings, especially multilayer coatings, can be decreased additionally if metrology validates that an intermediate product is suitable in the planned application.

2. Direct metrology techniques

The author briefly discusses and compares in the first part of this chapter several basic, among many others, direct metrology techniques: microinterferometry (as one of optical methods) [6], stylus (mechanical) profilometry [7], scanning electron microscopy (SEM) [8], and AFM [9]. Several examples of groove metrology results are presented and discussed further including those obtained very recently.

Any method for measuring the profile of a surface relief grating requires some calibration procedure [10]. The considered here methods are also widely used for surface microroughness determination on a nanometer or sub-nanometer level. Measuring main groove parameters of a grating, in particular, the actual groove depth or blaze angle, adds to the requirements for the specific metrology method. The depth of the groove profile, defined depending on the accuracy of a vertical calibration, basically determine the wavelength for the peak efficiency in a given optical mounting geometry (classical or conical) [11]. The common error of the order of a few percent in the lateral calibration can affect the prediction of the blaze wavelength that should be within hard tolerance for many practical applications. This is because the groove vertical geometry is often expressed relatively to the grating period, in dimensionless units. Any lateral error becomes vertical error in the respective topographical transformation. Fortunately, lateral errors can be fairly determined because the grating period is well known beforehand with high accuracy and, thus, the grating data itself gives a calibration factor to correct the lateral scale unit. The accurate lateral calibration is also required for rigorous efficiency modeling codes, in which the use of the average groove profile shape is very important to obtain exact efficiency data in all significant diffraction orders.

The microinterferometer is sometimes called as 'optical profilometer'. It is essentially an interferometric head on a microscope, where the reference arm of the interferometer views a small, highly polished reference plate [6]. Such a reference can be removed from the results of measurements on highly polished surfaces that is important for grating measurements because many state-of-the-art gratings, especially for X-ray and Extreme Ultraviolet (EUV) applications, have the root-mean-square (RMS) roughness of the same order as the best reference plates. A Phase Shift Instruments model MicroXAM [12] has been used in the discussed work [10]. It has variable magnification from 2× to 100×; values of range and resolution for the 50× magnification is listed in **Table 1**. The instrument uses the zero path difference calculations independently for each pixel from a series of images obtained during a vertical sweep. This increases the available vertical range and the available slope angle range substantially, however any microinterferometer has two lateral resolution-restricted factors, which are not limitations in the other

Instrument	Microinterferometer 50×	Stylus profilometer	Atomic-force microscope	Units
Vertical resolution	0.05	0.1	0.05	nm
Vertical range	100	130	~5	μm
Lateral resolution	~0.3	~0.3	0.015	μm
Lateral range	163 (more w/stitching)	> 25000	100	μm
Limiting factor(s) for lateral resolution	MTF, sampling, need for retroreflection over the whole profile	Tip radius & angle	Tip radius	
Upper slope limit	–	45	~70	deg.

Table 1.
Measurement characteristics of three surface profiling instruments (after [10]).

considered methods. Namely, the optical resolution is due to the diffraction limit and pixel sampling is due to different magnifications and focusing. The theoretical limit on lateral resolution in such instruments is a half of the working wavelength, or, typically, about 0.1–0.3 μm. Thus, this method generally is not suitable to measure high frequency (short period) and/or low depth diffraction grating.

The stylus profilometer has a diamond tip to brought into direct contact with the surface, with calibrated contact force. As the tip moves across the surface, the motion of the tip is amplified, filtered, and detected. The basic limits inherent to such metrology devices are well-discussed, e.g., in [7]. Care must be exercised to prevent indentations of the surface by the tip, depending on materials and forces used, as well as accounting the tip size. The model used in this work is a Tencor P-10 [13]. **Table 1** presents the basic lateral and vertical ranges and resolutions typical for the instrument. Typical measurement parameters are: the tip radius of 0.1 μm (in the plane of dispersion), the tip speed of 5 μm/s, the digital sampling of 2 kHz, the tip force of 0.25 mg, and profile lengths of at least 100 μm (depending on the groove period). Note that in the last model of this instrument, KLA-Tencor HRP-260, the tip radius can be up to 25 nm and it has a high-resolution stage that produces scanning results similar to an AFM device. In the recent investigation we have used another model, namely, XP-1 Stylus (Ambios Technology, USA) [14].

The AFM instrument model that has been used in the discussed earlier work was a Digital Instruments Nanoscope III [15]. The recent investigation was carried out using the atomic-force microscope model NT-MDT NTEGRA Aura [16]. NTEGRA Aura is a Scanning Probe Microscope for studies in the conditions of controlled environment and low vacuum. The Q-factor of the cantilever in vacuum increases, thus gaining the sensitivity, reliability and accuracy of 'probe-sample' light forces measurements. At that, the change from atmosphere pressure to 10^{-3} Torr vacuum provides the tenfold gain of Q-factor. By further vacuum pumping, Q-factor reaches its plateau and changes insignificantly. Thus, NTEGRA Aura comparing to the high-vacuum devices it needs much less time, about a minute, to get the vacuum that is needed for the tenfold Q-factor increase. NTEGRA Aura has built-in closed loop control for all the axes, optical system with 1 μm resolution and ability to work with more than 40 different AFM regimes.

We have used for the Si-grating technology investigation a flexible Carl Zeiss SUPRA 25 SEM system with a versatile analytical specimen chamber that can be easily expanded with a choice of optional detectors and a full range of accessories [17]. Utilizing the unique GEMINI field emission column, it delivers superb resolution over the complete high voltage range with the magnification of up to 500000. The large 5-axes motorized cartesian stage is particularly useful for handling a number of smaller specimens simultaneously. It is equally useful for accommodating bulky or irregular shaped specimens.

Table 1 summarizes the capabilities and limits of the three metrology devices, which have been characterized earlier for grating metrology. As one can see, the atomic-force microscope has the finest lateral and perfect vertical resolutions. The stylus profilometer and the microinterferometer have comparable vertical ranges, however, without a possibility to determine superfine (atomic-scale) structures, i.e., nanoroughness, and abrupt slopes (see, e.g., **Figures 3–5**). On the other hand, the stylus profilometer has significantly larger lateral range for probing to the millimeter spatial range. Also, the AFM data gives a typical example of non-linearity that should be accounted and described further.

In the groove profile experiment, a series of step height calibration standards [18] has been used [10]. The vertical axis was calibrated using one of the smallest steps of 10 nm. Then the rest of the step height series was measured. Small errors, up to 8%, were observed for heights much higher than that used to calibrate the atomic-force microscope. The fit to correct such nonlinearity was used when the nonlinearity gives a significant difference. The obtained results are summarized in **Table 2**.

In the manufacture and analysis of diffraction gratings, it is necessary to control certain of their parameters at each stage of the process. A SEM research [19] is permissible only at the stage of development of the manufacturing technology, because after each technological operation, see, e.g., [20] or Section 4.5, the sample of the Si-etched grating becomes less and less, since a fragment is separated from the sample to obtain a SEM image of a transverse cleavage (CS). In contrast to this, AFM studies are non-destructive; therefore, control of parameters in the manufacture of gratings is usually carried out with the help of AFM. We made a comparison between AFM (NTEGRA Aura microscope) topographies and SEM (SUPRA 25 system) images of Si-etched grating samples with the period of 2 μm. **Table 3** shows the results of AFM and SEM investigations of Si-etched gratings obtained at different stages of their fabrication. Table also presents the numerical comparing between the AFM and SEM results of the measured groove geometric parameters for the samples studied. The calculated value of the blaze (working facet) angle from the SEM studies was obtained from the sine determined by dividing the experimental values of the groove depth by the width of the working facet.

To measure the roughness of Si(100) plates etched through a DWL mask in KOH and intended for developing the technology of manufacturing Si-etched

Nominal height, nm	Microinterferometer	Stylus profilometer	Atomic-force microscope
8.7	7.92	8.1	—
25.8	—	—	25.7
42.7	43.4	42.4	—
530	—	—	520
1046	—	—	1005
1590	—	—	1469
960	—	—	—
Note No.	3, 4	2	1

1. *Nonlinear at ~8% at highest step when calibrated to a 10 nm step.*
2. *Using 0.1 μm tip, could not resolve depth of 3.3 μm period, AFM step height standard.*
3. *Used at 50× magnification.*
4. *At 100× did not have lateral resolution to see the 3 μm period samples tested using AFM.*

Table 2.
Step height data (after [10]).

Sample No.	Groove depth, nm		Working facet width, nm		Working facet angle, deg.	
	SEM	AFM	SEM	AFM	SEM	AFM
1	151 (47)*	133 (38)*	1571	—	—	—
2	149 (37)*	141 (44)*	—	—	—	—
3	111	121	1630	1710	3.90	4.05
4	111	111	1603	1594	3.97	—
5	105	114	1590	1580	3.89	4.13

Height of Si-nubs.

Table 3.
AFM and SEM data for blaze Si-etched gratings.

Sample No.	RMS roughness, nm			
	Non-etched field		Etched bottom	
	Profilometer	AFM	Profilometer	AFM
1	2.6	0.2	2.8	0.8
2	3.2	1.6	4.0	2.5
3	3.4	0.2	4.5	1.8
4	2.8	1.7	2.9	4.9
5	2.4	1.5	3.5	4.3

Table 4.
AFM and stylus Profilometry data for Si-etched plates.

gratings, several high quality samples were selected using white light optical microscopy. The roughness of the etched bottom and the non-etched area was measured by two compared methods: Stylus Profilometry (XP-1 Stylus profilometer) and AFM (NTEGRA Aura microscope). Our studies were carried out on an atomic force microscope in the semi-contact or tapping mode; all scans had 512×512 points. We used TipsNano [4] silicon cantilevers with a typical radius of tips ~6 nm. The results of roughness measurement by two methods on topological elements (stripes) of 50 μm wide are presented in **Table 4**. As follows from the presented data, the RMS roughness obtained by different methods may differ by more than an order of magnitude. This is due to the radius of the stylus and the scanning length, which in that case were 2 μm and 80 μm, respectively. However, this device is equipped with a stylus with a radius of 0.2 μm, which, in principle, allows one to measure low- and mid-frequency gratings with smaller roughness. Note that for the etched bottom, where the average roughness is several times higher, the scatter of results is much smaller and ranges from several tens of percent to several times.

3. AFM groove metrology problems

A cantilever tip convolution, which limits the resolution of both the atomic-force microscope and the stylus profilometer, has been much studied and various algorithms to account for this effect has been developed and intensively used (see, e.g., [21] and also in this book). In the results presented in **Table 2** such algorithms have not be used. However, the general used rule is that the known tip radius should be much less than the measured periods of gratings. Typically, the radius of a fresh AFM cantilever tip is about 5–15 nm; so, the rule of thumb is that for groove profiles of mid- and high-frequency gratings (say, periods of 100–300 nm and less) tip deconvolution algorithms should be used. In the vertical direction, the depth parameter is smaller, and, apparently, the groove profiles recorded somewhat non-correctly for high-frequency gratings only. However, it depends also on absolute values of the groove profile depth, which can vary in two orders of magnitude.

Another important and general AFM problem, in particular for fine-structure gratings with steep slopes and high aspect ratios of grooves, is the shape and the radius of AFM cantilever tips. Tip size has the major impact on the resolution of images obtained by any atomic-force microscope. The knowledge of the tip radius and shape is essential for the quantitative interpretation of nano-scale lateral steps, in particular, for roughness having short correlation lengths. Tip wear is therefore a key limitation in the application of AFM [22]. The results of nanoindentation experiments with diffraction gratings permanently confirm this conclusion. The measurement of the tip radius before and after measuring groove profiles of gratings was performed in Ref. [10], and they found that the radius to be in the range of 10–20 nm. One measurement found a fresh tip to be ~10 nm radius and a used one to be ~20 nm. Therefore, one should restrict an AFM-profiling work to gratings of period much longer than 10–20 nm, as it has been discussed above.

One more problem in AFM measurements of diffraction gratings is the grooves with steep facet slopes, which can be 80 degrees and more for echelle gratings [2]. This problem is similar to measurements of the rectangular (lamellar) groove profile in microelectronics [23]. To accurately measure such general trapezoidal profiles with steep or even negative sidewalls, a large change in the angle of inclination of a cantilever (or scanner, or sample) and/or special cantilever (tip) shape are required, as well as taking into account the aspect ratio of measured grooves [24]. Several studies applied to periodic structures demonstrate that some combination of the tilted probe, special orientations of AFM images and appropriate deconvolution algorithms allows the precise groove shape reconstruction at any aspect ratio [25]. An example of such problem successfully solved is the average groove profile (two grooves) of a 112/mm echelle R5 grating (blaze angle ~78°) derived from AFM images and presented in **Figure 2**.

In **Figure 3**, typical power spectral density (PSD) 1D functions for Si(111) substrate and Si-etched grating samples (see also Section 4.5) are shown. An estimator of the PSD function is factually the periodogram for any periodic, or quasi-periodic, or random profile, or some combination. Assuming the ergodicity of a stochastic process connected with a random generation of asperities, the PSD function can be found as the Fourier transform of the autocorrelation function [26]. Although these functions are mathematically equivalent, one can analyze easy any corrugations of the profile shape simultaneously, i.e., random roughness and groove depth variations, using the 1D or 2D PSD function. Then, the RMS roughness is directly calculated through PSD as the root square of the integral over an effective range of allowed spatial frequencies. Thus, a wide lateral scanning range may require for an AFM instrument to take into account in the evaluated RMS roughness all spatial frequencies (or correlation lengths). It is especially important for low-frequency (long-period) gratings having additionally large correlation lengths of random roughnesses. A good discussion related to this problem and devoted of the use of AFM and similar instruments for measurements of PSD functions of smooth

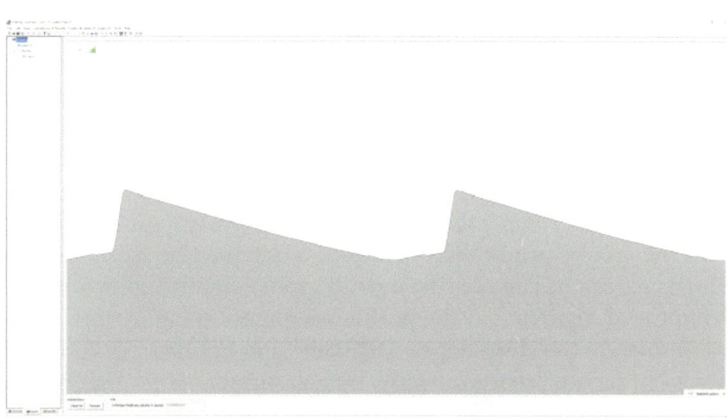

Figure 2.
Average AFM groove profile for 112/mm echelle R5 grating.

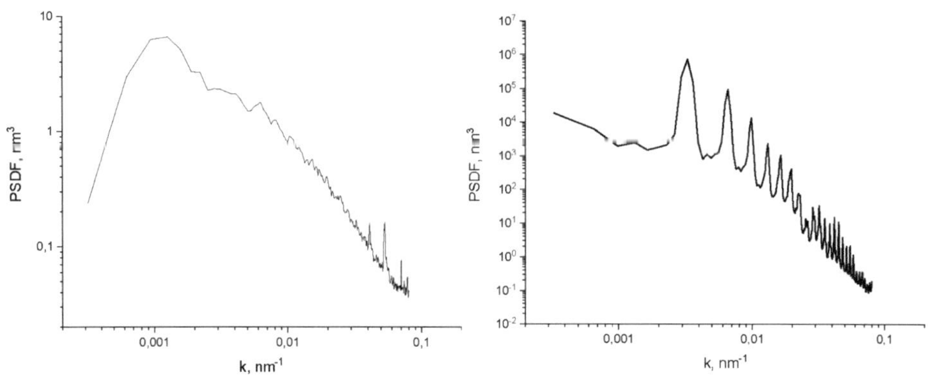

Figure 3.
PSD function: (left) for Si(111) substrate; (right) for Si-etched grating with 500/mm and 4° blaze angle.

mirrors for imaging systems working in the X-ray–EUV range can be found in [27]. So, if one need to use images with a scanning area of about 100 × 100 μm² then thermal drifts, hysteresis, and essential scanner nonlinearities should be accounted.

4. AFM groove metrology results

4.1 Space ruled grating for Visible: NIR

The abovementioned metrology techniques were applied to validate the efficiency of a chosen grating from an ordered grating set which is mounted in the Space Telescope Imaging Spectrograph (STIS) flown aboard the Hubble Space Telescope (HST) [28]. A − 1-order reflection grating with 67.556/mm blazed for 750 nm (1.44° nominal blaze angle) working in the range from 500 to 1000 nm at 8° incidence angle was chosen by us for a certification [29]. The pattern size was 1.5 inches by 1.5 inches, and the ruled area was 30 mm by 30 mm. A sister-replica to this grating, designated 'Ng41M' or by its manufacturers' (Richardson Gratings of Newport Corp.) serial number, 1528, is in use on the HST/STIS as a red survey grating (blazed in the red visible and near infrared range) [10]. In its flight application, this grating had a reflective overcoating of 100 nm Al plus 25 nm MgF_2. However, in these wavelengths the effect of the MgF_2 layer is minor and simulations have showed no valuable difference, within a small part of the accuracy in the measured diffraction efficiency) with such coating or without it. This grating was chosen as an example because: (1) high quality efficiency data exists for it, including rigorous efficiency calculus using the realistic groove profile shape; and (2) groove profiles can be measurable by the mentioned above three methods for a direct comparison.

Portion of a trace of grating No. 1528 taken with the microinterferometer is shown in **Figure 4**. Both the depth and the profile shape are somewhat distorted in compare with the groove profiles in **Figures 5, 6** obtained by the other considered methods. However, as one can see, the overall groove depth and profile are evident. It is clear from the all figures that the profile roughness is higher on the upper sloped portion than on the steep edges. Difficulties in holding the sample steady during 'flyback' prevented reproducibility of measurements for that microinterferometric study.

The groove profile was characterized in details AFM measurements. The tips used here were 10 or 20 nm in radius. An example of the typical groove profile of No. 1528 grating is presented in **Figure 6**. **Figure 6** shows an example of AFM data

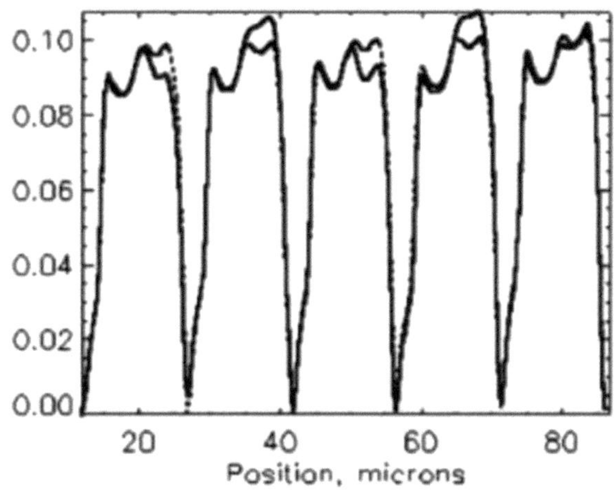

Figure 4.
Portion of a microinterferometer trace of ruled grating No. 1528 (after [10]).

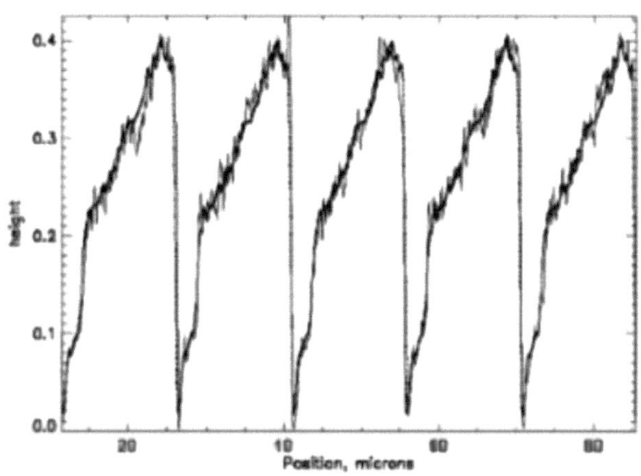

Figure 5.
Portion of a stylus profilometer trace of ruled grating No. 1528 (after [10]).

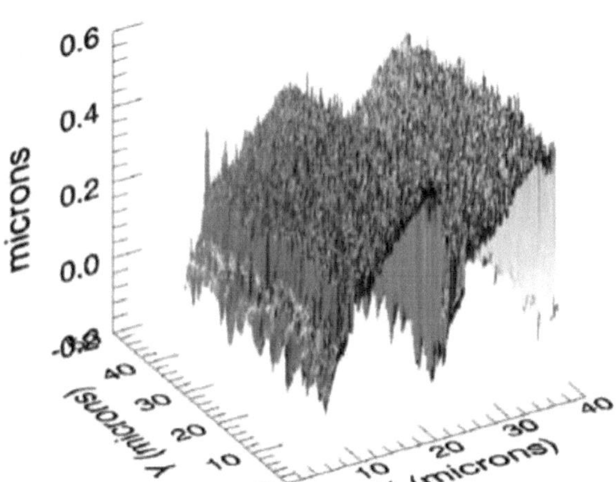

Figure 6.
AFM image of two grooves of ruled grating No. 1528 (after [10]).

for a portion of the surface of the investigated ruled grating. The basic groove profile shape is clearly evident, along with portions of the profile that are rougher than others, and some roughness along the grooves is indicated as well. **Figure 6** shows that the minimum of the grooves is clearly visible in the AFM image. If, as usually, one selects the bottom of the groove as the minimum value, there are two complete grooves in each scan.

The resulting average groove profile – with averaging performed both across the grooves and along as well – is shown in **Figure** 7. The solid line is based on the AFM data, and the dotted line is based on the stylus profilometer data: the groove tops are aligned for the purpose of this comparison; the relatively sharp groove bottom is not as well resolved by the stylus profilometer. The periodicity of the profile is shown by comparing a model of the averaged scan based on the average groove profile shape to the average scan. This is demonstrated by dotted lines plotted against the initial data in **Figure 4** (microinterferometer) and **Figure 5** (stylus profilometer). Once the average profile has been determined, the fitting routine finds the sawtooth and two-angle shape fits by the method of least squares. It is found in the considered case the blaze angle of 1.45° and the anti-blaze angle of 30° (**Figure 8**). The efficiency in general is fairly insensitive to the anti-blaze angle, and the fitting procedure does not fit it as consistently as it does a case of the blaze angle. Thus, the final average groove profile derived from AFM measurements for efficiency modeling purposes is shown in **Figure 8** with 100 discretization points [29].

4.2 Soft-X-ray: EUV blaze gratings

The surface of gratings, namely, the master [30] and replica [31] gratings, as examples of 'good' products, were characterized using a Topometrix Explorer Scanning Probe microscope [32], a type of atomic-force microscopes. The gratings have 2400/mm, a concave radius of curvature of 2.0 m, and a patterned area of size 45 mm by 35 mm. The master grating was fabricated by Spectrogon UK Limited (formerly Tayside Optical Technology). The groove pattern was developed in fused silica by a holographic technique using ion-beam etching to produce an approximately triangular, blazed groove profile. Ion-beam etching results in a groove profile much closer to triangular than the ideal blazed (sawtooth) profile with the apex angle of ~90°. The master grating was uncoated. The replica of the master grating was produced by Hyperfine, Inc. As a result of the replication process, the replica grating had an aluminum surface. A thin SiO_2 coating was applied to the Al surface

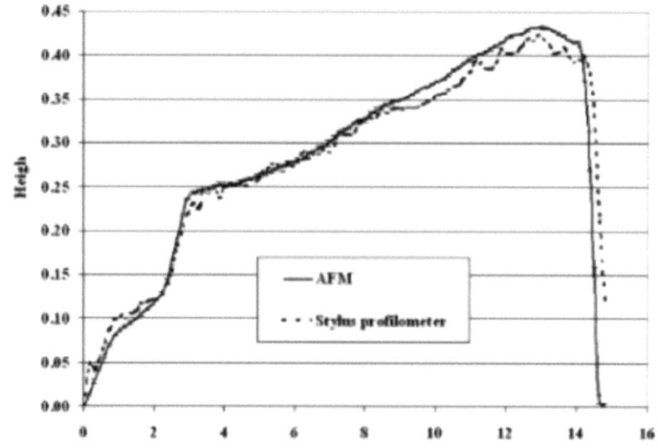

Figure 7.
Average groove profile for grating No. 1528 based on AFM and stylus profilometer data.

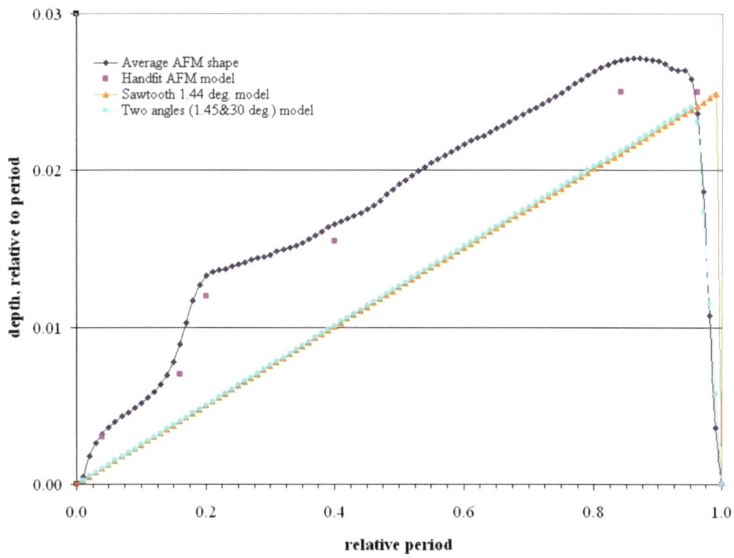

Figure 8.
Models of normalized to period groove profiles of No. 1528 grating.

for the purpose of reducing the nanoroughness and protecting the surface from an additional oxidation.

The AFM images typically had 500 × 500 pixels and a scan range of 1 to 20 μm (pixel size 20 to 400 Å). The silicon probe had a pyramid shape. The base of the pyramid was 3 to 6 μm in size, the height of the pyramid was 10 to 20 μm, and the height to base ratio was approximately 3. The tip of the pyramid had a radius of curvature 100 to 200 Å. The AFM scans were performed using the non-contact resonating mode, where the change in the oscillation amplitude of the probe is sensed by the instrument. A surface topology reference sample was used to optimize the AFM scanning parameters, to calibrate the height scaling of the instrument, and to evaluate the performance of the AFM. This was essential for the accurate characterization of the gratings. The surface topology reference sample consisted of an array of approximately square holes fabricated on the silicon dioxide surface of a silicon die by VLSI Standards, Inc. [18]. The top surface of the die was coated with a thin layer of Pt. The hole array had a pitch of 3 μm and a hole depth of 180 Å.

One typical AFM image of the master grating measured using 16-Å pixels is shown in **Figure 9**, where the vertical scale has been scaled to reveal the texture of the groove surface. The RMS roughness, determined by integrating the PSD function over 2–40 μm^{-1} range, was 3.2 Å. Most of the roughness is concentrated at low spatial frequencies as is apparent from the analysis of the PSD function. The central portion of the AFM image shown in **Figure 9** that covers one period of the grating pattern was selected for further investigations. An analysis program was written in the Interactive Display Language (IDL) for this purpose and it is discussed in detail in Ref. [31].

The histogram of the pixel heights, for one period of the grating pattern, is shown in **Figure 10**. The maxima at 10 Å and 85 Å in **Figure 10** are caused by rounding of the groove profile at the peaks and the troughs which is a result of the pattern fabrication process. An ideal groove profile, either sawtooth or triangular, would have a flat height histogram. The separation between the peaks in **Figure 10** represents the average groove height, approximately 75 Å. The local blaze angle at each pixel was determined by using a least squares algorithm to fit a linear curve to

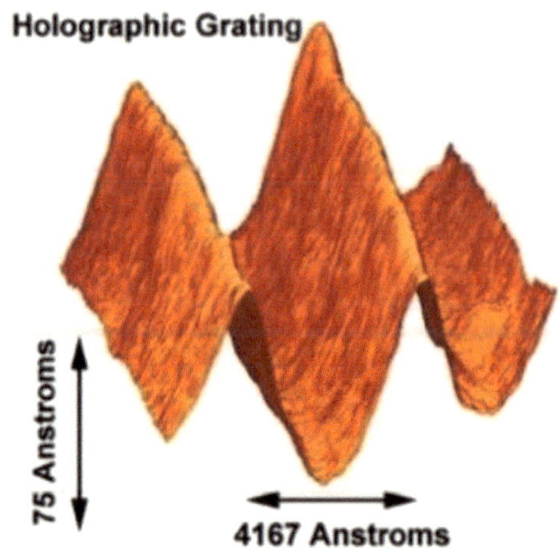

Figure 9.
AFM image of 2400/mm holographic (master) grating (after [30]).

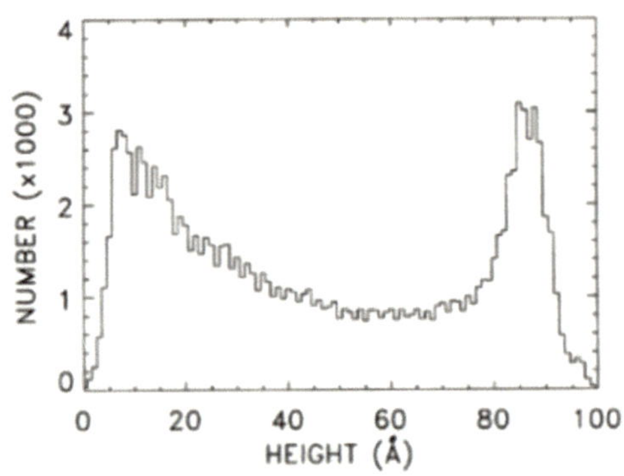

Figure 10.
Histogram of pixel heights from AFM image of 2400/mm holographic grating (after [30]).

the data points in a sliding window. The window was 25 pixels (400 Å) long in the direction perpendicular to the grooves and one pixel wide parallel to the grooves. The blaze angle is the arctangent of the fitted slope. The histogram of the blaze angles, for all rows of data in one period of the grating, is shown in **Figure 11**. The peak at 2.5 deg. represents the classical blaze angle, and the peak at 5.5 deg.

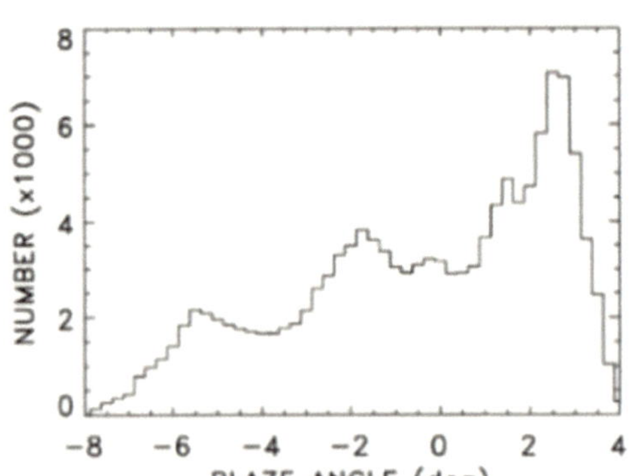

Figure 11.
Histogram of blaze angles from AFM image of 2400/mm holographic grating (after [30]).

represents the steep facet of the ideal sawtooth profile as modified by the ion-beam etching process. For a density of 2400/mm and for facet angles of 2.5 deg. and 5.5 deg., an ideal grating would have a groove height of 125 Å. However, the measured value of 75 Å (**Figure 10**) indicates a significant degree of rounding at the peaks and troughs of the groove profile. In addition, the measured ratio of the heights of the 2.5 deg. and 5.5 deg. features in the angle histogram (**Figure 11**) is approximately 3, greater than the ratio of approximately 2 that is expected based on the average facet angles.

The interpretation of the widths of the features in **Figure 11** is difficult because they are complicated functions of the surface roughness, the width of the sliding window, and the probe geometry. This is addressed in the publications [30, 33, 34]. The feature at −2 deg. in **Figure 11** results from the fits to the peaks and troughs of the groove profile, where the local slope is changing rapidly but has an average value near zero. Simulations show that the −2-deg offset of this feature from zero is a consequence of the unequal average blaze angles of the two facets. To provide a groove profile for the efficiency calculation, a representative AFM scan perpendicular to the grooves was chosen at random and scaled to the average groove height. The resulting groove profile is shown in **Figure 12**. This groove profile has 210 points.

An AFM image of two grooves of the replica grating is shown in **Figure 13**. The scan was performed across the grooves over a range of 1 μm (20-Å pixels). The vertical scale in **Figure 13** has been expanded to reveal the texture the texture of the grating surface. The PSD function derived from a 2 μm-size image spanning nearly 5 grooves is shown in **Figure 14**. The peak in the 2 to 3 μm^{-1} frequency range results from the 0.4167 μm groove period. The RMS roughness is 7 Å in the 4–40 μm^{-1}

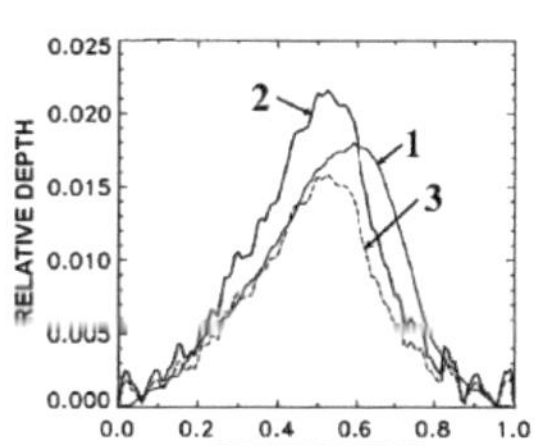

Figure 12.
Average groove profile from AFM image of 2400/mm grating: (1) 7.5-nm- deep master; (2) 9.0-nm-deep replica; (3) 6.6-nm-deep scaled replica (after [35]).

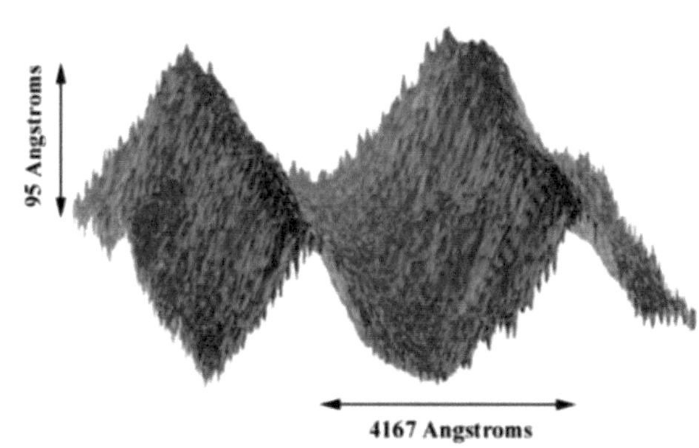

Figure 13.
AFM image of 2400/mm replica grating (after [31]).

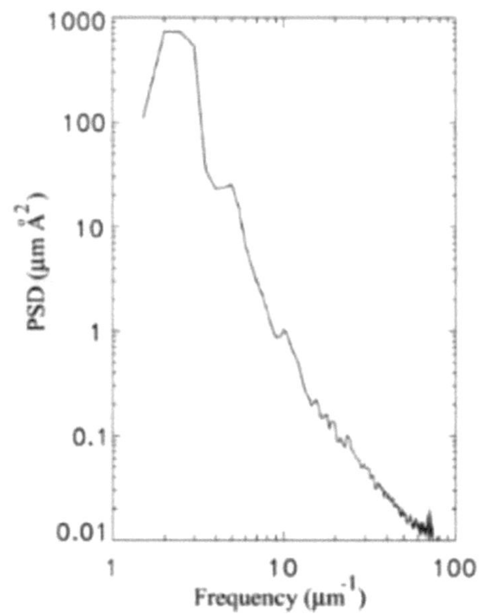

Figure 14.
PSD function of 2400/mm replica grating from AFM image (after [31]).

frequency range. By comparison, the RMS roughness of the master grating measured by the same type of AFM instrument was 3.2 Å, and this implies that the replica grating is significantly rougher than the master grating. This may result from the replication process, which for a concave grating is at least a two-step process. Furthermore, the master grating was fabricated on a fused silica surface by a holographic technique and was ion-beam polished, while the Al surface of the replica grating may contribute to its larger nanoroughness. The replica grating without the SiO_2 coating was not characterized by AFM. Typical average groove profile derived from the AFM image (1 μm in size) of the replica grating is shown in **Figure 15**. The groove profile is approximately triangular in shape with rounded corners and troughs and with facet angles of 3.4 deg. and 6.2 deg. The average groove depths derived from the AFM images are in the range 85 to 95 Å. These values of the facet angles and the groove depth are larger than the corresponding values for the master grating, 2.5 deg. and 5.5 deg. facet angles and 75 Å average groove depth (**Figure 12**). Thus, the grooves of the replica grating are deeper and the facet angles are steeper compared to those of the master grating.

Multilayer gratings were produced by application of Mo_4Ru_6/Be multilayer coatings [35, 36] to two replicas of the described holographic master grating.

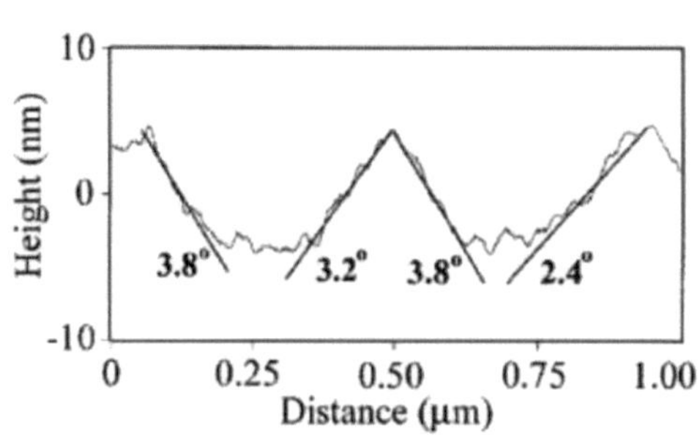

Figure 15.
Average groove profile from AFM image of 2400/mm Mo_4Ru_6/Be grating (after [36]).

Beryllium-based multilayer coatings can provide substantial reflectance at wavelengths near 11 nm. Such a Mo_4Ru_6/Be multilayer coating with 50 bi-layers was applied to the grating substrate. The coating was deposited by the magnetron-sputtering technique. Here we describe one of the multilayer gratings.

The surface of the multilayer grating was also characterized using the same Topometrix Explorer scanning probe microscope. The grating topography was measured merely for the master, replica, and multilayer gratings. The scan was performed across the grooves over a range of 1 μm (2-nm pixels). Typical groove profiles derived from the AFM image (1 μm in size) of the master, replica, and multilayer gratings are shown in **Figures 9, 13, 15**, respectively. These groove profiles have from 120 to 210 points. The groove profiles are approximately triangular in shape with rounded corners and troughs and with facet angles of 2.5° & 5.5°, 3.4° & 6.2°, and 3.0° & 4.1°, respectively. The average groove depths derived from the AFM images are in the range 7 to 8 nm, 8.5 to 9.5 nm, and 8 to 9 nm, respectively. Within the AFM groove-to-groove variation of the facet angles, the border shapes did not significantly change after multilayer coating. As determined above the average surface of the multilayer grating was characterized using a scaled replica AFM profile (**Figure 12**).

4.3 Space holographic gratings for VUV: NUV

The aforementioned AFM method was applied to simulate the efficiency of a 5870/mm G185M grating intended for operation at vacuum-ultraviolet (VUV) wavelengths below 200 nm [37]. This grating has the highest groove density and the shortest operational wavelength range of all Cosmic Origins Spectrograph (COS gratings planned for the last servicing mission to the HST) [38]. The G185M master grating was recorded holographically on 40 mm by 15 mm rectangular fused silica blank and the Pt coated at HORIBA Jobin Yvon Inc. [3]. An adhesive Cr coating, a working Al coating, and a protective (from oxidation) MgF_2 coating were deposited on Au-coated replica gratings at NASA/GSFC.

Resonance efficiency anomalies associated with waveguide funneling modes inside the MgF_2 dielectric layer degrading the G185M COS NUV grating performance were measured and qualitatively described at NASA/GSFC [39]. We used PCGrate-SX v. 6.1 [40] to model the efficiency of the G185M subwavelength grating with real boundary profiles (measured by AFM) and refractive indices (RIs) taken from different sources, including best fits of the calculated efficiency data to experimental ones [37].

The border profiles were characterized using AFM measurements. The profile of the G185M grating (replica C) intended for operation in the 170–200-nm range was AFM-measured before and after deposition of the $Cr/Al/MgF_2$ coating (**Figure 16**). As seen from the figure, after the deposition the profile depth decreased by about a

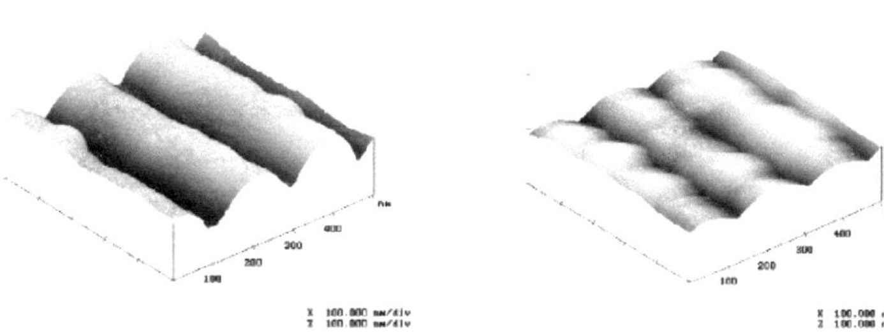

Figure 16.
G185M AFM-measured surfaces before (left) and after (right) coating Al plus MgF_2 (after [37]).

factor of 2.05 (46.4 nm against 22.6 nm), and the profile shape changed noticeably too, thus evidencing the case of nonconformal layering of the grating. For the reason that all G185M gratings were manufactured from the same master and by the same technology, one may suggest that all of them share before- and after-coating profiles. The average before-coating groove profile had 165 points and the average after-coating profile had 163 points.

To determine which of the two AFM-measured boundary profiles, MgF2 (border profile 1 measured after Cr/Al/MgF_2 coating) or (Cr)-Au (border profile 2 measured before Cr/Al/MgF_2 coating), is closer to the MgF_2-Al boundary, we started with modeling the non-polarized (NP) efficiency of a two-boundary grating. We assume a conformal MgF_2 layer (the lower MgF_2-Al boundary is identical in shape to the MgF_2 one) with the 40.1 nm thickness. The calculated efficiencies (**Figure 17**, pink curve) differ from the measured values in time throughout the whole wavelength range, thus implying invalidity of a model with a conformal layer. All calculated efficiency data presented in **Figure 17** were obtained with the RIs of Al and MgF_2 taken from the handbook of Palik [41]. Although hereinafter the experimental efficiency data of two grating replicas (A and B) are displayed, we will focus primarily on discussing the grating A data (solid dark blue squares in **Figure 17**), because replica A is the grating on which more measurements were performed.

The next step is to use two models with nonconformal layers, one with the lower boundary being the same as border 2 (**Figure 17**, yellow curve) and the other with the boundary scaled from border 2 at all points by a factor of 0.488 to the profile depth of border 1 (**Figure 17**, bright green curve). In both cases, a vertical displacement of one boundary with respect to the other (shift of the boundary reference levels) was 40.1 nm, as in the conformal model. As evident from **Figure 17**, the nonconformal model with unscaled lower boundary yields a noticeably superior qualitative agreement with experimental data. This suggests that the MgF_2-Al boundary more closely resembles border profile 2 than border profile 1. The model takes into account the fact that the thickness difference of 23.8 nm between the lower and upper boundaries should be added to the conformal vertical displacement (40.1 nm) to obtain an adequate vertical displacement for the nonconformal MgF_2

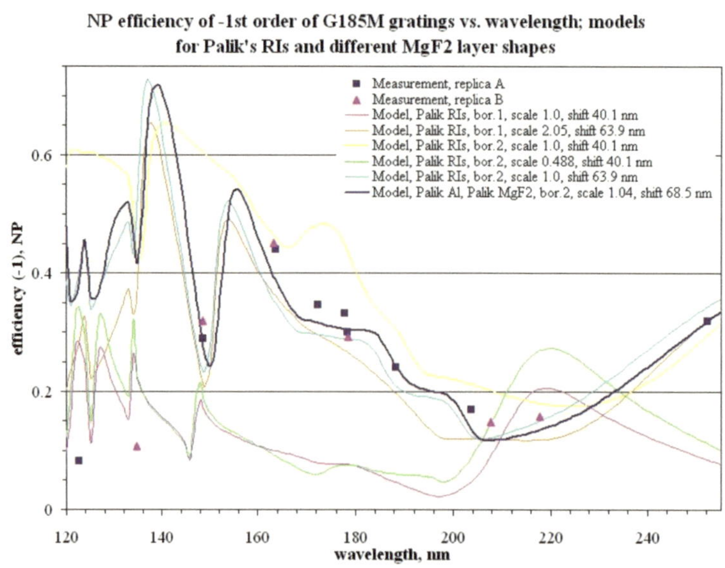

Figure 17.
G185M –1st-order NP efficiency measured and calculated for different layer shapes.

layer. In this way the period-averaged thickness of the nonconformal MgF_2 layer is kept approximately equal to 40.1 nm within the boundary shape distortion.

To determine the effect of profile shape, we set up models with equal depths and vertical shifts. The first one has border 1 scaled to the depth of border 2 (making it grater by a factor of 2.05) and a vertical displacement between the zero boundary levels equal to 63.9 nm. As seen from **Figure 17**, the efficiency of this model (orange curve) is close to that of another model with unscaled border 2 and a vertical shift of 63.9 nm (sky blue curve), while it is inferior by 40% or more as far as matching the experimental efficiencies. The latter suggests that, to set up an exact model, one has not only to determine the depth of the MgF_2-Al boundary but also to take into account the shape of its profile – see **Figure 18**.

Having determined the type of the MgF_2-Al boundary profile, we have to refine it by scaling the shape in depth and then comparing the efficiencies obtained for each model with experimental data. Another fitting parameter is the vertical displacement of the boundaries. By automatic modeling of the efficiency over a small-meshed grid of these two parameters and wavelength, one can determine the average thickness of the MgF_2 layer from the best fit between the calculated and the experimental efficiencies. Even slight changes (with a few nanometers) in profile depth and vertical displacement give a noticeable rise to the efficiency at fixed wavelengths, particularly in resonance regions. **Figure 17** presents an efficiency curve (heavy dark blue) for the model with a lower-boundary scaling factor of 1.04 and a vertical displacement of 68.5 nm. The model with these parameters of the layer geometry provides the better least-squares fit (not worse than 20%) of calculated efficiency to experimental data, both in the medium and in the long-wavelength ranges. As to the short-wavelength part, no variations in the lower boundary profile chosen within our approach yield theoretical values of the efficiency close enough to the measured ones.

What only remains is to check whether the average-thickness parameters of the MgF_2 nonconformal layer used in the final model provide a better fit between the calculated and experimental values of efficiency throughout the wavelength range with a new MgF_2 RI library (Keski-Kuha–Goray) [37]. To do this, we scale the vertical displacement and boundary parameters for the final model. Graphical results of this three-parameter optimization (scale, shift, and wavelength) are displayed in **Figure 19**. The final geometrical model of border shapes and layer thicknesses is demonstrated in **Figure 18**. The optimization procedure using different thicknesses for all the layers accounted has been applied using the least-square method. An analysis of these results shows that the parameters of the final model do

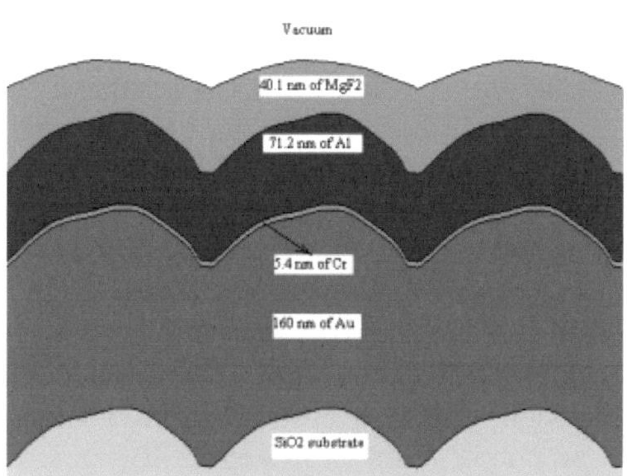

Figure 18.
Average G185M AFM border profiles before and after coating Cr/Al/MgF_2.

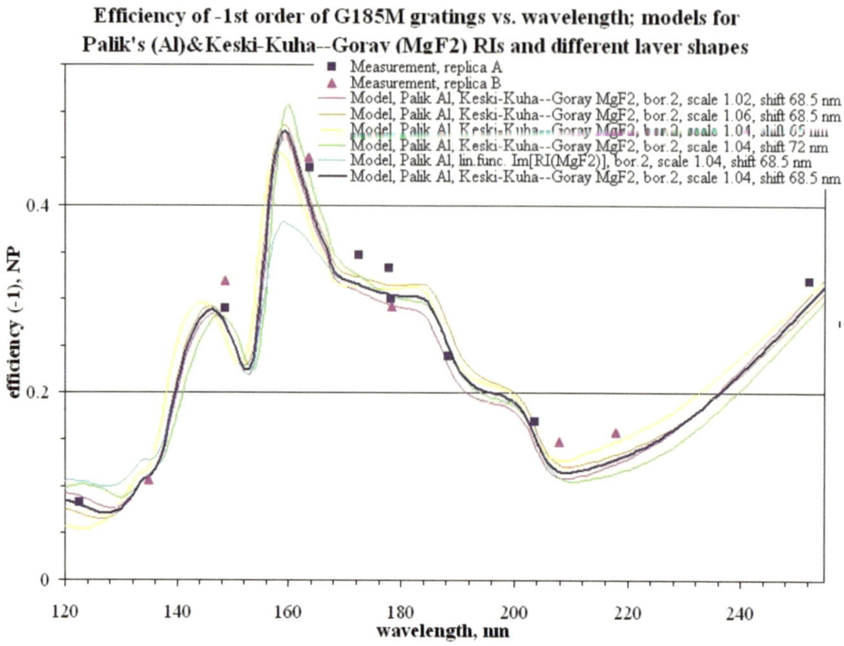

Figure 19.
Five-boundary G185M grating model. Horizontal and vertical scales are different (after [37]).

indeed provide the best agreement between the measured and calculated values of efficiency throughout the wavelength range. The relative deviation of experiment from theory for all wavelengths at which grating A was studied does not exceed 10% throughout the wavelength range. **Figure 21** presents also an efficiency curve (sky blue curve) calculated by use of the approximate values of the MgF_2 absorption index; all other parameters of the final model remain intact. A comparison of the curve efficiencies based on scaled (sky blue curve) and exactly calculated (heavy dark blue curve) values of absorption shows that the efficiency changes at the wavelengths where the RI imaginary values scale only slightly are indeed appreciable.

4.4 Imprinted off-plane blaze grating for soft X-rays

Grazing-incidence off-plane gratings have been suggested for the International X-ray Observatory (IXO) [42]. Compared with gratings in the classical in-plane mount, X-ray gratings in the off-plane mount have the potential for superior resolution and efficiency for the IXO mission [43]. The results of efficiency calculations for such a 5000/mm gold-blazed soft-X-ray grating in a conical (off-plane) mount using the average groove profile derived from AFM measurements was presented in [44].

An AFM study of the grooved area confirmed the larger than expected blaze angle. The AFM scans across the grooves near the center of the grating are shown in **Figure 20(a)**, where each scan is displaced vertically by 1 nm for ease of viewing. The standard deviation of the data points from the average scan curve is 0.89 nm and is a measure of the roughness of the groove profile. The histogram of the angles between each pair of scan points is shown in **Figure 20(b)**, where a Gaussian curve is fitted to the angle distribution. The top corners of the groove profiles are rounded, and this results in a rather broad distribution of angles with a centroid value of 13°.

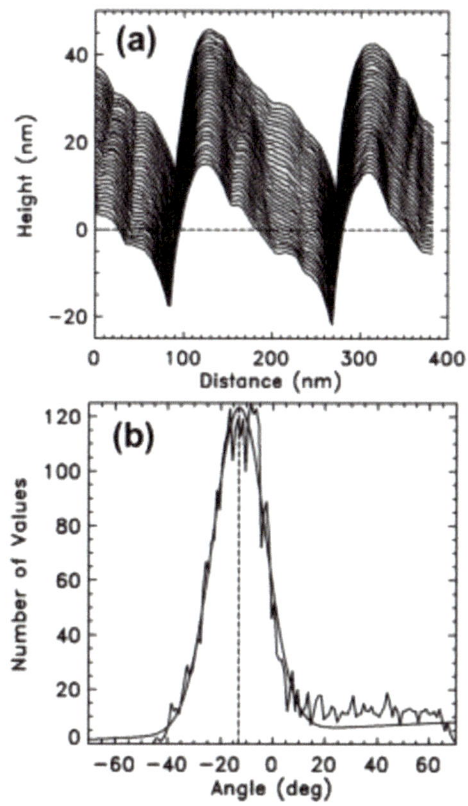

Figure 20.
(a) AFM scans across the grooves near the center of the grating; (b) histogram of the angles of pairs of points on the AFM scans giving a measure of the average blaze angle (after [43]).

The average values of the blaze angles measured at seven points distributed on the grooved area ranged from 8.9° to 15°, and the RMS roughness values ranged from 0.66 to 0.92 nm. Thus, there was considerable variation of the grooves over the 5 cm patterned area. AFM data that were taken before the titanium and gold coating of the imprinted grating showed RMS roughness of approximately 0.2 nm and blaze angles of around 8°, which indicate that deposition of the metal films onto the polymer-based imprint resist led to the observed changes in groove profile [45]. High diffraction efficiencies of the Au-imprinted 5000/mm grating using the average groove profile with 123 nodes of the polygonal groove profile derived from the AFM measurements (**Figure 21**) are demonstrated in Refs [44, 46].

4.5 Si-etched blaze gratings for X-rays: EUV

For medium- and high-frequency diffraction gratings, classical (in-plane) diffraction gives acceptable values of the efficiency of working orders only in the soft X-ray and EUV ranges [47]. However, grazing conical (off-plane) diffraction schemes have great advantages in efficiency when such gratings operate in short-wavelength regions of the X-ray spectrum (hard X-rays and tender X-rays), including in high orders and to obtain high dispersion and resolution. With such a mount, record efficiency, close to that of a respective mirror, can be obtained for sawtooth gratings with blaze angles of several degrees, which are much easier to manufacture. For a theoretical analysis of the diffraction efficiency of such gratings, the use of rigorous electromagnetic theories is required [48, 49].

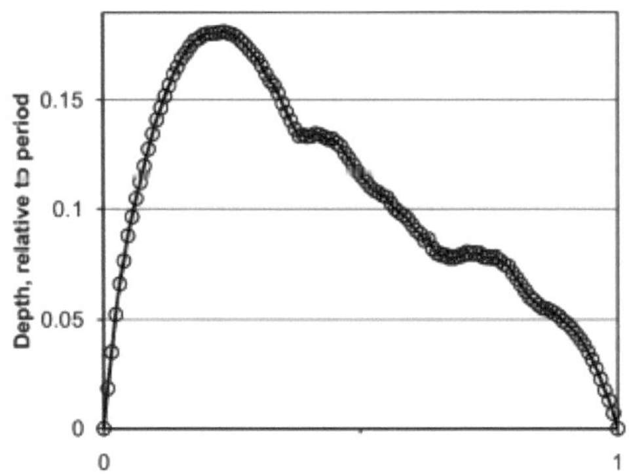

Figure 21.
Normalized average groove profile of an Au-imprinted 5000/mm grating measured by AFM (after [44]).

The manufacturing process of a reflective Si-etched grating of a triangular groove profile (sawtooth or blaze) can be conventionally divided into four main steps: (1) obtaining a pattern of a protective mask for etching grooves (DWL or EBL, in our case); (2) anisotropic etching of grooves in a solution of potassium hydroxide (KOH); (3) etching to smooth the grating profile and polish the surface of the reflective (working) facets; (4) coating to increase reflectivity. In turn, each step consists of several operations that should be controlled using AFM and, if possible, SEM. Some AFM results (NTEGRA Aura microscope) obtained during the grating manufacturing process are considered further in detail.

To transfer the grating pattern directly to a silicon wafer (stage 2), it is etched in KOH with various concentrations at a temperature from room temperature to 50°C with vigorous stirring of the solution [20, 50, 51]. KOH etches the {111} planes more slowly than the rest of silicon, which leads to angular facets with a facet tilt determined by the orientation of the {111} planes relative to the surface plane (i.e., vicinal Si(111) plates). Therefore, KOH etches the pattern of the grooves in the Si while simultaneously setting the blaze angle of grating facets. The author uses here the results of our original Si-etched grating production technique, however, with references to the similar methods for mastering such gratings.

In our AFM studies, the following was performed: measurement of the surface roughness of the working facet on an area of 1×1 and $10 \times 10\ \mu m^2$ and measurement of the grating profile, etching depth and blaze angle of the working facet when scanning $10 \times 10\ \mu m^2$. The measurements were made in the tapping mode using scans of 512×512 pixels. We used TipsNano [4] silicon cantilevers with a typical radius of tips ~6 nm. Examples of the AFM topography of Si-etched grating samples with a smoothed profile on the area of 1×1 and $10 \times 10\ \mu m^2$ are shown in **Figure 22a, b** for sample No. 5.

Figure 23a shows the topography profile of specimen No. 5 along line 1 (black curve) and the blaze angle of the working facet (blue curve). The angle is calculated as the arctan of the coordinate derivative and converted to degrees. **Figure 23b** shows the profile of the slope of the non-working facet along line 1 for sample No. 3/1.

The results of AFM studies of the geometrical groove parameters of the samples of Si-etched gratings with a period of 2 μm are presented in **Table 5**. In the results presented in **Table 5** the deconvolution algorithms have been used, although we evaluated mid-frequency gratings. The histogram (normalized density of probability) of blaze, anti-blaze, and apex angles of grooves of the grating with 500/mm and 4° blaze angle is demonstrated in **Figure 24** (left). The three peaks on this curve are clearly associated with the corresponding working and non-working facet angles, as well as with the angle of the smoothed top of the groove profile. The average groove

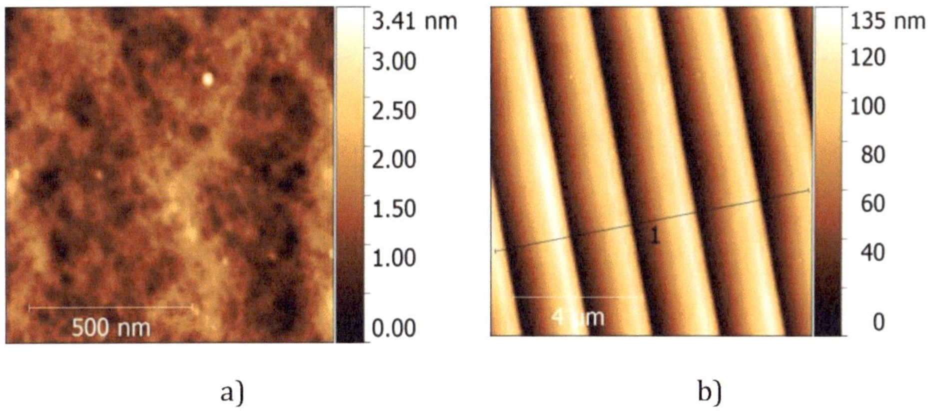

Figure 22.
Surface topography obtained by AFM scanning of area of sample No. 5: a) 1 × 1 µm²; b) 10 × 10 µm².

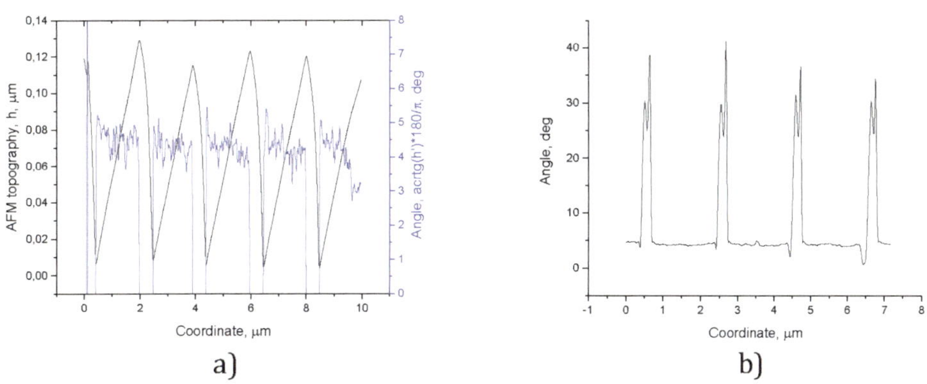

Figure 23.
Profile topography obtained by AFM scanning of area of 10 × 10 µm: a) surface and blaze angle of the working facet, sample No. 5; b) anti-blaze angle of the non-working facet, sample No. 3/1.

Sample No.	Groove depth/Si-nub height, nm	Working/non-working facet width, nm	Working facet RMS roughness, nm	Blaze angle, deg.	Anti-blaze angle, deg.
1	95/38	1512/340	0.462	—	—
2	97/44	1544/340	0.345	—	—
3/1	121	1710	0.278	4.05	20
3/2	111	1594	0.340	—	—
5	114	1580	0.337	4.13	20

Table 5.
Groove geometrical parameters of Si-etched grating samples according to AFM.

profile topography and the respective angles one can see in **Figure 24** (right). The peak corresponding to the blaze angle is pronounced and indicates a high quality of the developed sawtooth grating. The average groove profile derived from AFM data for one grating was used then for rigorous calculus of 3D diffraction efficiencies of orders vs. incidence angle and wavelength in the soft-X-ray–EUV range and classical mount (**Figure 25**). The other AFM groove profile data for similar Si-etched

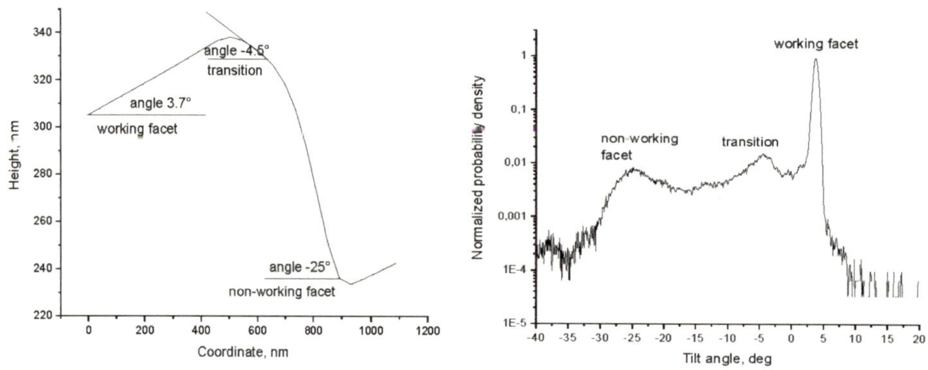

Figure 24.
AFM groove parameters of 500/mm and 4° blaze grating: (left) histogram of groove angles including smoothed groove top ('transition'); (right) average groove topography and respective angles.

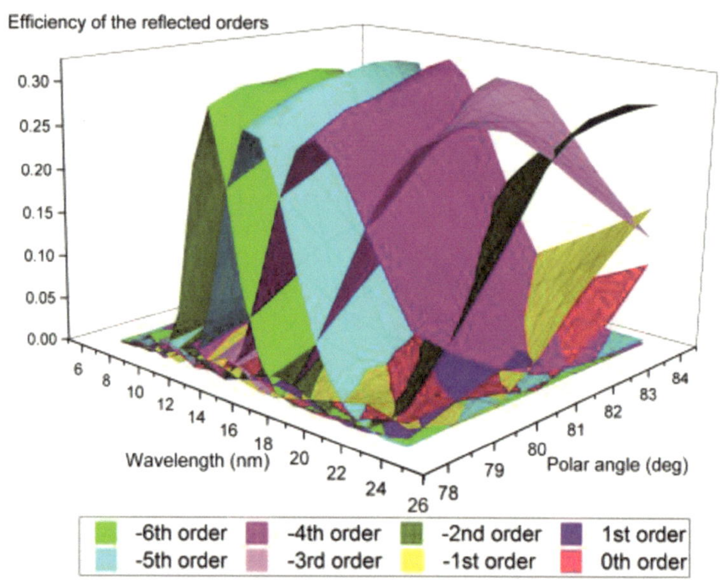

Figure 25.
3D diffraction efficiency in principal orders of 500/mm Au-coated Si-etched grating rigorously calculated using the realistic groove profile vs. incidence angle and wavelength.

gratings produced by DWL, or EBL, or holographic recording can be found in [52, 53] and references there in.

In order to reduce the roughness of the grating surface, the authors of [54] use a nine-cycle RCA-1/HF etching procedure to remove any irregularities and roughness, i.e., perform both smoothing and polishing etching; and they report submicron roughness. To reduce the roughness of the working facet at the polishing stage, several etchants have been tested, including tetramethylammonium hydroxide (TMAH) and the isotropic silicon etchant HF: HNO_3: H2O. **Table 6** shows the AFM results of processing in different etchants of the surface of samples, punctured from the same grating immediately after anisotropic etching in KOH. As one can see from **Table 6**, the RMS roughness of working facets can be reduced to <0.3 nm for a few etching processes. The initial RMS roughness (before a polishing process) was ~1.2 nm (compare with results in **Table 5**).

Sample No.	Working facet RMS roughness, nm	Polishing etchant/etching time, s
1	0.269	Isotropic/30 s
2	0.244	Isotropic/20 s
3	0.271	Isotropic, using HF before/60 s
4	0.315	Isotropic/60 s
5	0.246	TMAH/2 min
6	0.291	TMAH/4 min
7	0.336	TMAH/6 min
8	0.312	TMAH/8 min
76KDB Si(111)4°-substrate, ∅76.2 mm	0.149	No process

Table 6.
Groove roughness of Si-grating samples according to AFM after polishing.

5. Conclusion

In the chapter, some earlier and recent results of the use of AFM & STM methods for groove metrology of various surface relief (ruled, holographic, lithographic, imprinted) diffraction gratings, mostly intended for short wavelengths, were described and discussed. Examples of a few comparisons with the other widely-used direct metrology techniques, such as SEM, stylus profilometry and microinterferometry, were also demonstrated and compared. In addition, the most critical problems connected with AFM methods for groove metrology of bulk, thin-film-coated and multilayer-coated gratings were discussed, such as: the tip deconvolution and its radius; groove shape and abrupt groove slopes; RMS nano-roughness and PSD functions

The detailed AFM groove metrology results were presented by the author for several important grating samples: the Space Telescope Imaging Spectrograph grating flown aboard the HST and working in the Visible–NIR; the similar master, replica and multilayer soft-X-ray–EUV blaze gratings; the Cosmic Origins Spectrograph grating used in the last servicing mission to the HST and working in the VUV–NUV; imprinted off-plane blaze grating planned for the International X-ray Observatory and working in the soft X-rays; and recently developed Si-etched blaze diffraction gratings indented to work in the X-rays–EUV at high efficiency and a very low level of scattering light. These gratings were chosen because high quality efficiency data exists, in particular, for space gratings or/and X-ray gratings characterized by synchrotron radiation sources; and their groove profiles, together with random nanoroughness, were measured by AFM to be included in rigorous efficiency and scattered light intensity calculus.

The rigorous calculation accounts for the real profile of the grooves and their thickness as well as suitable refractive indices. It was not possible earlier to achieve such good agreements between measured and calculated efficiencies of high- and mid-frequency gratings working in the short spectral ranges due to the lack of realistic, i.e., measured using the AFM technique, groove profile shapes, as it has been demonstrated in the present study. Today, using an appropriate AFM instrument and the respective method one has a possibility to determine with a superfine (atomic-scale) spatial resolution grating-like structures, i.e., their groove profiles including abrupt slopes and random nanoroughness. Moreover, such non-destructive AFM analysis is the only suitable one to apply to current production and evaluation of such complicated and expensive devices like as most of X-ray diffraction gratings are.

Acknowledgements

I thank David A. Content, John F. Seely, Tamara N. Berezovskaya, Vladislav A. Sharov for the information provided.

This work was partially supported by the Russian Foundation for Basic Research (RFBR) (Grant No. 20-02-00326) and the Russian Science Foundation (RSF) (Grant No. 19-12-00270) in the theoretical part.

Author details

Leonid I. Goray[1,2]

1 Alferov University, Saint Petersburg, Russia

2 Institute for Analytical Instrumentation, Saint Petersburg, Russia

*Address all correspondence to: lig@pcgrate.com

References

[1] Yusupov IY, Mikhailov MD, Herke RR, Goray LI, Mamedov SB, Yakovuk OA. Investigation of the arsenic sulphide films for relief-phase holograms. In: International Seminar on 3D Holography (Three-Dimensional Holography: Science, Culture, Education); 1989, Kiev, Ukraine; Proc SPIE 1238; 1991. p. 240–247. DOI: 10.1117/12.19398

[2] Palmer C. Diffraction Grating Handbook. 8th ed. MKS Instruments, Inc.; 2020. 250 p.

[3] Horiba Corp. page. [Internet]. 2021. Available from: https://www.horiba.com/de/scientific/products/diffraction-gratings/for-scientific-applications/ [Accessed: 2021-02-20]

[4] TipsNano Co page. [Internet]. 2021. Available from: https://tipsnano.com/catalog/calibration/ [Accessed: 2021-02-20]

[5] Helmholtz-Zentrum Berlin page. [Internet]. 2021. Available from: https://www.helmholtz-berlin.de/forschung/oe/wi/praezisionsgitter/index_en.html [Accessed: 2021-02-20]

[6] Bennett J, Mattson L. Introduction to Surface Roughness and Scattering. 2nd ed. Washington, D.C.: OSA; 1999. 130 p.

[7] Stout KJ, Blunt L. Three-Dimensional Surface Topography. 2nd ed. Penton Press; 2000. 320 p.

[8] Scanning Electron Microscopy. Elsevier BV ScienceDirect page. [Internet]. 2021. Available from: https://www.sciencedirect.com/topics/engineering/scanning-electron-microscopy [Accessed: 2021-02-20]

[9] Binnig G, Quate CF, Gerber Ch. Atomic Force Microscope. Physical Review Letters. 1986;56(9):930–933 DOI: 10.1103/PhysRevLett.56.930

[10] Content DA, Arsenovic P, Kuznetsov IG, Hadjimichael T. In: International Symposium on Optical Science and Technology (Optical Spectroscopic Techniques, Remote Sensing, and Instrumentation for Atmospheric and Space Research IV); 2001, San Diego, CA, United States; Proc SPIE 4485; 2002. p. 405–416. DOI: 10.1117/12.454276

[11] Goray L, Jark W, Eichert D. Rigorous calculations and synchrotron radiation measurements of diffraction efficiencies for tender X-ray lamellar gratings: conical versus classical diffraction. J Synchrotron Rad. 2018;25:1683–1693. DOI: 10.1107/S1600577518012419

[12] MicroXAM. Phase Shift Technology, Inc. instrument specification. [Internet]. 2002. Available from: https://www.ccmr.cornell.edu/wp-content/uploads/sites/2/2015/11/MicroXam-specs.pdf [Accessed: 2021-02-20]

[13] KLA-Tencor P-10 Surface Profiler. ClassOne Equipment, Inc. page [Internet]. 2021. Available from: https://classoneequipment.com/product/kla-tencor-p-10-surface-profiler-2/ [Accessed: 2021-02-20]

[14] XP-S Stulus Profiler. Ambios Technology, Inc. instrument specification. [Internet]. 2021. Available from: http://cloudfront.eas.caltech.edu/aphms/research/facilities/micro_nano/documents/micro-nano-profilometer-user-manual.pdf [Accessed: 2021-02-20]

[15] NanoScope III SPM System. Digital Instruments, Inc. instrument specification. [Internet]. 2021. Available from: http://www.photonicmicrodevices.com/files/Digital_Instruments_Nanoscope_General_Description_and_Specs.pdf [Accessed: 2021-02-20]

[16] NTEGRA-Aura SPM. NT-MDT BV page. [Internet]. 2021. Available from:

https://ntmdt.nl/home/products/ntegra-aura/ [Accessed: 2021-02-20]

[17] Carl-Zeiss Supra 25 SEM. EVISA page. [Internet]. 2010. Available from: http://www.speciation.net/Database/Instruments/Carl-Zeiss-AG/Supra-25-;i665 [Accessed: 2021-02-20]

[18] VLSI standards. VLSI page. [Internet]. 2021. Available from: https://www.vlsistandards.com/products/overview.asp [Accessed: 2021-02-20]

[19] Herada T, Taira E, Kita T, Itou M. Groove Profile Measurement of Diffraction Gratings Using Scanning Electron Microscope. In: 31st Annual Technical Symposium on Optical and Optoelectronic Applied Sciences and Engineering (Application and Theory of Periodic Structures, Diffraction Gratings, and Moire Phenomena III); 1987, San Diego, CA, United States; Proc SPIE 0815; 1987. DOI: 10.1117/12.941742

[20] Voronov DL, Ahn M., Anderson EH, Cambie R, Chang C-H, Goray LI, Gullikson EM, Heilmann RK, Salmassi F, Schattenburg ML, Warwick T, Yashchuk VV, Padmore HA. High efficiency multilayer blazed gratings for EUV and soft X-rays: Recent developments. In: SPIE Optical Engineering + Applications (Advances in X-Ray/EUV Optics and Components V); 2010, San Diego, CA, United States; Proc SPIE 7802; 2010. p. 780207. DOI: 10.1117/12.861287

[21] Canet-Ferrer J, Coronado E, Forment-Aliaga A, Pinilla E. Correction of the tip convolution effects in the imaging of nanostructures studied through scanning force microscopy. Nanotechnology. 2014;25(39):395703. DOI: 10.1088/0957-4484/25/39/395703

[22] Vorselen D, Kooreman E, Wuite G, Roos WH. Controlled tip wear on high roughness surfaces yields gradual broadening and rounding of cantilever tips. Sci Rep. 2016;6:36972 DOI: 10.1038/srep36972

[23] Andrle A, Hönicke P, Gwalt G, Schneider P-I, Kayser Y, Siewert Frank, Soltwisch V. Shape- and element-sensitive reconstruction of periodic nanostructures with grazing incidence X-ray fluorescence analysis and machine learning. arXiv.org. 2021, 12 Feb; arXiv:2102.06600v1 [physics.app-ph]

[24] Garnaes P-E, Hansen NA, Holm J, Borsetto F, Kühle A. Profiles of a high-aspect-ratio grating determined by spectroscopic scatterometry and atomic-force microscopy. Applied Optics. 2006; 45(14):3201–3212.

[25] Kondratov AV, Rogov OY, Gainutdinov RV. AFM reconstruction of complex-shaped chiral plasmonic nanostructures. Ultramicroscopy. 2017; 181,81–85. DOI: doi.org/10.1016/j.ultramic.2017.05.013

[26] Goray L, Lubov M. Analysis of mirror soft-x-ray–EUV scattering using generalized continuous growth model of multiscale reliefs. Optics Express. 2015; 23(8):10703–10713

[27] Church EL, Takacs PZ. Surface Scattering. In: Bass M, editor. Handbook of Optics: Volume I – Geometrical and Physical Optics, Polarized Light, Components and Instruments. 3rd ed. McGraw-Hill; 2010. Ch. 8

[28] Content DA, Boucarut RA, Bowers CW, Madison TJ, Wright GA, Lindler DJ, Huang LK, Puc BP, Standley C, Norton TA. Development and testing of diffraction gratings for the Space Telescope Imaging Spectrograph. In: SPIE's 1996 International Symposium on Optical Science, Engineering, and Instrumentation (Space Telescopes and Instruments IV); 1996, Denver, CO, United States; Proc SPIE 28074; 1996. p. 267. DOI: 10.1117/12.255105

[29] International Intellectual Group, Inc page. [Internet]. 2021. Available from:

https://www.pcgrate.com/etestlab/certific/index_html#9 [Accessed: 2021-02-20]

[30] Kowalski MP, Seely JF, Goray LI, Hunter WR, Rife JC. Applied Optics. 1997;36(34):8939–8943. DOI: 10.1364/AO.36.008939

[31] Seely JF, Goray LI, Hunter WR, Rife JC. Applied Optics. 1999;38(7): 1251–1258. DOI: 10.1364/AO.38.001251

[32] Topometrics Explorer Scanning Probe Microscope. MIT page. [Internet]. 1997. Available from: http://electron.mit.edu/~gsteele/mirrors/elchem.kaist.ac.kr/jhkwak/TopometrixWeb/Explore r.htm [Accessed: 2021-02-20]

[33] Seely JF, Kowalski MP, Cruddace RG, Heidemann KF, Heinzmann U, Kleineberg U, Osterried K, Menke D, Rife JC, Hunter WR. Multilayer-coated laminar grating with 16% normal-incidence efficiency in the 150-Å wavelength region. Applied Optics. 1997:36(31) 8206–8213. DOI: 10.1364/AO.36.008206

[34] Kowalski MP, Barbee TW, Heidermann KF, Gursky H, Rife JC, Hunter WR, Fritz GG, Cruddace RG. Efficiency calibration of the first multilayer-coated holographic ion-etched flight grating for a sounding rocket high-resolution spectrometer. Applied Optics. 1999;38(31):6487–6493 (1999). DOI: 10.1364/AO.38.006487

[35] Goray LI, Seely JF. Efficiencies of master, replica, and multilayer gratings for the soft-x-ray-extreme-ultraviolet range: modeling based on the modified integral method and comparisons with measurements. Applied Optics. 2002;41 (7):1434–1445. DOI: 10.1364/AO.41.001434

[36] Seely JF, Montcalm C, Baker S, Bajt S. High-efficiency MoRu–Be multilayer-coated gratings operating near normal incidence in the 11.1–12.0-nm wavelength range. Applied Optics. 2001;40(31):5565–5574. DOI: 10.1364/AO.40.005565

[37] Goray LI, Keznetsov IG, Sadov SYu, Content DA. Multilayer resonant subwavelength gratings: effects of waveguide modes and real groove profiles. JOSA A. 2006;23(1):155 –165. DOI: 10.1364/JOSAA.23.000155

[38] Green JC. Cosmic origins spectrograph. In: Astronomical Telescopes and Instrumentation (UV, Optical, and IR Space Telescopes and Instruments); 2000, Munich, Germany; Proc SPIE 4013; 2000. p. 352. DOI: 10.1117/12.394017

[39] Kuznetsov IG, Wilkinson E, Content DA, Boucarut RA, Madison TJ, In: Optical Science and Technology, SPIE's 48th Annual Meeting (Optical Modeling and Performance Predictions); 2003, San Diego, California, United States; Proc SPIE 5178; 2004. p. 267. DOI: 10.1117/12.507946

[40] International Intellectual Group, Inc page. [Internet]. 2021. Available from: https://www.pcgrate.com/load purc/download/ [Accessed: 2021-02-20]

[41] Palik E, editor. Handbook of optical constant of solids, II, III. New York: Academic Press; 1985, 1991, 1998.

[42] NASA IXO page [Internet]. 2021. Available from: https://asd.gsfc.nasa.gov/archive/ixo/technology/xgs.html [Accessed: 2021-02-20]

[43] Seely JF, Goray LI, Kjornrattanawanich B, Laming JM, Holland GE, Flanagan KA, Heilmann RK, Chang C-H, Schattenburg ML, Rasmussen AP. Efficiency of a grazing incidence off-plane grating in the soft x-ray region. Applied Optics. 2006;45(8):1680–1687. DOI: 10.1364/AO.45.001680

[44] Goray LI, Schmidt G. Solving conical diffraction grating problems with integral equations. J. Opt. Soc. Am. A. 2010;27(3): 585–597. DOI: 10.1364/AO.45.001680

[45] Chang C-H, Montoya JC, Akilian M, Lapsa A, Heilmann RK, Schattenburg ML, Li M, Flanagan KA, Rasmussen AP, Seely JF, Laming JM, Kjornrattanawanich B, Goray LI. High fidelity blazed grating replication using nanoimprint lithography. J Vac Sci Technol B. 2004;22:3260–3264. DOI: 10.1116/1.1809614

[46] Goray LI, "Off-plane grazing-incidence fan-groove blazed grating to serve as a high-efficiency spectral purity filter for EUV lithography. In: SPIE Optics + Photonics, Components, and Applications (Advances in X-Ray/EUV Optics); 2006, San Diego, CA, United States; Proc SPIE 6317; 2006. p. 63170O. DOI: 10.1117/12.678151

[47] Goray L, Jark W, Eichert D. Rigorous calculations and synchrotron radiation measurements of diffraction efficiencies for tender X-ray lamellar gratings: conical versus classical diffraction. J Synchrotron Rad. 2018;25: 1683–1693. DOI: 10.1107/S1600577518012419

[48] Popov E, editor. Gratings: Theory and Numeric Applications. 2nd rev. ed. AMU Press; 2014. Ch. 12

[49] Voronov DL, Goray LI, Warwick T, Yashchuk VV, Padmore HA. High-order multilayer coated blazed gratings for high resolution soft x-ray spectroscopy. Optics Express. 2015;23(4):4771–4790. DOI: 10.1364/OE.23.004771

[50] Voronov DL, Anderson EH, Cambie R, Salmassi F, Gullikson EM, Yashchuk VV, Padmore HA, Ahn M, Chang C, Heilmann RK, Schattenburg ML. 5000 groove/mm multilayer-coated blazed grating with 33% efficiency in the 3rd order in the EUV wavelength range. In: SPIE Optical Engineering + Applications (Advances in X-Ray/EUV Optics and Components IV); 2009, San Diego, California, United States; Proc SPIE 7448; 2009. p. 74480J. DOI: 10.1117/12.826921

[51] Ahn M, Heilmann RK, Schattenburg ML. Fabrication of ultrahigh aspect ratio freestanding gratings on silicon-on-insulator wafers. J Vac Sci Technol B. 2007;25(6):2593–2597. DOI: 10.1116/1.2779048

[52] Voronov DL, E.H. Anderson, Cambie R, Cabrini S, Dhuey SD, Goray LI, Gullikson EM, Salmassi F, Warwick T, Yashchuk VV, Padmore HA. A 10,000 groove/mm multilayer coated grating for EUV spectroscopy. Optics Express. 2011:19 (7):6320–6325. DOI: 10.1364/OE.19.006320

[53] Voronov DL, Anderson EH, Cambie R, Goray LI, Gawlitza P, Gullikson EM, Salmassi F, Warwick T, Yashchuk VV, Padmore HA. Development of near atomically perfect diffraction gratings for EUV and soft x-rays with very high efficiency and resolving power. Journal of Physics: C. 2013;425:152006. DOI: 10.1088/1742-6596/425/15/152006

[54] Golub L, Cheimets P, DeLuca EE, Madsen CA, Reeves KK, Samra J, Savage S, Winebarger A, Bruccoleri AR. EUV imaging and spectroscopy for improved space weather forecasting. J Space Weather Space Clim. 2020;10:37 DOI: 10.1051/swsc/2020040

Chapter 4

Atomic Force Microscopy Reveals the Role of Vascular Smooth Muscle Cell Elasticity in Hypertension

Yi Zhu

Abstract

The vascular smooth muscle cell (VSMC) mechanical properties not only provide intrinsic cellular functions, but also influence many vascular and circulation functions in physiology. In this report, the VSMCs of thoracic aorta from 16-18 week age Wistar-Kyoto normotensive rats (WKY) and spontaneously hypertensive rats (SHR) were used as research subjects to reveal hypertension mechanism at a single cell level using atomic force microscopy (AFM). The apparent elastic modulus was significantly increased in VSMCs from SHRs compared to those from WKYs. Treatment with cytochalasin D (CD), ML7, Y27632 and lysophosphatidic acid (LPA) modulated VSMC stiffness of WKYs and SHRs. A spectral analysis approach was applied to further investigate the time- dependent change in VSMC elasticity of WKYs and SHRs. This report demonstrated the efficacy of real-time analysis of VSMC elasticity by AFM nano-indentation, and revealed real-time functional differences in biomechanical characteristics of VSMCs with drug treatments.

Keywords: atomic force microscopy, vascular smooth muscle cell, elasticity, Wistar-kyoto normotensive rat (WKY), spontaneously hypertensive rat (SHR)

1. Introduction

Vascular smooth muscle cells (VSMCs) locate blood vessel medial layer as a main component and bear mechanical stress and pressure from blood flow, and sustain vascular tone and resistance. A number of recent studies have demonstrated that changes in a cell's elastic characteristics can affect its response to the external mechanical force [1, 2]. The single-cell mechanical property and behavior of VSMC is chiefly considered to play a crucial role in the development of vascular diseases, and atomic force microscopy (AFM) is currently the most wonderful tools for determining this interaction [3–5].

The VSMC intrinsic properties not only perform a normal cellular function to sustain and support vascular geometric architecture, but also take some important actions to participate the regulation of biophysical and biochemical properties for blood vessel [6–8]. With the development and applications of AFM technology, people gradually concentrate their research on reconstituted tissues and single cell detections [9–11]. Hypertension is a common age-related vascular disease, and many factors can induce age-related vascular dysfunctions and diseases [12]. However, the detailed mechanisms that induce hypertension still need to be elucidated. Currently, people attempt to analyze and reveal the hypertension mechanism in single molecule and single cell level [13–15]. The cytoskeleton contents, the polymerization and arrangement of actin filaments were directly responsible for the VSMC elasticity [16, 17]. The investigation in single cell level can supplement studies on complicated living organisms or an intact tissue to determine the related pathways that regulate cell elasticity and adhesion [18, 19]. Furthermore, cells are in micro-scales and easy to break, and AFM provides a probability to manipulate VSMC at an individual cell level due to its nano-sensitivity under liquid environment [4, 20, 21]. The experimental medicines are administered in micro-volume by a pipette and ensured drugs to diffuse and aim the measured cells, and people can fully and perfectly employ AFM to perform a continuous real-time measurement in the absence and presence of drugs on a single cell. The single spectral analysis is an approach to reveal mathematical decomposition of the elasticity waveform and further demonstrates the underlying molecular mechanism [22–24]. The drug cytochalasin D (CD) depolymerizes and breaks apart actin filaments, and the drug ML7 dephosphorylates myosin light chain to inhibit the establishment of actin binding with myosin [8]. Additionally, the drug Y27632 inhibits the Rho-associated protein kinase (ROCK) and the drug lysophosphatidic acid (LPA) enhances integrin proteins to adhere the extracellular matrix (ECM) and activates Rho kinase to phosphorylate myosin light chain kinase (MLCK) [25, 26]. In this report we chose these drugs using AFM to measure the stiffness of thoracic aortic VSMCs *in vitro*. An investigation was taken to study and reveal the real-time record of single VSMC mechanical property and behavior. Moreover, we analyzed and interpreted the oscillatory waveforms of VSMC elasticity for various drug treatments to reveal the underlying cellular and molecular mechanisms of VSMC stiffness in hypertension by a spectral analysis approach.

2. Materials and methods

2.1 Vascular smooth muscle cell isolation, cell culture, and treatments

Male WKYs and SHRs at 16–18 weeks of age were utilized in this study. All animal procedures were done under the *Guide for the Care and Use of Laboratory Animals* (NIH 85–23, revised 2011). Primary VSMCs from thoracic aorta of three experimental WKY and SHR rats were enzymatically isolated and cultured in Dulbecco's Modified Eagle's *medium* (DMEM) with 10% fetal bovine serum (FBS), 10 mmol/L HEPES, 2 mmol/L L-glutamine, 1 mmol/L sodium pyruvate, 100 U/mL penicillin, 100 μg/mL streptomycin, and 0.25 μg/mL amphotericin B and used at passages 2 to 4 [4].

2.2 VSMC image and stiffness measured by AFM

Single VSMC image and lively measurements of cell elasticity were operated in contact mode by an AFM instrument, which is a Bioscope System (Model IVa, Veeco Mertrology Inc., Santa Barbara, CA) mounted on an Olympus IX81 microscope (Olympus Inc., NY). The employed AFM probes were silicon nitride microlevers (Model micro lever cantilever, Veeco Mertrology Inc., Santa Barbara, CA; spring constant ranging 10–30 pN/nm) and purchased from Veeco Mertrology Inc. (Santa Barbara, CA). The VSMC image scanning and stiffness measurement were operated using contact mode in colorless DMEM solution without serum at room temperature. The radius of AFM tip for the VSMC image scanning is 15 ± 5 nm, and the radius curvature of the probe is 7.5 ± 1 nm. The apparent dimensions of the samples obtained by AFM are usually much larger than the real dimensions because of the radius curvature of the probe, that is, the "magnification effect". However, the size of measured VSMC is far larger than the AFM tip radius size, thus there are not any magnification effects in the VSMC images [27]. The AFM probe was put in the mid-site between VSMC margin and the nucleus for nano-indentation to measure WKY and SHR elasticity. The AFM probe was continuously indented 2 minute to collect force curves for determining the mean stiffness of individual WKY and SHR VSMC, and the experimental VSMCs from three rats were assessed then averaged together for the stiffness of WKYs and SHRs. The force curves were interpreted using proprietary software NForceR (registration number TXu1–328-659), and the VSMC elastic modulus was translated from these force curves into Young's modulus using a modified Hertz model. The calculation of the elastic modulus was Eq. (1) as follow:

$$F = \frac{2E\delta^2}{\pi\left(1-\upsilon^2\right)}\tan(\alpha) \tag{1}$$

where the indentation force (F) was stated and described using Hooke's law ($F = \kappa\Delta x$, κ and Δx denote the AFM probe's spring constant and the probe's apparent deflection). The indentation depth (δ) is identified from the difference in the AFM piezo movement in z direction and the AFM probe deflection. E is the Young's modulus of experimental cell as the value of elasticity, and ν denotes 0.5 for cell as the Poisson ratio. The numerical α is the semi-included angle of the cone for a pyramidal tipped probe and determined by the probe shape.

2.3 Dynamic stiffness in single VSMC measurement by AFM

The experimental VSMCs were nano-indented for the duration of 30 minute to examine the temporal characteristics of the cell stiffness, and then VSMCs were treated in micro-volume by a pipette with CD (10 μmol/L; Sigma, St. Louis, MO), with ML7 (10 μmol/L; Sigma, St. Louis, MO), with Y27632 (5 μmol/L; Sigma, St. Louis, MO), with LPA (2 μmol/L; Sigma, St. Louis, MO) for another 30 minute continuous AFM investigation. The curves were continuously recorded and collected during the whole measuring procedure, and applied to determine elastic stiffness, absence and presence of drugs. A spectral analysis procedure was exploited for analysis and following translation of the oscillation waveforms for elasticity data, and linear trends were evaluated and subtracted from each series ahead of a spectral analysis. To reveal the average group behavior of the oscillations, three values of amplitude, frequency and phase for every experimental subject were further investigated and averaged: phases (φ) as a simple mean; frequencies (f) were converted to periods (1/f) ahead of averaging; amplitudes (A) were log10-transformed before averaging the mean. The mean period and mean log-amplitude were then transformed back to frequency and amplitude. A composite time series for each treatment set was constructed as Eq. (2):

$$y(t) = \bar{A}_1 \sin\left(2\pi \bar{f}_1 t + \bar{\phi}_1\right) + \bar{A}_2 \sin\left(2\pi \bar{f}_2 t + \bar{\phi}_2\right) + \bar{A}_3 \sin\left(2\pi \bar{f}_3 t + \bar{\phi}_3\right) + \bar{b}_1 t + \bar{b}_0 \tag{2}$$

where b_1 and b_0 denote respectively the slope and intercept of the linear trend, and the bar above each component indicates the average value [13, 28]. A brief explanation for the singular spectrum analysis equation and application was provided in this report, and we detail stated and described the measurement of dynamic stiffness by AFM in single VSMC and the analysis of oscillation waveform by singular spectrum analysis.

2.4 Statistical analysis

Data are expressed as mean ± SEM for the number of samples reported in this report. Statistically significant differences between WKYs and SHRs were determined by Student's t-test. A value of $P < 0.05$ was considered a significant difference.

3. Results and discussion

3.1 VSMC AFM image and topography

The SHR developed from WKY rat as an animal model for specific studies of hypertension, thus we analyzed and compared heights and topographic images (**Figure 1A**) of VSMCs isolated from thoracic aorta of WKYs (n = 4, from 3 rats) and SHRs (n = 5, from 3 rats). The VSMC surface areas of WKYs vs. SHRs were 10695 ± 339 μm^2 vs. 12380 ± 483 μm^2, and the VSMC surface area of SHRs was significantly larger than WKYs ($p < 0.05$). The AFM height images obtained from the Z sensor signal are shown in **Figure 1A**. These images describe the vertical displacement of the cantilever tip when scanning the topography of the VSMC samples. The images portray a cytoskeletal structure at the VSMC surface of the WKY and SHR rodent animals. From **Figure 1A** images, it can also be reflected that the geometric characteristics of VSMC and the configuration of the entire model vary with the size and form of hypertension. For the height measurement, we set AFM to predetermine the line across the cell and take a 30 second period reading. Waited 600 seconds (10 minutes), and started recording line scans of height again for another 30-second period. We repeated this procedure for five times to obtain the VSMC height (**Figure 1B**). The VSMC topography and shape of SHRs showed to be larger and higher than WKYs ($p < 0.05$) due to α-SMA over production and

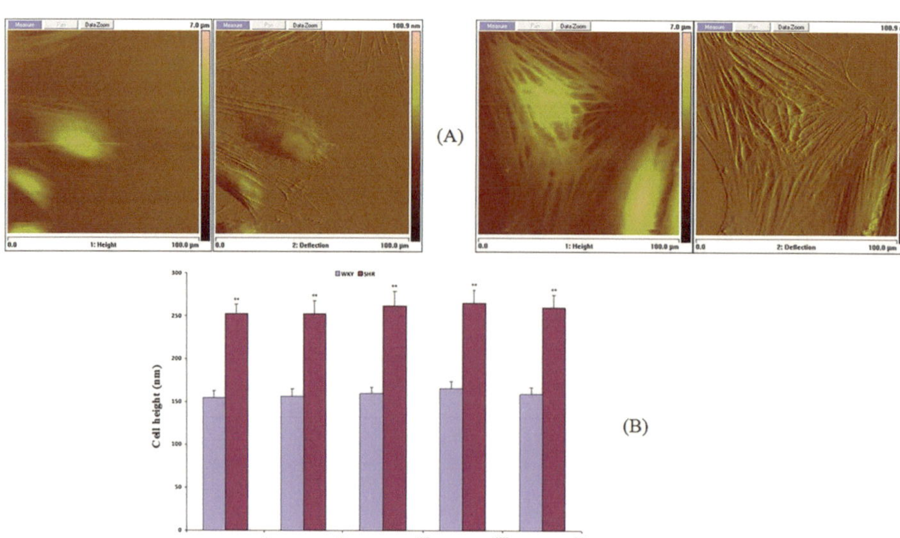

Figure 1.
*Topographic characterizations of WKY and SHR VSMCs. (A) Example of WKY TA (left) and SHR TA VSMC (right) AFM deflection and height image. (B) Scanning height data of WHK (n = 5, from 3 rats) and SHR (n = 5, from 3 rats). **p < 0.01 (SHR TA VSMC compared to WKY TA VSMC in different time regions).*

F-actin over assembly. The expression of cytoskeletal actin in SHRs is obviously higher than ($p < 0.05$) WKYs at the same age and the denser actin filaments make a lot of crosslinking polymers inside VSMCs [29].

3.2 Drugs effect on VSMC elasticity

The VSMC elasticity data are consistent with the above topographical observations, and indicate that the intrinsic property of a single cell reflects its mechanical characteristics. By CD and ML7 evaluations, the elasticities of VSMCs were dramatically reduced and there were no significant differences between WKYs and SHRs [29]. The drug Y27632 (5 μmol/L) was performed to treat WKY and SHR VSMCs, VSMC elasticity of SHRs showed a higher value in presence of Y27632 in comparison to that of WKYs ($p < 0.005$, **Figure 2A**), whereas the drug lysophosphatidic acid (LPA) (2 μmol/L) increased VSMC elasticity in both WKYs and SHRs, but to a larger extent in SHR ($p < 0.001$, **Figure 2A**).

The time series behavior in VSMC elasticity of WKYs and SHRs with 10 μmol/L CD, 10 μmol/L ML7 (**Figure 2B**), 5 μmol/L Y27632 (**Figure 2C**) and 2 μmol/L LPA treatments in single cell level was further investigated by a spectral analysis approach. After 10 μmol/L CD treatment there were not any significant differences in three components of VSMC elasticity oscillatory behaviors between WKYs and SHRs. Interestingly, in the second component the amplitude of WKY was higher than that of SHR ($p < 0.05$) by CD treatment (**Figure 3A**). Possibly CD depolymerizes and disrupts actin filaments in SHR cells, and shows a lower amplitude in the second component. The mechanism will be further revealed. Additionally, ML7 is a drug to inhibit myosin light chain phosphorylation, and is applied to VSMC to

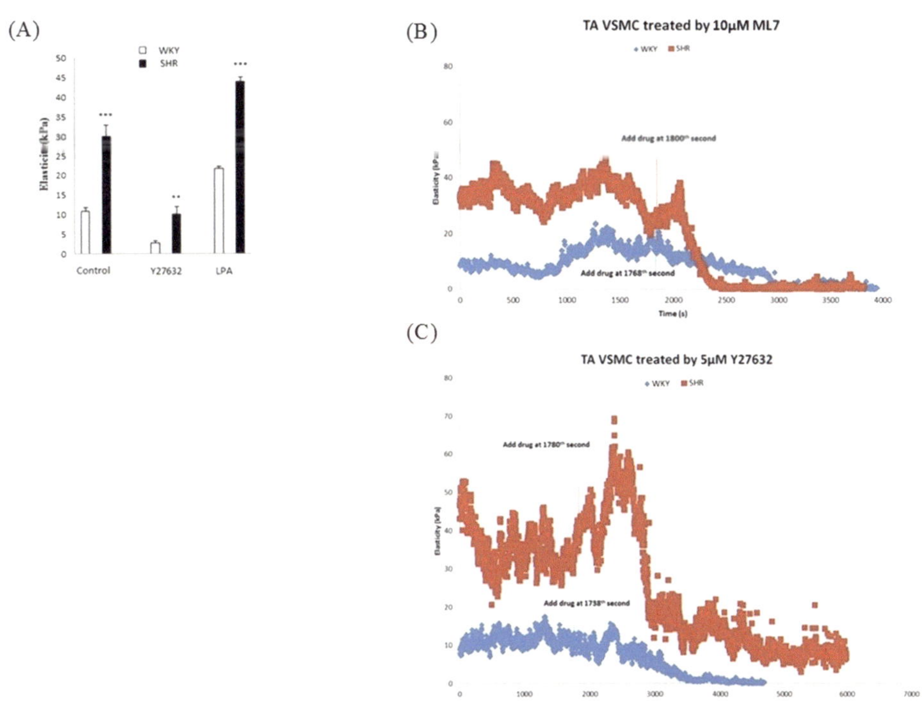

Figure 2.
*(A) VSMCs were treated with 5 μmol/L Y27632 to inhibit the ROCK, with 2 μmol/L LPA to increase integrin adhesion to ECM for 30 minute measurement. **P < 0.005 (WKY vs. SHR), ***P < 0.001(WKY vs. SHR). (B) Examples of real-time cell elastic modulus for typical WKY (blue) and SHR (red) vascular smooth muscle cells using 10 μmol/L ML7 treatment and (C) 5 μmol/L Y27632 treatment are shown.*

test its effect on SHR and WKY VSMC elasticity. After 10 μmol/L ML7 treatment, there were also not any significant differences between WKY and SHR VSMCs in three components of oscillatory behaviors. In three principle components, the frequencies and amplitudes of both WKYs and SHRs were not significantly different ($p > 0.05$) (**Figure 3B**).

From an individual cell point of view, SHRs strongly and vehemently responded to both LPA and Y27632 treatments. The spectral analysis clearly demonstrated the dynamic oscillatory behaviors in VSMC elasticity that are driven by actin–ECM interactions at the absence and presence of these two drugs. The drugs activate or inactivate VSMC elastic characters through a series of cascade responses, thus after drug treatments the frequencies and amplitudes of first spectral component (large visible oscillation, **Figure 3C** and **D**) between SHRs and WKYs existed significant differences ($p < 0.05$). Moreover, the amplitudes of the second spectral component between SHRs and WKYs existed significant differences ($p < 0.05$) (**Figure 3C** and **D**).

Rho kinase acts as a signaling molecule to influence VSMC stiffness via the Ca^{2+}-CaM-MLCK pathway and with some cascade cycles by phosphorylation and dephosphorylation. Rho kinase also regulates aortic VSMC stiffness via actin/serum response factor (SRF)/myocardin in hypertension [19]. The WKYs and SHRs are at 16–18 weeks of age, from the prior reports the basal expression level of myosin light chain (MLC) in both WKYs and SHRs was closed, but the expression of pMLC was found to be increased in SHRs [29]. The drug Y27632 inhibits Rho kinase to phosphorylate MLCK and dephosphorylates pMLC, indirectly keeps actin away from binding with myosin to depolymerize the establishment of actin-myosin complex and eliminates the VSMC elasticity [30]. In addition, the ECM-integrin-cytoskeletal axis is an important pathway to influence VSMC stiffness by regulating α-SMA expression, and the coordinate ability of ECM-integrin-actin is attenuated to produce hypertension [13, 15]. The cytoskeletal α-smooth muscle actin (α-SMA) importantly responds mechanical forces through ECM-integrin-cytoskeletal axis to mediate VSMC stiffness, and it is over expressed to be a decisive factor leading to an increase in aortic stiffness for inducing hypertension. The α-SMA expression

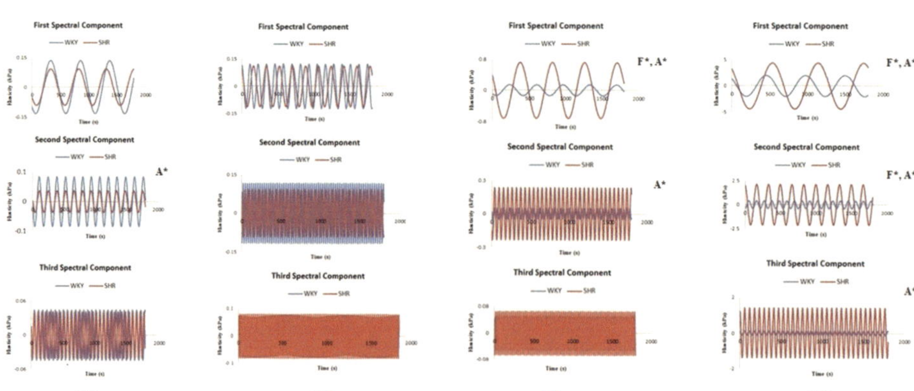

Figure 3.
Over a 30 minute period, mathematical analysis of the elastic modulus waveform in the presence of drugs indicated three principle components of oscillation by spectral analysis for WKY and SHR thoracic aorta vascular smooth muscle cells. (A) WKYs (5 cells from 3 animals) and SHRs (4 cells from 3 animals) treated by 10 μM CD. (B) WKYs (9 cells from 3 animals) and SHRs (5 cells from 3 animals) treated by 10 μMML7. (C) WKYs (5 cells from 3 animals) and SHRs (6 cells from 3 animals) treated by 5 μM Y27632. (D) WKYs (5 cells from 3 animals) and SHRs (5 cells from 3 animals) treated by 2 μM LPA. F and A* indicate that the frequency and amplitude of the component were significantly different from WKYs and SHRs in the presence of drugs, $P < 0.05$.*

and polymerization in SHR VSMCs are obviously higher than WKYs at the same age [29]. Attenuating the stiffness of VSMC via several pathways, Y27632 gently damaged the crosslinking of α-SMA filaments and the contractility of myosin production in comparison to the drug CD and ML7, consequently VSMC elasticity of SHRs showed a higher value in presence of Y27632 in comparison to that of WKYs.

The drug lysophosphatidic acid (LPA) activates Rho kinase to phosphorylate MLCK and promote α-SMA polymerization [25]. Meanwhile, LPA enhances integrin to adhere to the ECM and activate Rho kinase. Increasing adhesion interaction between α5β1integrin and fibronectin (one component of ECM) is related to increasing cell stiffness via ECM-integrin-cytoskeletal axis, and MLCK was also found to be over expressed in VSMCs of SHRs [23, 29, 31, 32]. LPA increases α5β1integrin to adhere to ECM for promoting actin expression and polymerization. The interaction between α5β1 integrin and FN is specific and important in the mechanical transduction of VSMC, and α5β1 integrin is the major receptor for FN [33–35]. The α5β1integrin provides the bio-mechanical linkage between α-SMA and fibronectin (FN) in extracellular space, and α-SMA responds the bio-mechanical forces through integrin-mediated cell-ECM interactions to alternate cytoskeleton system of VSMCs [36]. The ECM-integrin-cytoskeletal axis and the contractility of myosin production are two independent pathways to regulate the VSMC stiffness [14]. The α-SMA was highly expressed in the VSMC of SHRs, moreover, MLCK was also found to be over expressed in SHRs to stiffen the VSMC due to enhancing the VSMC contractile process [37], thus this analysis showed VSMC elastic moduli of SHRs were greater than ($p < 0.05$) those of WKYs.

Previous studies have shown that both internal and external biomechanical forces can act through the cytoskeleton, thereby affecting local elasticity and cell behavior [38, 39]. The differences in stiffness and time-dependent oscillations were largely influenced by actin cytoskeletal dynamics. Dynamical alternation of α-SMA constructs different high-level linkage structures in VSMCs, affecting cell elasticity and cellular stress relaxation behavior [40–42]. To further verify the internal characteristics of cells that reflect the mechanical properties of cells, various drug treatments that affect the cytoskeleton and corresponding vascular smooth muscle contraction mechanisms have been carried out on VSMCs for in-depth research. Three spectral components are determined in the oscillation mode, so it is reasonable to assume that more than one mechanism causes the spontaneous oscillation of cell elasticity, and therefore may play a role in the increased vascular stiffness observed in hypertension. The elastic oscillations of VSMCs represent the inherent characteristics of cells and involve the cytoskeleton structure responsible for the interaction of actin-ECM and actin-myosin. At the same time, the oscillation of VSMC elasticity reveals the polymerization and depolymerization of α-SMA. Different pharmacological mechanisms produced the different individual cell elastic behavior after the drug treatments. Spectral analysis showed that compared with WKY rats, SHRs usually have lower frequencies and larger amplitudes. The general pattern of slow, larger oscillations in SHRs and faster, smaller oscillations in WKY VSMCs [29]. After Y27632 treatment, the frequency of the first wave component is significantly reduced in SHR VSMCs, whereas the amplitudes of the first and second wave components are increased in SHR VSMCs. The frequency and amplitude showed not to be a significant difference between WKY and SHR VSMCs in the third wave component. All in all, the spectral analysis indicated that Y27632 gently attenuated VSMC stiffness. The drug LPA polymerizes α-SMA to increase VSMC stiffness, from the spectral analysis the amplitudes of all three wave components are enhanced in SHR VSMCs, and the frequencies of the first and second wave components are significantly reduced in comparison to WKY VSMCs. The drug CD breaks apart the actin cytoskeletal network and the drug ML7 dephosphorylates myosin light chain to block the interaction between actin and myosin. These two drugs CD and ML7 strongly and irreversibly destroy the cross-linking of actin filaments and the contractility produced by myosin. Both WKYs and SHRs completely lose their elasticity, and therefore exhibit inactivity in the three components of the oscillation.

In summary, cellular mechanisms underlying differences in VSMC stiffness were investigated using AFM. For decades, pharmacists have developed many drugs for the treatment of certain vascular diseases based on the role of the actin-integrin axis in the mechanical properties of VSMCs, and AFM provides a way to manipulate an individual VSMC due to its nano-sensitivity under physiological condition and liquid environment. At present, people have used AFM in many applications to study the mechanical properties of a single VSMC, and the intrinsic changes of VSMC enable people to open up new therapeutic ways for the treatment of multiple diseases and update our understanding of vascular biology [30, 43]. The future trends of employing AFM tip coating techniques for adhesive assessment and super-resolution fluorescence microscopy for cytoskeletal tracking will further resolve individual VSMC elasticity and its role in physiological process of living organisms [44]. The *in vivo* mechanism of how individual VSMC elasticity to regulate vascular processes is still unknown, thus the AFM detection *in vitro* combined with the investigation *in vivo* by other techniques will also provide a perspective view to describe the VSMC elastic characters in the coming researches [45, 46].

List of abbreviations

AFM	atomic force microscopy
α-SMA	α-smooth muscle actin
CD	Cytochalasin D
DMEM	Dulbecco's Modified Eagle's *medium*
ECM	extracellular matrix
FBS	fetal bovine serum
HEPES	4-(2-hydroxyethyl)-1-piperazineethanesulfonic acid
LPA	lysophosphatidic acid
ML7	Hexahydro-1-[(5-iodo-1-naphthalenyl) sulfonyl]-1*H*-1, 4-diazepine
MLCK	myosin light chain kinase
ROCK	Rho-associated protein kinase
SHR	spontaneously hypertensive rat
VSMC	vascular smooth muscle cell
WKY	wistar-kyoto normotensive rat
Y27632	(1R, 4r)-4-((R)-1-aminoethyl)-N-(pyridin-4-yl) cyclohexanecarboxamide

Author details

Yi Zhu[1,2]

1 Department of Medicine, University of Alabama-Birmingham, Birmingham, AL, United States

2 Dalton Cardiovascular Research Center, University of Missouri-Columbia, Columbia, MO, United States

*Address all correspondence to: yalezhuphd@gmail.com

References

[1] Dhar S., Sun Z., Meininger G.A., Hill M.A. (2017). Nonenzymatic glycation interferes with fibronectin-integrin interactions in vascular smooth muscle cells. *Microcirculation*, 24(3), e12347 (1-15).

[2] Hill M.A., Meininger G.A. (2016). Small artery mechanobiology: roles of cellular and non-cellular elements. Microcirculation, 23(8), 611-613.

[3] Leloup A.J.A., Van Hove C.E., De Moudt S., De Meyer G.R.Y., De Keulenaer G.W., Fransen P. (2019). Vascular smooth muscle cell contraction and relaxation in the isolated aorta: a critical regulator of large artery compliance. *Physiological Reports*, 7(4), e13934 (1-13).

[4] Sanyour H.J., Li N., Rickel A.P., Torres H.M., Anderson R.H., Miles M.R., Childs J.D., Francis K.R., Tao J.N., Hong Z.K. (2020). Statin mediated cholesterol depletion exerts coordinated effects on the alterations in rat vascular smooth muscle cell biomechanics and migration. The Journal of Physiology, 598(8), 1505-1522.

[5] Zhu W., Kim B.C., Wang M., Huang J., Isak A., Bexiga N.M., Monticone R., Ha T., Lakatta E.G., An S.S. (2018). TGFβ1 reinforces arterial aging in the vascular smooth muscle cell through a long-range regulation of the cytoskeletal stiffness. *Scientific Report*, 8(1), a2668 (1-12).

[6] Jia G., Habibi J., DeMarco V.G., Martinez-Lemus L.A., Ma L., Whaley-Connell A.T., Aroor A.R., Domeier T.L., Zhu Y., Meininger G.A., Mueller K.B., Jaffe I.Z., Sowers J.R. (2015). Endothelial Mineralocorticoid Receptor Deletion Prevents Diet-Induced Cardiac Diastolic Dysfunction in Females. Hypertension, 66(6), 1159-1167.

[7] Yang, J., Jin, K., Xiao, J., Ma, J., and Ma, D. (2017). Endogenous tissue factor pathway inhibitor in vascular smooth muscle cells inhibits arterial thrombosis. Frontiers of Medicine, 11, 403-409.

[8] Zhang J., Zhao X., Vatner D.E., McNulty T., Bishop S., Sun Z., Shen Y.T., Chen L., Meininger G.A., Vatner S.F. (2016). Extracellular Matrix Disarray as a Mechanism for Greater Abdominal Versus Thoracic Aortic Stiffness With Aging in Primates. Arteriosclerosis, Thrombosis, and Vascular Biology, 36(4), 700-706.

[9] Li N., Sanyour H., Remund T., Kelly P., Hong Z. (2018). Vascular extracellular matrix and fibroblasts-coculture directed differentiation of human mesenchymal stem cells toward smooth muscle-like cells for vascular tissue engineering. Materials Science and Engineering: C, 93(1), 61-69.

[10] Zhou Z., Qu J., He L., Yang S., Zhang F., Guo T., Peng H., Chen P., Zhou Y. (2020). Stiff matrix instigates type I collagen biogenesis by mammalian cleavage factor I complex-mediated alternative polyadenylation. *JCI insight*, 5(3), e133972 (1-15).

[11] Nance M.E., Whitfield J.T., Zhu Y., Gibson A.K., Hanft L.M., Campbell K.S., Meininger G.A., McDonald K.S., Segal S.S., Domeier T.L. (2015). Attenuated sarcomere lengthening of the aged murine left ventricle observed using two-photon fluorescence microscopy. American Journal of Physiology-Heart and Circulatory Physiology, 309(5), H918-H925.

[12] Touyz R.M., Alves-Lopes R., Rios F.J., Camargo L.L., Anagnostopoulou A., Arner A., Montezano A.C. (2018). Vascular smooth muscle contraction in hypertension. Cardiovascular Research, 114(4), 529-539.

[13] Zhu Y., He L., Qu J., Zhou Y. (2018). Regulation of Vascular Smooth Muscle Cell Stiffness and Adhesion by $[Ca^{2+}]_i$: an Atomic Force Microscopy-Based Study. Microscopy and Microanalysis, 24(6), 708-712.

[14] Huang H., Sun Z., Hill M.A., Meininger G.A. (2018). A calcium medicated mechanism coordinating vascular smooth muscle cell adhesion during KCl activation. Frontiers in Physiology, 9, a1810 (1-13).

[15] Zhu Y., Qu J., Zhang F., Zhou Z, Yang S., Zhou Y. (2019). Calcium in Vascular Smooth Muscle Cell Elasticity and Adhesion: Novel Insights into the Mechanism of Action. Frontiers in Physiology, 10, a00852 (1-8).

[16] Rickel A.P., Sanyour H. J., Leyda N.A., Hong Z.H. (2020). Extracellular Matrix Proteins and Substrate Stiffness Synergistically Regulate Vascular Smooth Muscle Cell Migration and Cortical Cytoskeleton Organization. ACS Applied Bio Materials, 3(4), 2360-2369.

[17] Shen K., Kenche H., Zhao H., Li J., Stone J. (2019). The role of extracellular matrix stiffness in regulating cytoskeletal remodeling via vinculin in synthetic smooth muscle cells. Biochemical and Biophysical Research Communications, 508(1), 302-307.

[18] Zhou N., Lee J.J., Stoll S., Ma B., Wiener R., Wang C., Costa K.D., Qiu H. (2017a). Inhibition of SRF/myocardin reduces aortic stiffness by targeting vascular smooth muscle cell stiffening in hypertension. Cardiovascular Research, 113: 171-182.

[19] Zhou N., Lee J.-J., Stoll S., Ma B., Costa K.D., Qiu H. (2017b). Rho Kinase Regulates Aortic Vascular Smooth Muscle Cell Stiffness Via Actin/SRF/Myocardin in Hypertension. Cellular Physiology and Biochemistry, 44, 701-715.

[20] Sanyour H., Childs J., Meininger G.A., Hong Z. (2018). Spontaneous oscillation in cell adhesion and stiffness measured using atomic force microscopy. *Scientific Report*, 8(1), a2899 (1-10).

[21] Sanyour H.J., Li N., Rickel A.P., Childs J.D., Kinser C.N., Hong Z.K. (2019). Membrane cholesterol and substrate stiffness coordinate to induce the remodeling of the cytoskeleton and the alteration in the biomechanics of vascular smooth muscle cells. Cardiovascular Research, 115(8), 1369-1380.

[22] Hong Z., Reeves K.J., Sun Z., Li Z., Brown N.J., Meininger G.A. (2015). Vascular smooth muscle cell stiffness and adhesion to collagen I modified by vasoactive agonists. PloS One, 10(3), e0119533.

[23] Sehgel N.L., Sun Z., Hong Z., Hunter W.C., Hill M.A., Vatner D.E., Vatner S.F., Meininger G.A. (2015a). Augmented vascular smooth muscle cell stiffness and adhesion when hypertension is superimposed on aging. Hypertension, 65(2), 370-377.

[24] Sehgel N.L., Vatner S.F., Meininger G.A. (2015b). "Smooth Muscle Cell Stiffness Syndrome"—Revisiting the Structural Basis of Arterial Stiffness. Frontiers in Physiology, 6, a00335 (1-15).

[25] Staiculescu M.C., Ramirez-Perez F.I., Castorena-Gonzalez J.A., Hong Z., Sun Z., Meininger G.A., Martinez-Lemus L.A. (2014). Lysophosphatidic acid induces integrin activation in vascular smooth muscle and alters arteriolar myogenic vasoconstriction. Frontiers in Physiology, 5, a00413 (1-12).

[26] Turner C.J., Badu-Nkansah K., Crowley D., van der Flier A., Hynes R.O. (2015). α5 and αv integrins cooperate to regulate vascular smooth muscle

and neural crest functions in vivo. Development, 142(4), 797-808.

[27] Nettikadan S., Tokumasu F., Takeyasu K. (1996). Quantitative Analysis of the tran-scription factor Ap2 binding to DNA by atomic force microscopy. Biochemical and Biophysical Research Communications, 226 (3), 645-649.

[28] Zhu Y., Qiu H., Trzeciakowski J.P., Sun Z., Li Z., Hong Z., Hill M.A., Hunter W.C., Vatner D.E., Vatner S.F., Meininger G.A. (2012). Temporal analysis of vascular smooth muscle cell elasticity and adhesion reveals oscillation waveforms that differ with aging. Aging Cell, 11(5), 741-750.

[29] Sehgel N. L., Zhu Y., Sun Z., Trzeciakowski J.P., Hong Z.K., Hunter W. C., Vatner D. E., Meininger G. A., Vatner S. F. (2013). Increased vascular smooth muscle cell stiffness: a novel mechanism for aortic stiffness in hypertension. American Journal of Physiology - Heart and Circulatory Physiology, 305(9), H1281-H1287.

[30] Pierce G.L. (2017). Mechanisms and subclinical consequences of aortic stiffness. Hypertension, 70(5), 848-853.

[31] Hong Z., Sun Z., Li Z., Mesquitta W.T., Trzeciakowski J.P., Meininger G.A. (2012). Coordination of fibronectin adhesion with contraction and relaxation in microvascular smooth muscle. Cardiovascular Research, 96(1), 73-80.

[32] Hong Z., Sun Z., Li M., Li Z., Bunyak F., Ersoy I., Trzeciakowski J.P., Staiculescu M.C., Jin M., Martinez-Le mus L., Hill M.A., Palaniappan K., Meininger G.A. (2014). Vasoactive agonists exert dynamic and coordinated effects on vascular smooth muscle cell elasticity, cytoskeletal remodelling and adhesion. Journal of Physiology, 592(6), 1249-1266.

[33] Sun Z., Martinez-Lemus L.A., Trache A., Trzeciakowski J.P., Davis G.E., Pohl U., Meininger G.A. (2005). Mechanical properties of the interaction between fibronectin and α5β1-integrin on vascular smooth muscle cells studied using atomic force microscopy. American Journal of Physiology -Heart and Circulatory Physiology, 289 (6), H2526-H2535.

[34] Wu X., Mogford J.E., Platts S.H., Davis G.E., Meininger G.A., Davis M.J. (1998). Modulation of calcium current in arteriolar smooth muscle by ανβ3 and α5β1 integrin ligands. Journal of Cell Biology, 143(1), 241-252.

[35] Wu X., Davis G.E., Meininger G.A., Wilson E., Davis M.J. (2001). Regulation of the L-type Calcium Channel by α5β1 Integrin Requires Signaling between Focal Adhesion Proteins. Journal of Biological Chemistry, 276 (32), 30285-30292.

[36] Hays T.T., Ma B., Zhou N., Stoll S., Pearce W.J., Qiu H. (2018). Vascular smooth muscle cells direct extracellular dysregulation in aortic stiffening of hypertensive rats. *Aging Cell*, 17(3), e12748 (1-13).

[37] Rodenbeck S.D., Zarse C.A., McKenney-Drake M.L., Bruning R.S., Sturek M., Chen N.X., Moe S.M. (2017). Intracellular calcium increases in vascular smooth muscle cells with progression of chronic kidney disease in a rat model. Nephrology Dialysis Transplantation, 32(3), 450-458.

[38] Fletcher D.A., Mullins R.D. (2010). Cell mechanics and cytoskeleton. Nature, 463(7280), 485-492.

[39] Qiu H., Zhu Y., Sun Z., Trzeciakowski J.P., Gansner M., Depre C., Resuello R.R., Natividad F.F.,

Hunter W.C., Genin G.M., Elson E.L., Vatner D.E., Meininger G.A., Vatner S.F. (2010). Short communication: vascular smooth muscle cell stiffness as a mechanism for increased aortic stiffness with aging. Circulation Research, 107(5), 615-619.

[40] Sun Z., Martinez-Lemus L.A., Hill M.A., Meininger G.A. (2008). Extracellular matrix-specific focal adhesions in vascular smooth muscle produce mechanically active adhesion sites. American Journal of Physiology-Cell Physiology, 295 (1), C268-C278.

[41] Sun Z., Li Z., Meininger G.A. (2012). Mechanotransduction through fibronectin-integrin focal adhesion in microvascular smooth muscle cells: is calcium essential? American Journal of Physiology - Heart and Circulatory Physiology, 302(10), H1965-H1973.

[42] Wu X., Sun Z., Foskett A., Trzeciakowski J.P., Meininger G.A., Muthuchamy M. (2010). Cardiomyocyte contractile status is associated with differences in fibronectin and integrin interactions. American Journal of Physiology -Heart and Circulatory Physiology, 298 (6), H2071-H2081.

[43] Martinez-Lemus L.A., Hill M.A., Meininger G.A. (2009). The plastic nature of the vascular wall: a continuum of remodeling events contributing to control of arteriolar diameter and structure. Physiology (Bethesda), 24 (1), 45-57.

[44] Ella S.R., Yang Y., Clifford P.S., Gulia J., Dora K.A., Meininger G.A., Davis M.J. , Hill M.A. (2010). Development of an Image-Based System for Measurement of Membrane Potential, Intracellular Ca^{2+} and Contraction in Arteriolar Smooth Muscle Cells. Microcirculation, 17(8), 629-640.

[45] Lacolley P., Regnault V., Segers P., Laurent S. (2017). Vascular smooth muscle cells and arterial stiffening: relevance in development, aging, and disease. American Journal of Physiology –Physiological Reviews, 97(4), 1555-1617.

[46] Lacolley P., Regnault V., Avolio A.P. (2018). Smooth muscle cell and arterial aging: basic and clinical aspects. Cardiovascular Research, 114(4), 513-528.

Section 2

Raman Spectroscopy and Its Applications

Chapter 5

Deep Learning Approach for Raman Spectroscopy

M.H. Wathsala N. Jinadasa, Amila C. Kahawalage, Maths Halstensen, Nils-Olav Skeie and Klaus-Joachim Jens

Abstract

Raman spectroscopy is a widely used technique for organic and inorganic chemical material identification. Throughout the last century, major improvements in lasers, spectrometers, detectors, and holographic optical components have uplifted Raman spectroscopy as an effective device for a variety of different applications including fundamental chemical and material research, medical diagnostics, bio-science, in-situ process monitoring and planetary investigations. Undoubtedly, mathematical data analysis has been playing a vital role to speed up the migration of Raman spectroscopy to explore different applications. It supports researchers to customize spectral interpretation and overcome the limitations of the physical components in the Raman instrument. However, large, and complex datasets, interferences from instrumentation noise and sample properties which mask the true features of samples still make Raman spectroscopy as a challenging tool. Deep learning is a powerful machine learning strategy to build exploratory and predictive models from large raw datasets and has gained more attention in chemical research over recent years. This chapter demonstrates the application of deep learning techniques for Raman signal-extraction, feature-learning and modelling complex relationships as a support to researchers to overcome the challenges in Raman based chemical analysis.

Keywords: machine learning, deep learning, neural networks, fluorescence, *data analysis*

1. Introduction

Spectroscopy is an ubiquitous method in natural sciences and engineering for e.g. characterization of materials, molecules or mechanisms, kinetics and thermodynamics of chemical reactions. It is the study of the interaction between electromagnetic radiation and molecules/particles which involves either absorption, emission, or scattering. In Raman spectroscopy, it is the interaction of light with matter which is generating the Raman effect. This effect is the scattering of incoming radiation leading to a change of wavelength or frequency. A Raman spectrum is composed of peaks which show the intensity and wavelength of the Raman scattered light which is due to radiation interaction with individual chemical bond vibrations. These peaks are used to detect, identify, and quantify information about atoms and molecules. Raman spectroscopy is a prominent choice among other spectroscopic techniques, particularly in chemical systems containing water and/or polar solvents [1]. Weak Raman scattering of water enables in-situ analysis in aqueous chemical systems and in vitro and in vivo analysis in human and sensitive biological systems. While many analytical techniques require sample preparation (such as grinding, glass formation, or tablet pressing) before measurement, Raman analysis can be made on 'as received' samples. A measurement can be made within few seconds in a non-destructive, non-contact manner and therefore samples can be retained for other analysis if necessary. Raman scattering of light by molecules was first predicted using classical quantum theory by Smekal in 1923 [2] and experimentally observed by Raman and Krishnan in 1928 [3, 4]. After a century of first ever discovery of Raman fundamentals, today, different types of Raman spectroscopies have been developed such as time-resolved Raman spectroscopy, high pressure Raman spectroscopy, matrix-isolation Raman spectroscopy, Surface-Enhanced Raman Spectroscopy, Raman microscopy and Raman Imaging spectrometry [5]. Throughout the last century, major improvements in lasers, spectrometers, detectors, and holographic optical components have yielded Raman spectroscopy as a dominant tool for molecular verification in a wide range of scientific disciplines.

A primary role of scientists is the extraction of new knowledge from experimental data. Spectroscopic techniques produce profiles containing a high amount of data. It can take significant time and effort to read, interpret and model these data. Cozzolino [6] mentions that the three critical pillars that support the development and implementation of vibrational spectroscopy including Raman, are as the sample (e.g., sampling, methodology), the spectra and the mathematics (e.g., spectral analysis, algorithms, pre-processing, data interpretation, etc.). Thus, data analysis becomes the only flexible option that can be adjusted to assess data extracted from a specific application (i.e. sample) using a given spectroscopic method. Spectroscopic techniques are only as powerful as the information that can be extracted from the resulting spectral data. Simultaneous development of spectroscopy hardware components and data analysis throughout the last five decades, made a radical change for the propagation of spectroscopic techniques in different fields. In case of Raman spectroscopy, instead of having a spectrometer whose volume fills up an entire room including a group of scientists manually reading the spectra, today we have miniature spectroscopic analyzers supported by a computer and software which automatically read, treat, interpret, and summarize measurements within seconds (**Figures 1** and **2**).

The objective of this chapter is to show the role of data analysis to raise and expand the awareness of Raman spectroscopy. The chapter reviews key deep learning strategies under machine learning perspectives, that have been already applied in different Raman applications emphasizing how these strategies have contributed to solving Raman spectroscopic data challenges. The objective is to strengthen the role of data analysis to uplift the capability and standard of Raman spectroscopy.

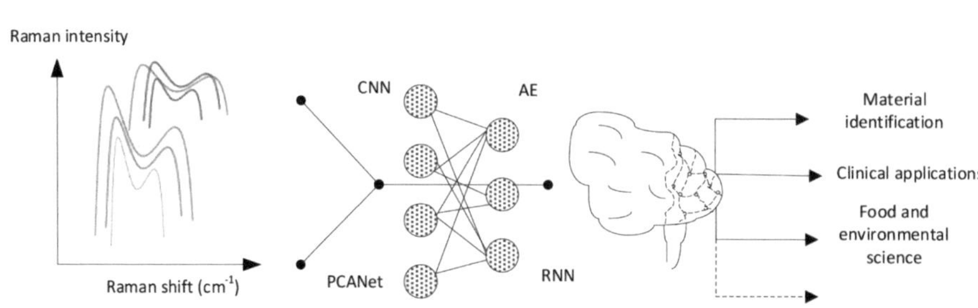

Figure 1.
Raman spectroscopy combined with machine learning methods in different applications.

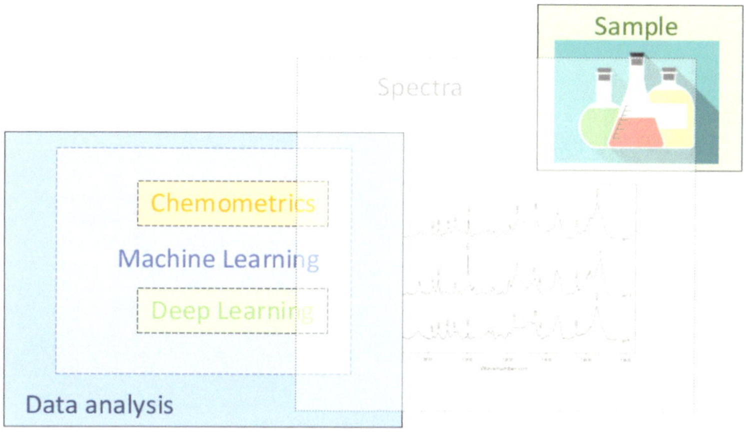

Figure 2.
Integration between sample, spectra, and data analysis to develop Raman applications.

"Deep learning" is a subset of machine learning in artificial intelligence. Many spectroscopists have a background in chemometrics and statistics for chemical analysis, but so far only a few are taking advantage of the potential provided by machine learning. There are many synergies and common concepts applied in between the areas of machine learning and spectroscopy which enhances productive inter-communication. The chapter provides comprehensions by showing how deep learning algorithms increase analytical insight into Raman spectra.

2. Deep learning

In a simplest way, deep learning can be introduced as a method which teaches computers to do a task. Very often this task is difficult to carry out by human brain due to limited brain capacities and limited time. It is a subset of Machine learning (ML) which is further a subset of artificial intelligence (AI).

Integrating data, information, machines, sensors, and software is a component of transforming conventional ways of human-oriented methods into more digitalized roots. It can level up efficiency and performance of an individual system and its related components by giving more deep insights. Artificial intelligence and machine learning persist to support this transformation.

Artificial intelligence (AI) compasses the science and engineering of making intelligent machines specially computer programs. Machine learning (ML) which is a subset of artificial intelligence, uses algorithms to optimize a certain task by using examples or experience and support AI to learn with explicit programming. Deep learning (DL) is a sub class of machine learning algorithms which consist of learning methods based on artificial neural networks (ANNs). **Figure 3** shows the interconnection of AI, ML, DL and chemometrics. ML algorithms that are not deep learning are referred as shallow learning. A simple explanation to understand the difference between a shallow learning and deep learning algorithm is shown in **Figure 4(a)**. It shows that in shallow learning, feature (useful patterns) extraction and classification are performed in two different stages. For instance, a general practice of a chemist who obtains a vibrational spectrum of an unknown chemical sample, starts with mapping individual peaks. Typically, this is performed by combining the knowledge of chemical vibrational modes and sample. There can be peaks that are not originated from sample chemical properties such as instrument noise or stray lights. The chemist will only utilize the peaks that reveal required

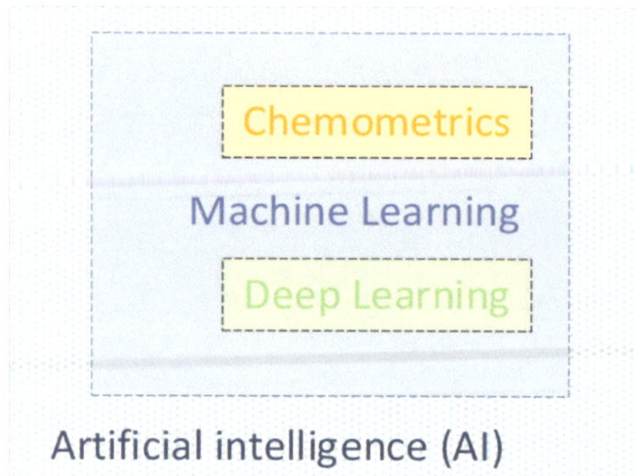

Figure 3.
Artificial intelligence, machine learning, deep learning and chemometrics.

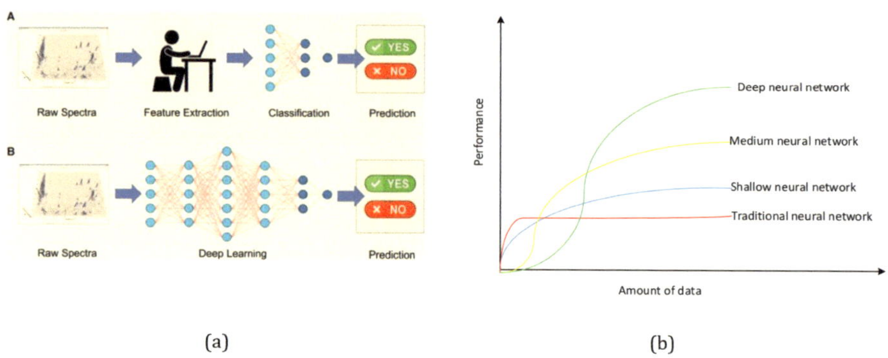

Figure 4.
(a) Comparison between machine learning and deep learning [7]; (b) performance vs. amount of data for traditional ML and DL algorithms.

information about the sample. This process is called feature extraction in ML language. Next, the chemist will proceed to the next step of analysis such as regression or classification. Conversely, in deep learning, feature extraction and subsequent analysis are performed automatically inside the single boundary of DL algorithm.

2.1 Chemometrics

For scientists utilizing spectroscopy, chemometrics is a very familiar term linked to data analysis. Chemometrics was established at the beginning of the 1970s by Svante Wold, Bruce L. Kowalski, and D.L. Massart [8, 9]. It is a chemical discipline that uses mathematical, statistical, and other methods employing formal logic to design or select optimal measurement procedures and experiments, and to provide maximum relevant chemical information by analyzing chemical data [10]. Throughout the past 50 years chemometrics revolutionized in the field of spectroscopy through the applications of multivariate calibration, (re)activity modeling, pattern recognition, classification, discriminant analysis, and multivariate process modeling and monitoring [11]. Wold and Sjöström [11] point out two strong trends where the future success of chemometrics remains; 1). Ability of chemometrics to handle the number of 'objects' observations, cases, or samples which is fairly small, and tends to become even smaller with time and 2). Ability of chemometrics to handle big data sets and those which continuously updated with more addition of data in the future. Data sets often remain smaller when experimentation is demanding more resources like time, personnel, laboratory space, instrumentation, chemicals, solvents and hence is becoming more and more expensive. Big data sets are generated when more samples are measured or several experimental runs are propagated over time for examples in combinatorial chemistry and process monitoring. Vogt [8] explains that maintaining chemometrics as an active and widely recognized research field, requires opening new research areas for chemometricians and without the power of parallel computation, many new and exciting avenues will remain unfeasible. For instance, limiting chemometrics to linear methodologies imposes restrictions because many chemical systems are nonlinear. Chemometrics has its main territory is analytical and measurement science, however fundamentally it can also be considered as a subset of machine learning. The understanding of chemical systems, and the respective underlying behavior, mechanisms, and dynamics, can be facilitated by the development of descriptive, interpretative, and predictive models. Common examples of chemometric techniques which develop such models are principal component analysis (PCA), partial least squares (PLS), linear discriminant analysis (LDA) and support vector machine (SVM). Studies showing the possibility of combining routine chemometrics methods with machine learning algorithms influence to break the stagnancy of chemometrics tools in the chemical laboratory.

In spectral data analysis, the amount of data plays a decisive role. DL algorithms give better performance for big data sets and as more data are being added. On the other hand, performance of a system which is analyzed by human brain or conventional machine learning algorithms, is limited after a certain size and scale of data. **Figure 4(b)** shows the performance curve for traditional machine learning algorithms and deep learning algorithms. Performance curve for traditional algorithms is saturated after a certain number of data because they are based on handcrafted rules. Creating many rules manually is an erroneous task. For instance, linear regression and random forests (which are traditional ML), tend to plateau at large data volumes. On the contrary, deep learning uses more than one level of non-linear feature transformation and therefore the performance keeps increasing with added data.

Shallow machine learning methods, such as shallow neural networks [12], support vector machines [13, 14], or kernel methods [15], have been applied to Raman spectroscopy with higher success, for instance, for the prediction of the physical, chemical, or biological properties of systems. More complex models and deep machine learning methods become useful as more data becomes available and more complex problems are experienced. It allows users to make decisions as data are collected, without human-in-the-loop processing [16]. Different type of data can be input to a DL algorithm such as sound, text, images, time series and video. Raman spectroscopy generates time series data such as in resonance Raman and image data such as in Raman image microscope. DL can be applied for machine perception including classification, clustering, and predictions and also a preferred choice for unstructured data like images where manual feature extraction are difficult.

2.2 Neural networks

Neural networks (NN) make up the backbone of deep learning algorithms and therefore, it is important to understand common terms in a neural network such as layers, weights and activation functions. **Figure 5(a)** shows a representation of an artificial neuron. The first layers are called input layers which passes incoming data (x_1, x_2, x_3,.....x_n) into other layers. Output layer is the last layer of neurons that produces given outputs (y) for the program.

All layers in between are called hidden layers. Weights (w_1, w_2, w_3,.....,w_n) are the parameters within a neural network that transforms input data within the network's hidden layers. A layer is the highest-level building block in deep learning and is a

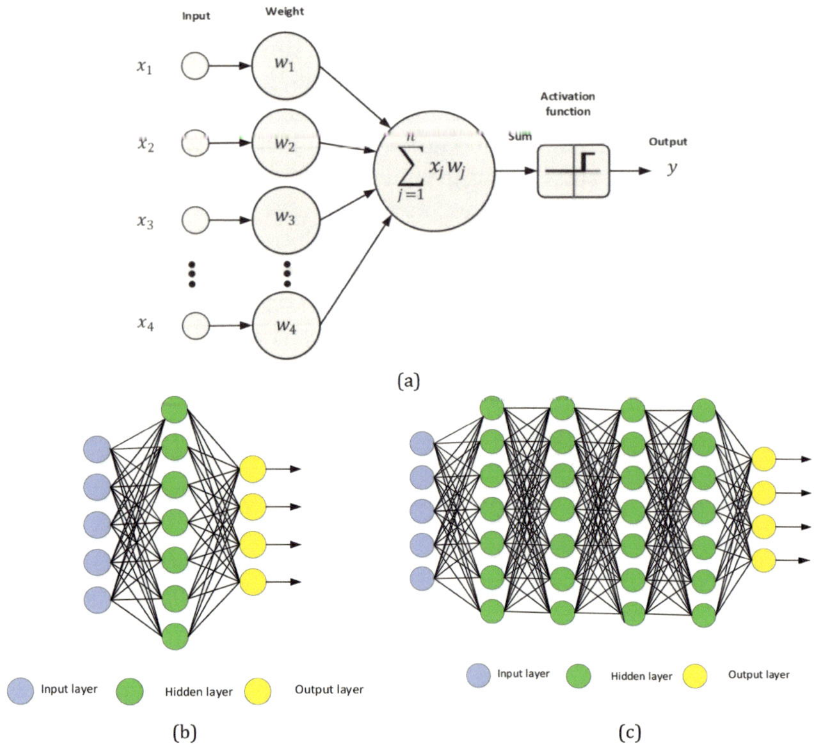

Figure 5.
Schematic representation of an artificial neuron (a), and a simple neural network displaying three basic elements: Input, hidden and output layers (b), and a deep neural network showing at least two hidden layers, or nodes (c).

container that usually receives weighted input, transforms it with a set of mostly non-linear functions and then passes these values as output to the next layer. An activation function takes in weighted data ($x_j w_j$ - matrix multiplication between input data and weights) and outputs a non-linear transformation of the data [17]. In generally, an activation function is a function that is added into an artificial neural network to help the network to learn complex patterns in the data. The most important feature in an activation function is its ability to add non-linearity into the network. Activation functions are applied after every layer in deep neural networks and they should be computationally inexpensive to be calculated. Sigmoid, Softmax, Tanh and ReLU are examples for activation function. **Figure 5(b)** shows a simple neural network which has a one hidden layer and **Figure 5(c)** shows a deep neural network which has at least two hidden layers. The neural network calculation is performed through the connections, which contain the input data, the pre-assigned weights, and the paths defined by the activation function. If the result is far from expected, the weights of the connections are recalibrated, and the analysis continues, until the outcome is as accurate as possible. Examples for neural networks are perceptron, feed forward neural network, multilayer perceptron, convolutional neural network, radial basis functional neural network, recurrent neural network, LSTM – long short-term memory, sequence to sequence models, modular neural network [18].

2.3 Deep learning algorithms

The objective of this chapter is to give an understanding about the possibilities of deep learnings in the field of Raman spectroscopy. Not many publications

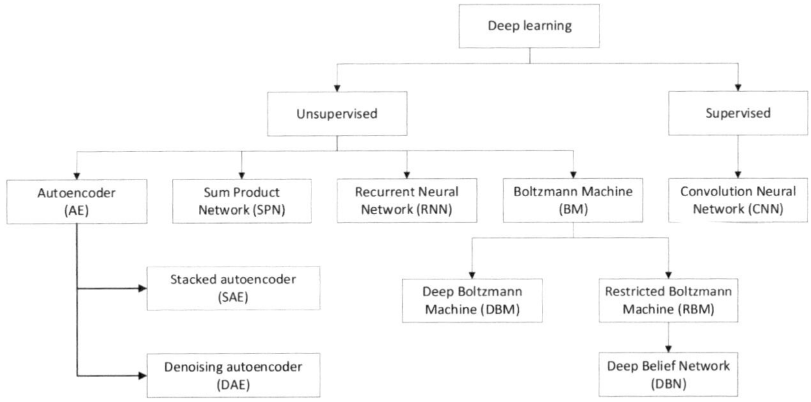

Figure 6.
Unsupervised and supervised algorithms commonly used in deep learning applications for Raman spectra.

can be found since the connection between Raman data and machine learning is still under the development stage. **Figure 6** shows some algorithms which have been applied for previous Raman data which the reader will find in the rest of the chapter. They are categorized under supervised, unsupervised and hybrid learning methods [19]. In supervised learning algorithms, we try to model relationships and dependencies between the target prediction output and the input features. The goal is to predict the output values for new data based on those relationships which it learned from the previous data sets. Therefore supervised algorithms are task driven. Supervised learning carries out tasks like regression and classification. A very common example for a supervised deep learning method is convolution neural network.

Unsupervised learning is a machine learning technique in which models are not supervised using training dataset. Instead, models itself find the hidden patterns and insights from the given unlabeled data. It can be compared to learning which takes place in the human brain while learning new things. It allows users to perform more complex processing tasks compared to supervised learning and is called a data driven approach. Dimensionality reduction, clustering and association are some tasks than an unsupervised machine learning platform can deliver. The ability to apply deep learning algorithms for unsupervised learning tasks is an important benefit because in big data sets unlabeled data are more abundant than the labeled data. Autoencoder, sum product network, recurrent neural network and Boltzmann machine can be considered as unsupervised deep learning algorithms. Supervised learning algorithms seek to answer the questions like "Based on the Raman fingerprint of this new sample I have just collected, which class in my database does it (most likely) belong to?" and/or "What is the level of purity this substance has?". Meanwhile unsupervised learning algorithms seek to answer the questions like "How similar to one another are these samples based on their Raman fingerprints?"

3. Can deep learning contribute to the development of Raman spectroscopy?

Raman scattering use a technique to interrogate chemical samples in question in a fast and non-destructive way. However, it is a weak scattering and therefore not always give straightforward results. As highlighted in Section 1 three success pillars of a spectral data analysis depends on "the type of sample to be measured", "the quality of the spectra "and "the choice of data analysis method". If any of these pillars fails, the final result will be weak in sensitivity, repeatability and reproducibility. Since we are interested about deep learning methods in this chapter, lets focus on issues related to spectral analysis and merge the contribution of deep learning to overcome those issues. Given below are four challenges that researchers have been experiencing when they analyze Raman fingerprints.

i. Assigning correct vibrational modes

Multicomponent chemical samples can contain vibrational peaks which look similar in shape and distribution over the Raman wavelength region. For instance, biological samples are composed of biochemicals such as lipids, proteins, nucleic acids, and carbohydrates. All the vibrations from these biochemicals are manifested in the Raman spectra of a biological sample making them convoluted and complex. Specially, for a fresh researcher, these spectra may appear very similar if analyzed by an untrained eye. Researchers working with Raman spectra of cells, tissues and bacteria also encounter the same problem. There are also incidents that different Raman spectrometers exhibit a small magnitude of change of Raman shift for the same component. This change can also be a significant problem if the spectra is crowded with several closely packed peaks.

ii. Analyte is influenced by the background

Weak Raman-active samples can be only analyzed if there is high spectral resolution, low spectral background, and high sensitivity. The relative intensities of the Raman bands of analytes change with solvents and are correlated with the absorption peak shift [20]. Occurrence of peaks from the matrix is true in many biomolecular Raman applications. For instance, paraffin fixed tissue may show a similar peak to a C–H stretch. Differentiating the actual spectra from the matrix therefore, becomes an equally important part before analysis.

iii. Fluorescence problem

One of the greatest challenge in Raman spectroscopy is that it is influenced by the turbidity, color, and fluorescence of the sample [21]. In spite of obvious advantages of Raman spectroscopy, the strong fluorescence background has so far restricted its use in many otherwise potential applications, for example, in the agricultural, food and oil industries, security control and crime investigations, for example. Marquardt [22] mentions that Raman biotech applications are currently is the most challenging because of the complex biological matrices and the associated fluorescence. Raman spectra are typically masked by a strong fluorescence background in most potential applications. This type of fluorescence intensity is normally several orders of magnitude larger than the Raman scattering signal, especially in biological samples. This is due to the fact that the probability of Raman scattering (cross-section) is much lower than that of fluorescence [23]. A strong fluorescence background gives rise to two problems. Firstly, it becomes the dominant element in the photon shot noise and thus detracts from the SNR (signal-to-noise ratio), and secondly, even if the Raman bands are narrow and the fluorescence has quite a smooth, featureless spectrum, errors in the mathematical estimation and removal (background subtraction) of the fluorescence increase with increasing fluorescence levels and result in increasing errors in both material identification and concentration measurement applications [23]. Fluorescence can be dealt with a variety of techniques such as the utilization of confocal configuration, photobleaching and the deployment of laser excitation at longer wavelengths. These techniques could be generally grouped into time domain, frequency-domain, wavelength-domain, and computational methods [24]. **Figure 7** shows three Raman spectra obtained from 514.5 nm laser, where the fluorescence effect of the original spectrum, *a*, could be reduced by increasing the irradiation effect as shown by spectra *b* and *c*. However, a key constraint to consider is the interaction between the laser wavelength and the sample, due to the effects of phototoxicity [25].

Computational methods can play a significant role for unmolding chemical Raman spectra from fluorescence spectra. Examples for such methods are polynomial fitting wavelet transform, and derivatives. Wei, Chen [24] describe pros and cons of polynomial fitting and derivative of Raman spectra. They mention that the optimal choice of order for polynomial fitting varies and the performance depends on the user's experience. The derivative of a measured Raman spectrum will eliminate the background components irrespective of their magnitudes and thus enhance the sharp Raman signal. However, high-frequency noises are often amplified by this method as well and the spectrum can be distorted because of the derivative process.

iv. Selection of optimum data processing technique

Understanding the system under study and making an informed judgment based on the experiments and correlating it with the available data is crucial for scientists. Selecting the correct signal processing method is a contributing factor towards understanding of the system. Improving the existing data analysis methods in Raman spectroscopy is a leading challenge. Preprocessing methods are very important to reduce inherent disturbances of a Raman spectrum such as baseline variation. Currently spectroscopists are limited to traditional chemometrics based preprocessing methods.

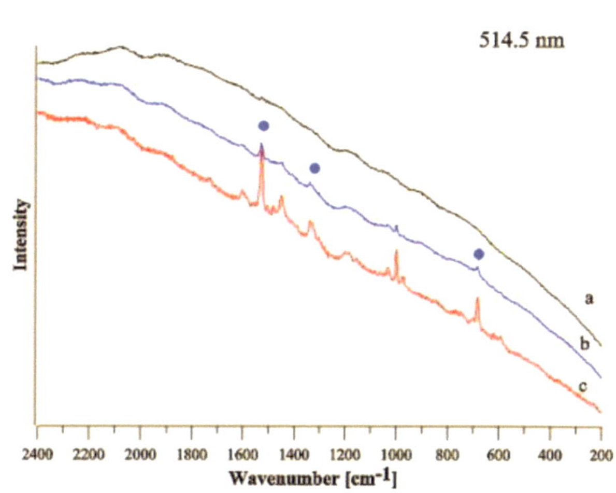

Figure 7.
Three Raman spectra of metallic paint showing the effect of fluorescence [7] (a) in normal measurement conditions (only fluorescence visible), (b) after irradiation for (b) 30 min and (c) 60 min at 50% laser power.

Models are calibrated using a fewer number of datasets, even in the situations where it is possible to use fairly a large calibration dataset. When the models are used for future large data sets these methods are limited in accuracy wise. For example, the instrument gives poor results when unknown interferences come with larger datasets such as spikes, cosmic rays and often require for re-calibration of the models time to time.

4. *Deep learning* algorithms in Raman applications

In this section, four deep learning algorithms and their derivations for different applications of Raman spectroscopy are described to provide an understanding of deployment of these methods as a means of strengthening computational methods and data analysis methods for Raman spectra.

4.1 Autoencoder

An autoencoder (AE) is an unsupervised type of artificial neural network used to learn efficient data coding. It consists of an encoder-decoder architecture as shown in **Figure 8**. Encoder consists of input data x while decorder includes output data x'. h usually referred to as code, latent variables, or latent representation and combines encoder and decorder. The aim of an autoencoder is to learn a representation (encoding) for a set of data, by training the network to ignore signal noise. Along with the reduction side of encoder, a reconstructing side is learned, where the autoencoder tries to generate from the reduced encoding a representation as close as possible to its original input x . This is done by training the AE to minimize the squared reconstruction errors $|x - x'|^2$. PCA is a linear transformation while auto-encoders are capable of modeling complex non linear functions (refer **Figure 8(b)**). PCA is faster and computationally cheaper than autoencoders. A single layered autoencoder with a linear activation function is very similar to PCA. The autoencoder weights are not equal to the principal components, and are generally not orthogonal, yet the principal components may be recovered from them using the singular value decomposition [26].

Advantages of autoencoder in Raman spectroscopy span in different areas such as dimensionality reduction, information retrieval, image processing and anomaly detection. Scientists have experimented several kinds of autoencoders such as convolution AE, denoising AE, sparse AE because they have different advantages. For

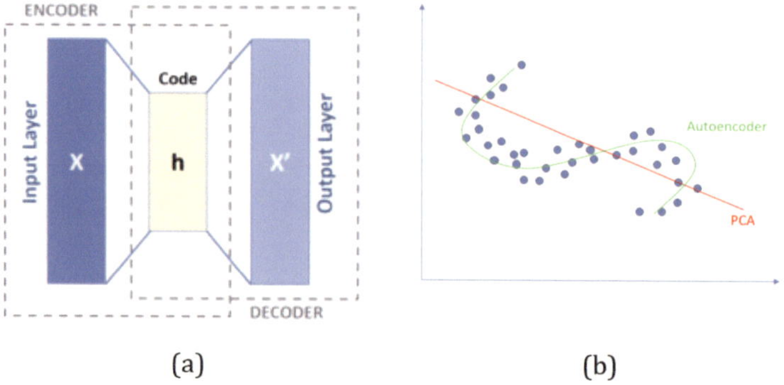

Figure 8.
Explanation of (a) autoencoder (AE) system and (b) simple demonstration of linear dimensionality reduction by PCA an non-linear dimensionality reduction by AE.

instance, sparse AE prevents overfitting. Convolutional AE is generally applied in the task of image reconstruction. If the network is trained on corrupted versions of the inputs with the goal of improving the robustness to noise, it is called a denoising autoencoder [27].

4.1.1 Anomaly detection without actually testing samples using an autoencoder network

Anomaly (outlier) detection has been an important research topic in data mining and machine learning while it also provides practical benefits in many real-world applications. Outlier detection has been used in spectroscopic data to detect and remove anomalous observations (if required). Most of the process analytical instruments implemented in industrial plants can also be converted to perform outlier detection in addition to their main task; for instance to detect a fault on a factory production line by constantly monitoring specific features of the products and comparing the real-time data with either the features of normal products or those for faults. Outliers arise due to mechanical faults, changes in system behavior, fraudulent behavior, human error, instrument error or simply through natural deviations in populations [28]. Modeling anomalies are not easy in real datasets as they appear irregularly and not often. Since abnormal data points appear rarely it is very costly to collect those data from real world [29]. Hodge and Austin [28] shows a survey on different techniques for outlier detection in machine learning. They highlighted that correct distribution model, correct attribute types, scalability, speed, any incremental capabilities to allow new exemplars to be stored and the modeling accuracy must be considered when selecting a suitable algorithm for outlier detection. In machine learning, multiclass or multinomial classification is the problem of classifying instances into one of three or more classes (classifying instances into one of two classes is called binary classification). In one-class classification which is also referred as class-modeling, whether a sample is compatible or not with the characteristics of a single class of interest is considered. The study by Hofer-Schmitz [30], presents an one-class anomaly detector based on autoencoder for Raman spectra for a biological application, where it's very costly to collect spectra of the outlier class. They use two chemical data sets with 10,000 samples and over 2000 samples for their evaluation. Bio-chemical approach to identify and characterize outliers takes months and therefore they measured normal class and trained one-class model using Autoencoder network to learn the normal classes' characteristics by minimizing the reconstruction error (score) with respect to the given loss function, similar to the learnt components of PCA. When using the learnt encodings to reconstruct irregular spectra, a sample's reconstruction was considered as anomaly if it exceeds a standard deviation threshold.

4.1.2 Sample classification using an autoencoder network

Houston [31] used six classification algorithms to identify whether a set of chemical samples contain chlorinated solvents or not, based on their Raman spectra. Dataset included 230 Raman spectra of solvents and solvent mixtures. An additional dataset comprising 24 Raman spectra of carbohydrates was compiled for use as examples of possible outlier data. k-Nearest Neighbors (kNN), Support Vector Machine, Decision Tree, Fully Connected Neural Network (FCNN), Gaussian Naïve Bayes, Locally Connected Neural Network (LCNN) were the algorithms used. The ability of the autoencoder models to correctly identify negative outliers were further demonstrated. Their results showed that a two-step process, combining an outlier detector and LCNN binary classifier, have better performance.

LCNN is quite the same as the Convolutional layer explained in Section 4.2. But has one (important) difference. In LCNN, there is a locally connected layer going from the inputs to the first hidden layer. In the Convolutional layer the filter is common among all output neurons. In Locally-Connected Layer, each neuron has its own filter. This type of layer let the network to learn different types of feature for different regions of the input, but if there is less number of data, it can also generate over-fitting.

4.1.3 Increasing signal-to-noise ratio (SNR) by convolutional denoising autoencoder (CDAE)

Obtaining the highest possible SNR and a good enough spectral resolution for a specific analysis are important factors while using Raman spectroscopy. The light of the Raman signal is refocused on a charge-coupled device (CCD) after dispersion by a diffraction grating which inevitably lower the signal. To obtain better Raman signals, generally, the excitation intensity is increased. However, this is not always practicable if the sample is sensitive to higher laser power. Physical and chemical properties of sensitive samples can be degraded by exposing to higher laser power. Therefore, in experimenting such, laser exposure times are extended while keeping a lower excitation intensity. As a result, stray light, environmental light, and the inherent interior noise of electronic or optical devices [13] result in noise adding up over longer integration time. These factors influence signal-to-noise ratio (SNR), thus further affecting the feature extraction of the valid signal. Fan, et al. [32] proposes a relevant automatic denoising method of convolutional denoising autoencoder (CDAE) to advance the SNR in Raman spectra without manual intervention. **Figure 9** shows the CDAE model which includes three layers of convolution and max-pooling (the encoder) and three layers of upsampling and convolution (the decoder) proposed by the authors Fan, Zeng [32]. The proposed CDAE model was implemented using Keras and Tensorflow. The authors show that the CDAE method outperforms other classical denoising methods such as Savitzky–Golay filter and wavelet transform.

4.1.4 Stacked sparse autoencoder (SSAE) to extract features from the unlabeled Raman data

Sparse autoencoder (SAE) may include more (rather than fewer) hidden units than inputs, but only a small number of the hidden units are allowed to be active at the same time. This sparsity constraint forces the model to respond to the unique statistical features of the training data. Sparse feature learning

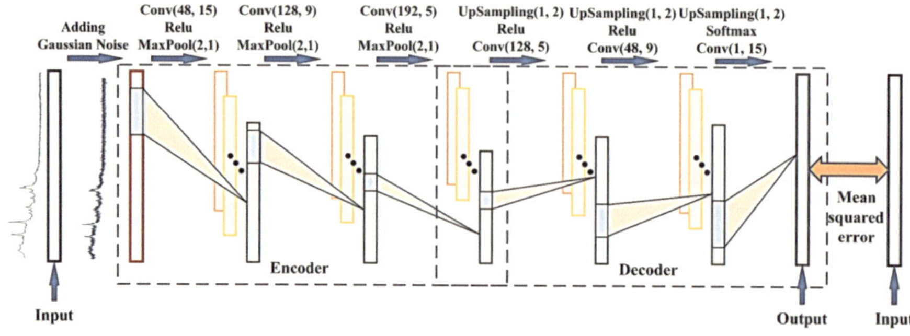

Figure 9.
The architecture of CDAE used to increase SNR in Raman spectra by fan, Zeng [32].

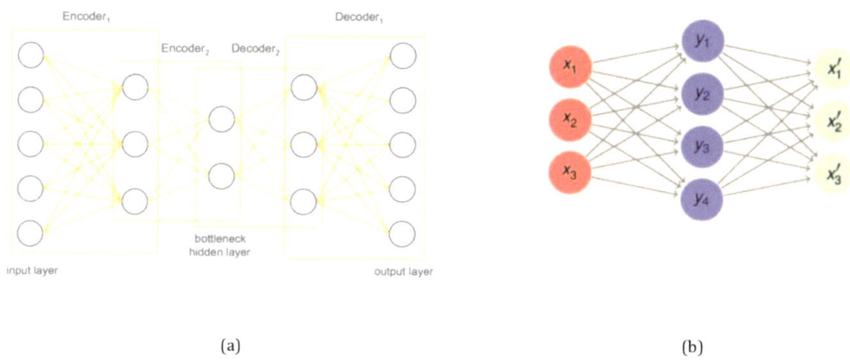

Figure 10.
Representation of (a) stacked autoencoder and (b) sparse autoencoder.

algorithms range from sparse coding approaches [33] to training neural networks with sparsity penalties. In the SAE, once the training process is performed, the decoder and reconstruction layer will be removed, and the features learned from the original data are preserved in the hidden layer. To extract high-level features, a stacked SAE (SSAE) is utilized. The SSAE consists of several SAEs, with the output of the previous SAE used as the input of a subsequent SAE (**Figure 10**) [34].

Feature extraction using a stacked sparse autoencoder integrated with a Softmax classifer (SMC) to extract the discriminative features from unlabeled Raman data of breath samples is proposed by Aslam [35]. They were successful to identify fifty peaks in each spectrum to distinguish the patients with gastric cancer and healthy persons. The architecture of this neural network comprises of two sparse autoencoder layers and the output of the stacked sparse autoencoder was wired into a Softmax layer as shown in **Figure 11**. This system reduces the distance between the input and output by learning the features and preserve the structure of the input data set of breath samples. The proposed deep stacked sparse autoencoder neural network architecture exhibits excellent results, with an overall accuracy of 98.7% for advanced gastric cancer classification and 97.3% for early gastric cancer detection using breath analysis.

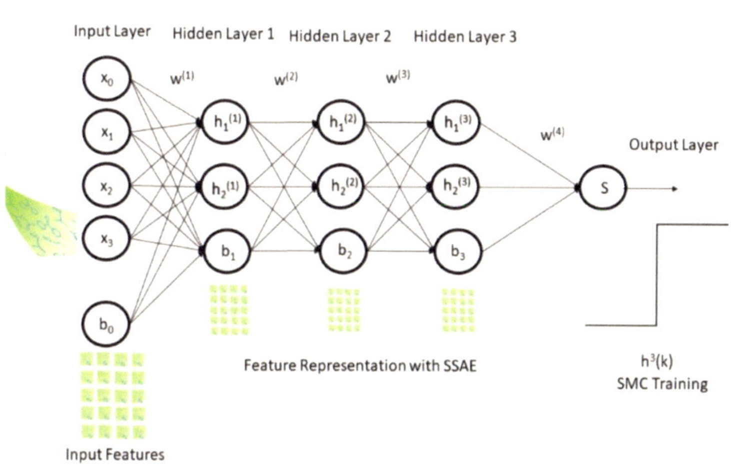

Figure 11.
Architecture of the stacked sparse autoencoder with a Softmax classifier [35].

4.2 Convolution neural network (CNN)

Several studies show that the convolution neural network (CNN) modeling method is potential to be used for spectral analysis. With the development of *deep learning*, CNN has become a major tool specially for image analysis. Independency from prior knowledge and human extraction is a feature of machine learning and CNN is one of best examples for that. In traditional algorithms, hand-engineered filters are used in preprocessing to understand the learning process, whereas in CNN, filters (or usually known as kernels) automatically learn the optimization without human intervention.

A typical CNN includes *convolutional, pooling* and *fully connected layers* as shown in **Figure 12**. Convolution is the process involving combination of two functions that produces the other function as a result. The convolution layers improve the performance of the network by shared weights and sparse connection. Pooling layer operates on each feature map independently. This reduces resolution of the feature map by reducing height and width of features maps, but retains features of the map required for classification. This is called Down-sampling. The output feature map(matrix) from pooling layer is converted into vector (one dimensional array) which is called flatten layer. Fully connected layer looks like a regular neural network. Soft-max is an activation layer normally applied to the last layer of network that acts as a classifier and it is used to map the non-normalized output of a network to a probability distribution.

4.2.1 CNN for predicting material properties and understanding composition-structure-property relationships

A CNN model was constructed by Umehara, et al. [37] in python using Keras package with Tensorflow backend to identify composition-property and composition-structure–property relationships that lead to fundamental materials insights through Raman spectra. They developed a model that could predict photoelectrochemical power density (*P*) of a solar fuels photoanode from materials parameters which were their composition and Raman signals of 1379 samples. **Figure 13** shows the CNN model structure used for this study where they used 10 different layers to derive model predictions. In addition to the prediction of material property, they also analyzed gradients in the trained model which was useful to reveal key data relationships that were not readily identified by human inspection or traditional statistical analyses. For instance, what is the impact of performance of the

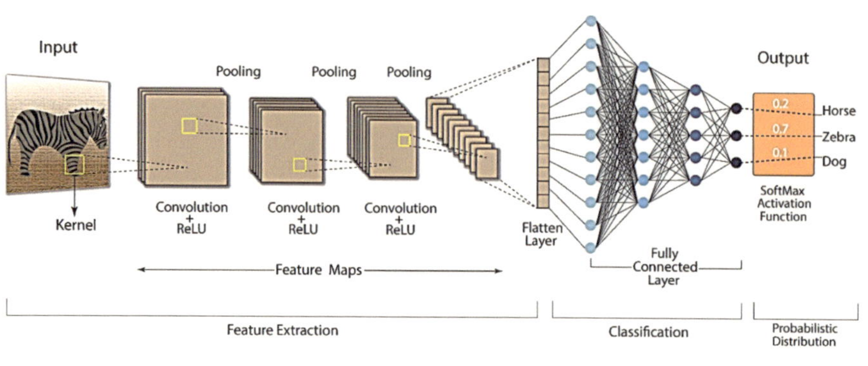

Figure 12.
Example showing how CNN works [36].

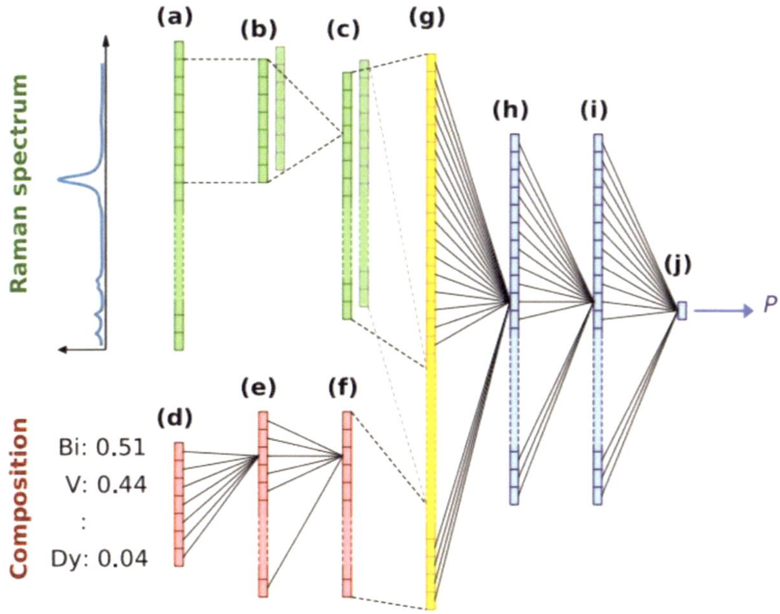

Figure 13.
Schematic of CNN model structure used by Umehara, stein [37]. The model takes the Raman spectrum and the composition –(Bi, V and Dy) as input to predict P = power density. Each of the 10 layers of the CNN model are labeled a to j. red: Dense layers acting on composition, green: Convolutional 1D layers acting on spectra, yellow: Flattening and concatenation layers, blue: Dense layers acting on both the composition and spectral data.

photoanode by changing the concentrations of alloying elements? Authors have also highlighted that the human interpretation of these key relationships produces the desired fundamental understanding, demonstrating a framework in which machine learning accelerates data interpretation by leveraging the expertise of the human scientist.

4.2.2 Identification of chemical species by CNN without preprocessing

Liu, et al. [38] describes an unified solution for the identification of chemical species. They used a trained convolutional neural network to automatically identify substances according to their Raman spectrum without preprocessing. Most of the Raman based regression procedures demand for preprocessing such as cosmic ray removal, smoothing and baseline correction. CNN combines preprocessing, feature extraction and classification in a single architecture which can be trained end-to-end with no manual tuning [38].

They evaluated their approach using the RRUFF spectral database [39], comprising mineral sample data and a superior classification performance was demonstrated compared with other frequently used *machine learning* algorithms including SVM, kNN, gradient boosting, CNN and random forest. In addition, 512 raw minerals spectra from the RRUFF database were used together with six widely-used baseline correction methods: modified polynomial fitting, rubber band, robust local regression estimation, iterative restricted least squares, asymmetric least squares smoothing and rolling ball. Results are shown in **Table 1** which indicates that the raw spectra on CNN achieved an accuracy of 93.3% and the second best method, kNN with rubber band baseline correction, achieved an accuracy of 82.5%.

Method	kNN (k = 1)	Gradient boosting	Random forest	SVM (linear)	SVM (Radial basis function)	CNN
Raw	0.429 ± 0.011	0.373 ± 0.019	0.394 ± 0.016	0.522 ± 0.011	0.434 ± 0.012	C.933 ± 0.007
Assymmetric least squares	0.817 ± 0.010	0.773 ± 0.009	0.731 ± 0.019	0.821 ± 0.012	0.629 ± 0.016	C.927 ± 0.008
Modified polynomial	0.778 ± 0.007	0.740 ± 0.016	0.650 ± 0.016	0.785 ± 0.014	0.629 ± 0.016	0.920 ± 0.008
Rolling ball	0.775 ± 0.009	0.737 ± 0.008	0.689 ± 0.018	0.795 ± 0.011	0.624 ± 0.013	0.918 ± 0.008
Rubber band	0.825 ± 0.007	0.792 ± 0.015	0.741 ± 0.009	0.806 ± 0.015	0.620 ± 0.010	0.911 ± 0.008
IRLS	0.772 ± 0.010	0.710 ± 0.008	0.675 ± 0.007	0.781 ± 0.011	0.614 ± 0.010	0.911 ± 0.008
Robust local regression	0.741 ± 0.009	0.694 ± 0.008	0.667 ± 0.0012	0.759 ± 0.013	0.600 ± 0.013	0 909 ± 0.007

Table 1.
Performance of different ML algorithms with and without baseline correction methods after applied on RRUFF Raman spectral data (of 512 minerals).

4.2.3 Tuning preprocessing of Raman spectra in one step by training a CNN model using simulated data

Wahl, et al. [40] show that a convolutional neural network can be trained using simulated data to handle several preprocessing steps automatically in a single step. These preprocessing methods include cosmic ray removal, signal smoothing, and baseline subtraction. Synthetic spectra were created by randomly adding peaks, baseline, mixing of peaks and baseline with background noise, and cosmic rays. Secondly, a CNN was trained on synthetic spectra and known peaks. Finally, a test set data which consisted of real Raman spectra of polyethylene, paraffin, and ethanol were used to evaluate the trained CNN model. The samples were placed on a polystyrene petri dish and their Raman measurements were taken so that the signals from the samples were mixed with signal from polystyrene. Measurements which only contained one cosmic ray were saved for the analysis. The performance of the CNN model was estimated by calculating the root mean squared error (RMSE). From 105 simulated observations, 91.4% predictions had smaller absolute error (RMSE). Authors also recommend that the similar simulation scheme for adaptations to problems with similar preprocessing challenges such as NIR, FT-IR, mass spectroscopy, and chromatograph and also take the benefit of the reduced computational time and time spent by an analyst in preparing data for the analysis. CNN preprocessing generated reliable results on measured Raman spectra from polyethylene, paraffin, and ethanol with background contamination from polystyrene.

4.2.4 CNN for bacterial detection, identification, and antibiotic susceptibility testing in a single step

Different bacterial phenotypes are characterized by unique molecular compositions. However they only lead to subtle differences in their corresponding Raman spectra. And due to the weak Raman scattering these subtle spectral differences are easily masked by background noise. Maintaining a higher signal-to-noise ratio by increasing the measurement time are often restricted in these types of samples. This challenge has been addressed by [41] using a trained convolutional neural network which can classify noisy bacterial spectra by using a very low measurement time of 1 second. The reference samples including bacterial and yeast isolates which generated 2000 spectra from a Raman microscope. Spectra were background corrected using a polynomial fit of order 5.

Figure 14 shows (a) spectral variation of Raman bacterial spectra and (b) the CNN architecture. CNN architecture used by these researchers consisted of an initial convolution layer followed by 6 residual layers and a final fully connected classification layer. Each residual layer contains 4 convolutional layers, and therefore the total depth of the network was 26 layers. The initial convolution layer has 64 convolutional filters, while each of the hidden layers has 100 filters. An identification accuracies of 99.7% was achieved by the researchers in this study when they validated the method using clinical samples.

4.3 PCANet

The principal component analysis network (PCANet), which is one of the recently proposed deep learning architectures, achieves the state-of-the-art classification accuracy in various databases [42]. It is also known as one of the simplest deep learning algorithms and can be adapted to small-scale data [43]. In the section below, application of PCANet deep learning for Raman spectroscopy is reviewed

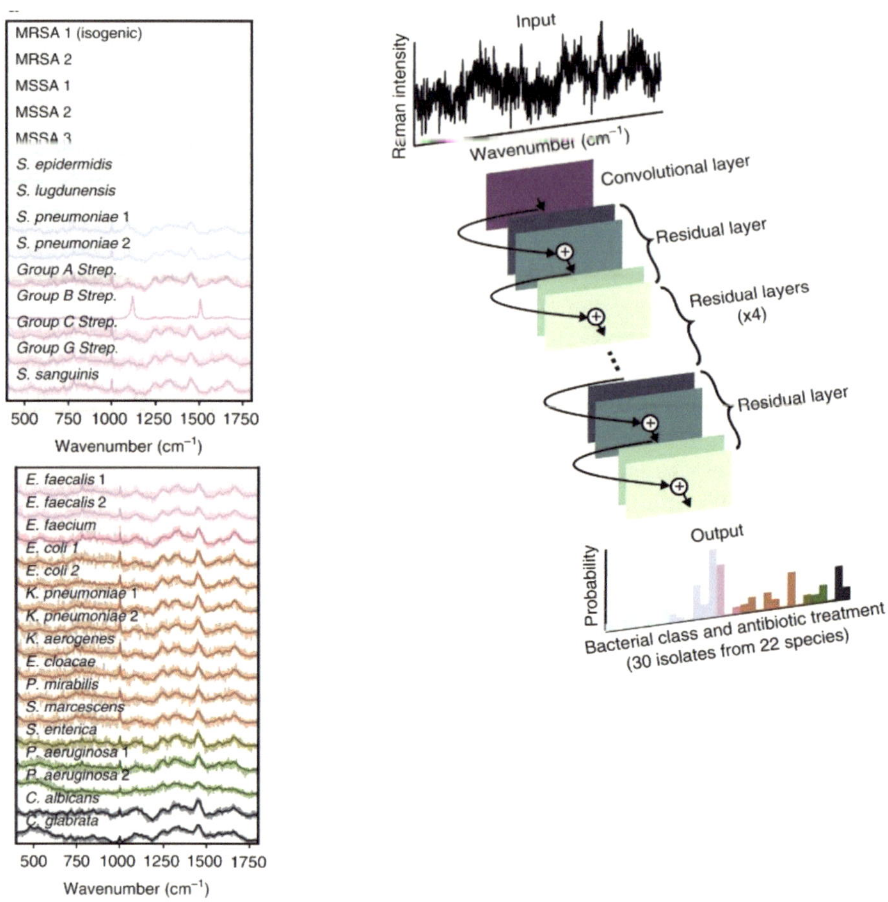

(a) (b)

Figure 14.
(a) Averages of 2000 spectra from 30 isolates in bold and overlaid on representative examples of noisy single spectra for each isolate. Raman spectra of bacterial species can be difficult to distinguish, and short integration times (1 s) lead to noisy spectra (SNR = 4.1). Spectra are color-grouped according to antibiotic treatment (b) low-signal Raman spectra are classified as one of 30 isolates, which are then grouped by empiric antibiotic treatment and one-dimensional residual network with 25 total convolutional layers [41].

using some of prominent research studies. Architecture of the PCANet is shown in **Figure 15**. It typically consists with only two convolutional layers.

The main algorithm used to learn the convolutional filters in PCANet is principal component analysis (PCA) algorithm. PCA is a linear transformation method which transforms original data to a new orthogonal coordinate system with less dimensionality. Eigenvalues and eigenvectors are calculated from the covariance matrix of the original dataset. Eigenvectors which have the highest eigenvalues are always selected while discarding that of small values. In the convolutional layer of PCANet, all local patches are convolved with the selected eigenvectors to create a new set of data which focus on the most important features of the input data. The main flow of PCANet can be divided into three stages. The function of the first two stages is similar, and the principal eigenvector of input matrix is obtained through the cascaded multiple-PCA filter in these two stages. In the last stage, the principal eigenvectors are performed by binary hash encoding and then processed to the composed block-wise histogram. Afterward, the histogram is combined with the classification algorithms to obtain the predicted data.

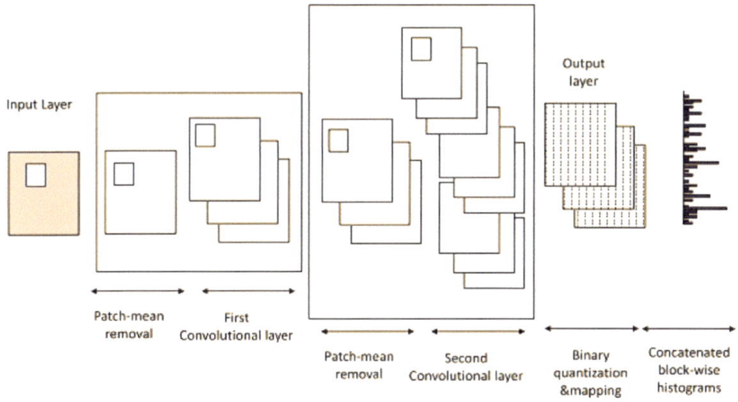

Figure 15.
PCANet architecture [44].

4.3.1 Recognition and quantitation of drugs in human urine by PCANet

Weng, et al. [45] shows that *deep learning* networks perform better than the common *machine learning* methods (including kNN, SVM, RF, LR, and PLS) and provide feasible alternatives for the recognition and quantitation of SERS. In their study, *deep learning* networks were used as fully connected networks, convolutional neural networks (CNN), fully convolutional networks (FCN), and principal component analysis networks (PCANet) to determine their abilities to recognize drugs in human urine and measure pirimiphos-methyl in wheat extract in the two input forms of a one-dimensional vector or a two-dimensional matrix.

4.3.2 Rapid detection of impurities using PCANet

Surface-enhanced Raman spectroscopy (SERS) has affected many areas in analytical detection, surface property investigation, biological event and marker sensing and imaging, and environment monitoring and its application in analytical science, food science, environmental sciences and biomedical sciences is enormous [46]. The study by Weng, et al. [47] proposes the suitability of SERS over NIR and FTIR for the automatic analysis of hazardous pesticide residues (acephate) in rice due to the significant interference from the aqueous phase. They used 82 contaminated rice samples for the model development and 14 contaminated rice samples were randomly selected as the prediction set. Finally, they combined the modeling methods in PCANet with the regression algorithms as PLSR, SVM, or RF ($PCANet_{PLSR}$, $PCANet_{RF}$, and $PCANet_{SVM}$) to obtain the residue level.

4.4 Recurrent neural network (RNN)

Recurrent Neural Network (RNN) is a tool in deep learning for problems that deal with sequential data [48]. Although, RNN was firstly designed to deal with sequential information, today it shows applications in time series data, natural language and converting non-sequencing data like images to sequences. The most used recurrent units are long short-term memory (LSTM) and gated recurrent unit (GRU). LSTM is a deep learning system that avoids the vanishing gradient problems in RNN [49]. The GRU is like a LSTM but it has fewer parameters than an LSTM [50]. Some results indicate that GRUs can outperform LSTMs while others show

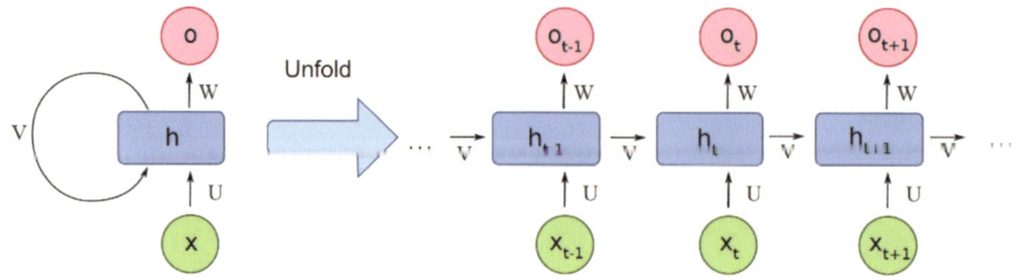

Figure 16.
Architecture of RNN; x: The input, O: The output, h: The main block of the RNN. It contains the weights and the activation functions of the network, V: Represents the communication from one time-step to the other [51].

the opposite results. The RNN models are trained with back propagation through time (BPTT) method. There are variants of RNN such as bidirectional RNN and deep RNN.

Possibility of processing input of any length, model size which is not affected with size of input, computational ability which takes into account historical information and weights which are shared across time which makes an efficient data handling are the advantages of RNN. On the other hand, it also has the drawback of having a slower computation, difficulty of accessing information from a long time ago and inability to consider any future input for the current state (**Figure 16**).

4.4.1 Species identification and model transfer using RNN

Species identification of human and animal blood is of critical importance in the areas of custom inspection, forensic science, wildlife preservation, and veterinary purpose. High-performance liquid chromatography (HPLC), mass spectroscopy (MS), nuclear magnetic resonance (NMR), polymerase chain reaction (PCR) are DNA profiling suitable methods, but they require experienced experts and professional laboratory. FTIR is also a promising candidate for this purpose but the presence in water makes the spectral analysis is challenging. Considering the interference of water and the risk of contact of pathogen, Wang, et al. [52], used Renishaw inVia confocal Raman spectrometer and a laboratory-built Raman spectrometer to find a method to discriminate of 20 kinds of blood species including human, poultry, wildlife, and experimental animals. The Raman spectra pre-processing methods included cosmic ray removal, Savitzky–Golay filter, baseline removal, normalization and standardization. The processed spectra were randomly grouped into training dataset (80%), validation dataset (10%) and testing dataset (10%). Data was input to different deep learning models such as RNN, GRU, LSTM and CNN and performance was compared. This study also proposes a solution for the wavenumber drift during long term use of instruments. Analyzing the blood samples are affected by the wavenumber drift and therefore instruments are required for immediate calibration. The usual RNN model could not function well for these unexpected drifts and therefore augmented Raman spectra with certain wavenumber drift were included intentionally in this study. Another speciality of this study is the migration learning of model transfer between Raman spectrometers with different performance. This was achieved by training a cross-instrument RNN model with spectra from 2 Raman spectrometers (1463 spectra from Renishaw Raman spectrometer and 1621 spectra from laboratory-built Raman spectrometer), which could be used for identification of blood species. This combined model showed accuracy is 98.2%.

4.4.2 Gated recurrent unit coupled with MCNN

The study [53] proposes the use of a gated recurrent unit (GRU) and multiscale fusion convolutional neural network (GRU-MCNN) to analyze Raman spectra of patients infected with hepatitis B virus (HBV). Current commonly used method for the detection of HBV is polymerase chain reaction, but the shortcomings of this method such as the possibility of cross-contamination of samples during the analysis which can generate false results and using a carcinogenic dying agent for the sample preparation can be eliminated using Raman spectroscopy non-invasive analysis. Unlike traditional methods for extracting spatial features, the MCNN first transforms the original data sets into a pyramid structure containing spatial information at multiple scales, and then automatically extracts high-level spatial features using multiscale training data sets [54]. GRU-MCNN model developed by [53] showed accuracy, precision, sensitivity and specificity over 0.97 for unprocessed data and it is even a higher value that was recorded for processed data.

4.5 Performing a deep learning analysis for data

Various deep learning tools are available in the market today, such as Neural Designer, H2O.ai, DeepLearningKit, Microsoft Cognitive Toolkit, Keras, ConvNetJS, Torch, Gensim, Deeplearning4j, Apache SINGA, Caffe, Theano, ND4J, and MXNet. Which one is the best, depends on the user and application. Many of these machine learning algorithms are available as free software modules and/or libraries for programming environments like Python, R, C++ and C# [55–58] to mention some. In the python programming environment, Keras and TensorFlow modules are popular for deep learning. Microsoft has the free ML.NET machine learning environment that is supported using the Visual Studio tools. Matlab and Python are widely used in academics and use a GUI interface enabling ML without writing the code by the user, however, some programming skills are needed. If the user wants to explore ML in depth and write his own code from scratch, R is often preferred, but there is really no agreed consensus on this matter. Python is a programming language which consists of a large standard library. One major advantage with Python is that it is free. Matlab is most highly regarded as not only a commercial numerical computing environment, but also as a programming language. Matlab has many functions for data processing and plotting. It also contains toolboxes such as Deep Learning toolbox. Toolboxes in Matlab usually comes with added cost. R is free, open-source software designed to run statistical analyses and output graphics.

5. Conclusion

Most common procedure employed in spectroscopic data analysis is selecting proper tools, validating them, and highlighting their use in real-world applications by a series of examples. Getting inspiration by the field of computer vision will surely accelerate the development of more robust methods in this process. The next generation of Raman data analysis will be using more advanced algorithms to further improve the analytical performance of spectral classification, regression, clustering, and rule mining. In supplementary, it will also be the key factor to break the limitations of Raman spectroscopic applications.

For instance, literature shows that molecular spectra predictions can be made instantly using deep learning at no further cost for the end user. Spectra with outliers are synthetically implemented and solved using autoencoders when such irregular spectra are costly or time consuming to obtain in reality. Scientists show that several Raman preprocessing steps can be performed using a single step by convolutional neural networks while in traditionally, combinations of preprocessing methods are performed as iterations to select the optimum preprocessing which demand time. Some DL algorithms show promising results by using raw spectra in entire wavelength as input for regression models region replacing monotonous variable selection methods. Classification problems in SERS and Raman spectra, have received the advantages of general image recognition deep learning methods which significantly improve selectivity and specificity over conventional classification methods. Unlabeled large Raman datasets which have been collected over years in clinical applications have been using to diagnose other diseases in addition for their main purpose where accuracy of the data interpretation are improved as dataset is being updated and heavier.

The classical linear methods of processing the extracted information from challenging Raman and SERS experiments no longer suffice. Deep learning is shaping up machine learning algorithms in many ways through carefully analyzing patterns and aberrations in those patterns. In analytical sciences, machine learning provides an unprecedented opportunity to extract information from complex datasets. Very often, the unfamiliarity of machine learning algorithms and definitions which is normally in the computer science domain, dictates the unpopularity of using them as tools in chemistry and analytical science. This chapter is aimed to elaborate the potential of deep learning methods with respect to its suitability in Raman spectral analysis. As these methods are applicable to other types of spectroscopies deep learning and artificial intelligence data processing in spectroscopy is bound to grow in the near future.

Acknowledgements

Authors would like to acknowledge the financial assistance from Faculty of Technology, University of South-Eastern Norway.

Conflict of interest

The authors declare no conflict of interest.

Author details

M.H. Wathsala N. Jinadasa*, Amila C. Kahawalage, Maths Halstensen, Nils-Olav Skeie and Klaus-Joachim Jens
Faculty of Technology, Natural Sciences and Maritime Sciences, University of South-Eastern Norway, Porsgrunn, Norway

*Address all correspondence to: wathsala.jinadasa@usn.no

References

[1] Ferraro, J.R., K. Nakamoto, and C.W. Brown, *Chapter 1 - Basic Theory*, in *Introductory Raman Spectroscopy (Second Edition)*, J.R. Ferraro, K. Nakamoto, and C.W. Brown, Editors. 2003, Academic Press: San Diego. p. 1-94.

[2] Smekal, A., *Zur Quantentheorie der Dispersion.* Naturwissenschaften, 1923. **11**(43): p. 873-875.

[3] Raman, C.V. and K.S. Krishnan, *A New Type of Secondary Radiation.* Nature, 1928. **121**(3048): p. 501-502.

[4] Das, R.S., Agrawal, Y. K., *Raman spectroscopy: Recent advancements, techniques and applications.* Vibrational Spectroscopy, 2011. **57**(2): p. 163-176.

[5] Ferraro, J.R., K. Nakamoto, and C.W. Brown, *Chapter 3 - Special Techniques*, in *Introductory Raman Spectroscopy (Second Edition)*, J.R. Ferraro, K. Nakamoto, and C.W. Brown, Editors. 2003, Academic Press: San Diego. p. 147-206.

[6] Cozzolino, D., *The Sample, the Spectra and the Maths—The Critical Pillars in the Development of Robust and Sound Applications of Vibrational Spectroscopy.* Molecules, 2020. **25**(16): p. 3674.

[7] Mendez, K.M., D.I. Broadhurst, and S.N. Reinke, *The application of artificial neural networks in metabolomics: a historical perspective.* Metabolomics, 2019. **15**(11): p. 142.

[8] Vogt, F., *Quo vadis, chemometrics?* Journal of Chemometrics, 2014. **28**(11): p. 785-788.

[9] Kowalski, B., S. Brown, and B. Vandeginste, *Editorial.* Journal of Chemometrics, 1987. **1**(1): p. 1-2.

[10] HÉBerger, K., *Chapter 7 - Chemoinformatics—multivariate mathematical–statistical methods for data evaluation*, in *Medical Applications of Mass Spectrometry*, K. Vékey, A. Telekes, and A. Vertes, Editors. 2008, Elsevier: Amsterdam. p. 141-169.

[11] Wold, S. and M. Sjöström, *Chemometrics, present and future success.* Chemometrics and Intelligent Laboratory Systems, 1998. **44**(1): p. 3-14.

[12] Mandrell, C.T., T.E. Holland, J.F. Wheeler, S.M.A. Esmaeili, K. Amar, et al., *Machine Learning Approach to Raman Spectrum Analysis of MIA PaCa-2 Pancreatic Cancer Tumor Repopulating Cells for Classification and Feature Analysis.* Life, 2020. **10**(9): p. 181.

[13] Ullah, R., S. Khan, S. Javaid, H. Ali, M. Bilal, et al., *Raman spectroscopy combined with a support vector machine for differentiating between feeding male and female infants mother's milk.* Biomedical optics express, 2018. **9**(2): p. 844-851.

[14] Sattlecker, M., C. Bessant, J. Smith, and N. Stone, *Investigation of support vector machines and Raman spectroscopy for lymph node diagnostics.* Analyst, 2010. **135**(5): p. 895-901.

[15] Yan, S., S. Wang, J. Qiu, M. Li, D. Li, et al., *Raman spectroscopy combined with machine learning for rapid detection of food-borne pathogens at the single-cell level.* Talanta, 2021. **226**: p. 122195.

[16] Glaws, A., R. King, and M. Sprague, *Deep learning for in situ data compression of large turbulent flow simulations.* Physical Review Fluids, 2020. **5**(11): p. 114602.

[17] A., S.K., *The Era of Artificial Intelligence, Machine Learning, and Data Science in the Pharmaceutical Industry*. 2021: Academic Press.

[18] Aggarwal, C.C., *Neural Networks and Deep Learning: A Textbook*. Springer.

[19] Flach, P.A., *Machine learning : the art and science of algorithms that make sense of data*. 2017, Cambridge ; New York: Cambridge University Press.

[20] Yamaguchi, T., Y. Kimura, and N. Hirota, *Solvent and Solvent Density Effects on the Spectral Shifts and the Bandwidths of the Absorption and the Resonance Raman Spectra of Phenol Blue.* The Journal of Physical Chemistry A, 1997. **101**(48): p. 9050-9060.

[21] Tobin, M.C., *Sample Problem in Raman Spectroscopy*, in *Developments in Applied Spectroscopy*, A. W.D., Editor. 1962, Springer: Boston, MA.

[22] Marquardt, B. *Addressing the Challenges of Process Raman Spectroscopy*. Spectroscopy April 2016 [cited 36; Available from: https://www.spectroscopyonline.com/view/addressing-challenges-process-raman-spectroscopy.

[23] Kostamovaara, J., J. Tenhunen, M. Kögler, I. Nissinen, J. Nissinen, et al., *Fluorescence suppression in Raman spectroscopy using a time-gated CMOS SPAD.* Optics Express, 2013. **21**(25): p. 31632-31645.

[24] Wei, D., S. Chen, and Q. Liu, *Review of Fluorescence Suppression Techniques in Raman Spectroscopy.* Applied Spectroscopy Reviews, 2015. **50**(5): p. 387-406.

[25] Butler, H.J., Ashton, L., Bird, B., Cinque, Gi., Curtis, K., Dorney, J., Esmonde-W. K., Fullwood, N. J., Gardner, B., Martin-H., P.e L., and M.J. Walsh, McAinsh, M. R., Stone, N., Martin, F. L., *Using Raman spectroscopy to characterize biological materials.* Nature Protocols, 2016. **11**(4): p. 664-687.

[26] Plaut, E., *From Principal Subspaces to Principal Components with Linear Autoencoders.* ArXiv, 2018. **abs/1804.10253**.

[27] Ithapu, V.K., Singh, V., Johnson, S. C., *Chapter 15 - Randomized Deep Learning Methods for Clinical Trial Enrichment and Design in Alzheimer's Disease*, in *Deep Learning for Medical Image Analysis*, S.K. Zhou, H. Greenspan, and D. Shen, Editors. 2017, Academic Press. p. 341-378.

[28] Hodge, V.J. and J. Austin, *A Survey of Outlier Detection Methodologies.* Artificial Intelligence Review, 2004. **22**(2): p. 85-126.

[29] Sharma, S., C. Bellinger, N. Japkowicz, R. Berg, and K. Ungar. *Anomaly detection in gamma ray spectra: A machine learning perspective*. in *2012 IEEE Symposium on Computational Intelligence for Security and Defence Applications*. 2012.

[30] Hofer-Schmitz, K., Nguyen, H., Berwanger, K. *One-class Autoencoder approach to classify Raman spectra outliers*. in *European Symposium on Artificial Neural Networks, Computational Intelligence and Machine Learning* 2018. Bruges (Belgium).

[31] Houston, J., Glavin, F. G., Madden, M. G., *Robust Classification of High-Dimensional Spectroscopy Data Using Deep Learning and Data Synthesis.* Journal of Chemical Information and Modeling, 2020. **60**(4): p. 1936-1954.

[32] Fan, X.-g., Y. Zeng, Y.-L. Zhi, T. Nie, Y.-j. Xu, et al., *Signal-to-noise ratio enhancement for Raman spectra based on optimized Raman spectrometer and convolutional denoising autoencoder.* Journal of Raman Spectroscopy, 2021. **52**(4): p. 890-900.

[33] Olshausen, B.A., Field, D. J., *Sparse coding with an overcomplete basis set: A strategy employed by V1?* Vision Research, 1997. **37**(23): p. 3311-3325.

[34] Idowu, O.P., J. Huang, Y. Zhao, O.W. Samuel, M. Yu, et al., *A stacked sparse auto-encoder and back propagation*

network model for sensory event detection via a flexible ECoG. Cognitive Neurodynamics, 2020. **14**(5): p. 591-607.

[35] Aslam, M.A., Xue, C., Chen, Y., Zhang, A., Liu, M., Wang, K. Cui, D., *Breath analysis based early gastric cancer classification from deep stacked sparse autoencoder neural network.* Scientific Reports, 2021. **11**(1): p. 4014.

[36] E., S.K. *Convolutional Neural Network | Deep Learning*. 2020; Available from: https://developersbreach.com/convolution-neural-network-deep-learning/.

[37] Umehara, M., H.S. Stein, D. Guevarra, P.F. Newhouse, D.A. Boyd, et al., *Analyzing machine learning models to accelerate generation of fundamental materials insights.* npj Computational Materials, 2019. **5**(1): p. 34.

[38] Liu, J., M. Osadchy, L. Ashton, M. Foster, C.J. Solomon, et al., *Deep convolutional neural networks for Raman spectrum recognition: a unified solution.* Analyst, 2017. **142**(21): p. 4067-4074.

[39] Lafuente, B., R.T. Downs, H. Yang, and N. Stone, *Highlights in Mineralogical Crystallography*, in *1. The power of databases: The RRUFF project*, A. Thomas and D. Rosa Micaela, Editors. 2015, De Gruyter (O). p. 1-30.

[40] Wahl, J., M. Sjödahl, and K. Ramser, *Single-Step Preprocessing of Raman Spectra Using Convolutional Neural Networks.* Appl Spectrosc, 2020. **74**(4): p. 427-438.

[41] Ho, C.-S., N. Jean, C.A. Hogan, L. Blackmon, S.S. Jeffrey, et al., *Rapid identification of pathogenic bacteria using Raman spectroscopy and deep learning.* Nature Communications, 2019. **10**(1): p. 4927.

[42] Wu, J., S. Qiu, Y. Kong, L. Jiang, Y. Chen, et al., *PCANet: An energy perspective.* Neurocomputing, 2018. **313**: p. 271-287.

[43] Chan, T., Jia, K., Gao, S., Lu, J., Zeng, Z., Ma, Y., *PCANet: A Simple Deep Learning Baseline for Image Classification?* IEEE Transactions on Image Processing, 2015. **24**(12): p. 5017-5032.

[44] Aly, S., Mohamed, A. , *Unknown-Length Handwritten Numeral String Recognition Using Cascade of PCA-SVMNet Classifiers.* IEEE Access, 2019. **7**: p. 52024-52034.

[45] Weng, S., H. Yuan, X. Zhang, P. Li, L. Zheng, et al., *Deep learning networks for the recognition and quantitation of surface-enhanced Raman spectroscopy.* Analyst, 2020. **145**(14): p. 4827-4835.

[46] Ouyang, L., W. Ren, L. Zhu, and J. Irudayaraj, *Prosperity to challenges: recent approaches in SERS substrate fabrication.* Reviews in Analytical Chemistry, 2017. **36**(1).

[47] Weng, S., W. Zhu, P. Li, H. Yuan, X. Zhang, et al., *Dynamic surface-enhanced Raman spectroscopy for the detection of acephate residue in rice by using gold nanorods modified with cysteamine and multivariant methods.* Food Chemistry, 2020. **310**: p. 125855.

[48] Britz, D. *Recurrent Neural Networks Tutorial, Part 1 – Introduction to RNNs*. 2015; Available from: http://www.wildml.com/2015/09/recurrent-neural-networks-tutorial-part-1-introduction-to-rnns/.

[49] Gers, F., N. Schraudolph, and J. Schmidhuber, *Learning Precise Timing with LSTM Recurrent Networks.* Journal of Machine Learning Research, 2002. **3**: p. 115-143.

[50] Britz, D. *Recurrent Neural Network Tutorial, Part 4 – Implementing a GRU/LSTM RNN with Python and Theano*. 2015; Available from: http://www.wildml.com/2015/10/recurrent-neural-network-tutorial-part-4-implementing-a-grulstm-rnn-with-python-and-theano/.

[51] Zhu, J., Z. Yang, M. Mourshed, Y. Guo, Y. Zhou, et al., *Electric Vehicle Charging Load Forecasting: A Comparative Study of Deep Learning Approaches.* Energies, 2019. **12**(14): p. 2692.

[52] Wang, P., L. Guo, Y. Tian, J. Chen, S. Huang, et al., *Discrimination of blood species using Raman spectroscopy combined with a recurrent neural network.* OSA Continuum, 2021. **4**(2): p. 672-687.

[53] Guo, Z., X. Lv, L. Yu, Z. Zhang, and S. Tian, *Identification of hepatitis B using Raman spectroscopy combined with gated recurrent unit and multiscale fusion convolutional neural network.* Spectroscopy Letters, 2020. **53**(4): p. 277-288.

[54] Zhao, W. and S. Du, *Learning multiscale and deep representations for classifying remotely sensed imagery.* ISPRS Journal of Photogrammetry and Remote Sensing, 2016. **113**: p. 155-165.

[55] Cole, M.R., *Hands-On Machine Learning with C# : Build smart, speedy, and reliable data-intensive applications using machine learning*. 1 ed. 2018: Packt Publishing.

[56] Kolodiazhnyi, K., *Hands-On Machine Learning with C++: Build, train, and deploy end-to-end machine learning and deep learning pipelines*. 2020: Packt Publishing.

[57] Raschka, S., *Python Machine Learning: Unlock deeper insights into Machine Leaning with this vital guide to cutting-edge predictive analytics*. 2015: Packt Publishing.

[58] Lantz, B., *Machine Learning with R - Second Edition: Expert techniques for predictive modeling to solve all your data analysis problems*. 2015: PAckt Publishing.

Chapter 6

High-Wavenumber Raman Analysis

Shan Yang

Abstract

Raman spectra are molecule specific, and their peaks in the fingerprint region (200-2000 cm^{-1}) are often sufficient for material identification. High-wavenumber signals (> 2000 cm^{-1}) are rare in inorganic material but rich in organic materials containing light hydrogen atoms. Reports on high-wavenumber (HW) Raman signals are far less than fingerprint signals. This could be partially attributed to the difficulty obtaining HW Raman signals, especially from biological materials containing fluorescent proteins. The development and the availability of InGaAs array and the near-infrared (NIR) laser enabled the acquisition of distinct HW Raman from bio-materials. In this chapter, we will introduce recent applications of HW Raman spectroscopy on different materials, especially on biological tissues. Raman instrumentation based on multiple lasers or multiple spectrometers will also be discussed.

Keywords: High-wavenumber Raman, High-fluorescent, Dental Hard tissues, Hydration, Water analysis

1. Introduction

Raman scattering is the inelastic scattering of photons by a matter; specifically, it results from the interactions between the photons and molecular vibrations in the matter. In a simple scenario, a molecule is composed of two atoms, and the vibrational energy of such a diatomic molecule can be approximately treated as a simple harmonic oscillator with the energy states of $E_n = \frac{\left(n+\frac{1}{2}\right)h}{2\pi}\sqrt{\frac{k}{\mu}}$, where n is the vibrational quantum number that can take values of 0, 1, 2, and so on; k is a spring constant, and $\mu = \frac{m_1 m_2}{m_1 + m_2}$ is the reduced mass of the two atoms. This approximation indicates the vibrational energy inversely proportional to the square root of the reduced mass. Therefore, diatomic molecules composed of light atoms have higher vibration energies; in other words, their Raman signals appear at higher wavenumber regions than those consisting of heavier atoms. For complex molecules composed of multiple atoms, a bond connecting two atoms will be influenced by other atoms nearby; however, bonds connecting lighter atoms still yield Raman signals at higher wavenumber regions in general. **Table 1** shows Raman signal positions of several common molecules and chemical bonds [1, 2].

Molecules or chemical bonds	Vibration modes	Raman frequency [cm^{-1}]
H_2	stretch	4395
H_2O	symmetric stretch	3657
CO	stretch	2170
HCl	stretch	2990
NO	stretch	1904
OH of hydroxyapatite	OH stretch	3570
PO_4 of hydroxyapatite	symmetric stretch	960
CH of CH4	symmetric stretch	2914

Table 1.
Raman frequencies of representative molecules and chemical bonds.

Detection of high-wavenumber (HW) Raman signals is as simple as other Raman signals for most of the inorganic materials (e.g., water, graphene) and nonliving organic material (e.g., alcohol, glycerol etc.). However, obtaining HWRaman signals from biological tissues, which have auto-fluorescent emissions under laser excitation, has been challenging. As an inelastic scattering, the Raman effect can happen with lasers at any wavelength from UV lights to near-infrared. However, excitation with lasers at a shorter wavelength is preferred because the efficiency of Raman scattering is inversely proportional to the fourth-order of the incident wavelength. Reduce the excitation, e.g., by a factor of 2, could easily lead to the increase of signal intensity by a factor of 16, which is significant to most Raman studies with low scattering efficiency (in the order of one part per million) [3]. On the other hand, the higher energy photons of the shorter wavelength are more capable of producing fluorescence from materials (especially biological tissues) which could overwhelm the weak Raman signals of interests [4]. Excitation with lasers at longer wavelength has been shown to be a major practice that could alleviate the strong fluorescence from biological tissues [5]. Therefore, for a given material with high auto-fluorescence, exploring various excitation wavelengths may be employed to find the optimal one [6–10].

In addition to the need to balance lower fluorescence interference and higher collecting efficiency, several factors in Raman instrumentations (e.g., the availability of lasers and detectors) should also be considered for detecting HWRaman signals. The most critical component in Raman instrumentation is the Raman spectrometer, which varies from simple, compact version to user-adjustable complex version. Research grade Raman spectrometers allow users to modify componential configurations such as replacing detectors, changing gratings, adjusting slit width, repositioning focusing mirror, etc. The combination of small slit width, high groove density grating, and long focal distance can allow research-grade spectrometers to achieve sub-wavenumber (cm^{-1}) spectral resolution [11]. Compact Raman spectrometers have fixed spectral range and resolution and usually allow none/limited configuration modifications for end users. However, utilizing high-dispersion and high-efficiency volume phase transmission gratings, as well as sensitivity enhancement (through back illumination and special coating) and noise reduction (through one or two stages of thermal electric cooling), the compact spectrometers can achieve comparable and even higher signal to noise ratio than research-grade spectrometers under similar experiment conditions (e.g., laser power, integration time, and spectral resolution). In addition, having no moving parts in the device, compact spectrometers are generally robust and appropriate for fieldwork or on-site tasks.

The small dimension of the detector sensor/chip determines that the spectral resolution must compromise with the covered spectral range, especially for compact spectrometers. Most commercial compact Raman spectrometers can only cover ~200-2000 cm^{-1} with a resolution in the range of 6-12 cm^{-1}. Majority of modern Raman spectrometers are based on silicon-based CCD detectors, which usually have optimal responses between 400 and 900 nm. Although this range may be slightly extended through UV or NIR enhanced technologies [12–15], CCD-detector based UV and NIR Raman studies out the range of 400-900 nm are scarce. As the result of the consideration of factors mentioned above, i.e., the need to balance Raman efficiency and autofluorescence and the availability of lasers and detectors, the combination of the 785 nm laser excitation and a NIR enhanced CCD detector has been chosen for most biologically related Raman studies. The results turned out to be satisfactory in many cases, especially signals in the fingerprint region (i.e., 200 – 2000 cm^{-1}).

The high-wavelength Raman signals from chemical bonds involving hydrogen atoms (e.g., CH and OH) will appear above 2900 cm^{-1} (see **Table 1**). With NIR enhanced silicon-based CCD detector, the CH groups appearing near 2900 cm^{-1} may be weakly observed, but the OH related signal expecting near 3500 cm^{-1} (corresponding to ~1090 nm) is hardly observable. Earlier literature reports on HW Raman signals from biological samples seemed primarily conducted by Fourier Transformed Raman spectroscopy (FT-Raman) [16–19]. FT-Raman is typically composed of Nd:Yag laser and germanium detector, it ghus has the great adavantages of fluorescence reduction and high resulution. However, because all the lights including the residuals of the laser simultaneously strikes the detector, FT-Raman is detector noise limited and typically requires much longer (up to 3 orders) integration time than dispersive Raman system [19, 20]. This explained why FT-Raman is still not very popular despite it has emerged over three decades.

Indium gallium arsenide (InGaAs) based arrays have emerged as a dispersive detector for spectroscopy devices around 2000 [21]. Unlike CCD, InGaAs has optimal sensitivity across 900-1700 nm, and has gained increasing interest for Raman applications based on 1064 nm laser as its great potential in further suppress the auto-fluorescence, especially from biological tissues. Unal et al. demonstrated that the InGaAs spectrometer designated for 1064 nm Raman spectroscopy can be combined with 852 nm laser to probe hydrations in bone tissues [22]. Yang and He et al. modified the system with a 866 nm laser and demonstrated water detection in other biological tissues, including animal skin, human teeth, and fruits [23–25]. For example, the Raman spectrum acquired with an InGaAs Raman spectrometer under 866/1064 nm dual excitation (red curve) shows strong signals from CH and OH bonds. Those signals were barely observable with Raman spectroscopy composing a 785 nm laser and a CCD detector (**Figure 1**).

2. HW Raman of non-living matters

Room temperature HW Raman signals typically result from chemical bonds formed by light atoms such as CO, CH, OH, CN. In contrast, the signals greater than 2500 cm^{-1} mostly result from chemical bonds containing hydrogen atoms (e.g., CH and OH groups), except for second-order processes, e.g., the 2D signal in graphene. HW Raman signals of the majority of inorganic compounds and non-living organic materials can be illuminated with visible light (e.g., 532 nm or 633 nm) without exciting problematic fluorescence interference. For example, Raman signals from water can be acquired with Raman spectroscopy at any wavelengths from UV to NIR as long as the detector responds. The HW Raman signals from hydroxyapatite crystal, the primary component in human mineral tissues, can also be obtained under the illumination of a green laser at 515 nm [2].

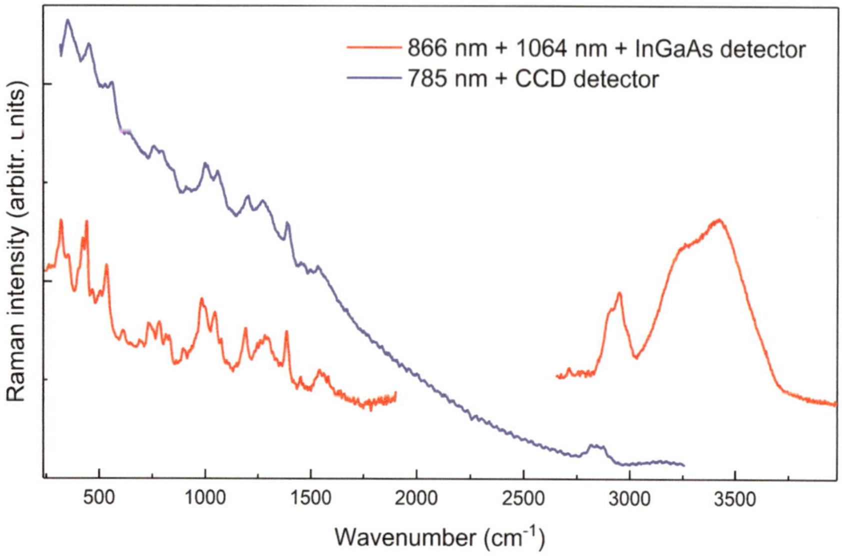

Figure 1.
Raman spectra of grape skin acquired with 785 nm (blue curve) excitation and CCD.

Simple and small molecule organic materials with a low visible light absorption rate can generally be examined with visible Raman spectroscopy. Thus their full range Raman spectra covering the HW Raman signals can be acquired as easily as fingerprint signals. Despite the fingerprint Raman signals of these material are sufficient for identification, HW Raman signals can provide additional information to understand the molecular dynamics. Nedić et al. studied water, methanol, ethanol dimer and trimers and found that both methanol and ethanol are better hydrogen bond donors and acceptors than water through investigating the red shif of HW OH signals [26]. In addition, Starciuc et al. were able to study the tranisitions of unclusered water to large water clusters through the analysis of highwavenumber and lowwave number Rmaan signals in mixed glycerol-water system [27].

3. HW Raman analysis on biological tissues using CCD detectors

Although the HW region Raman signals, including CH, NH, and OH bonds, are informative in analyzing biological tissues, as seen in many Fourier Transformed Infrared (FTIR) absorption spectrum, Raman studies in this region on biological tissues have been underperformed. Other than the few studies using FT-Raman spectrometer, there are several studies using CCD-based dispersive Raman spectrometers. Santos et al. investigated the effects of different optical fibers on 720 nm laser based Raman spectroscopy and demonstrated the system was capable of acquireing HW Ramang signals from sliced porce brain tissues [28]. Carvalho et al. demonstrated HW Raman signals may be obtained with 532 nm laser from oral cells immersed in distilled water [29]. Barroso et al. studied the difference of HW Raman signals between healthy and cancer oral tissues and found that water contents may be used as a biomarker for cancer diagnosis [30]. It is worth to note that all these biological tissues that was able to stuied by the visible light do not produce intense fluorescence that overwhelm the Raman signals. Skin tissue is also one type of tisse that does not have problematic fluorescenc under Raman study. Therefore, skin tissues especially the stratum corneum, have also been well studied in terms of HW Raman region. By using a confocal Raman micro-spectrometer to exclude out of plane fluorescence

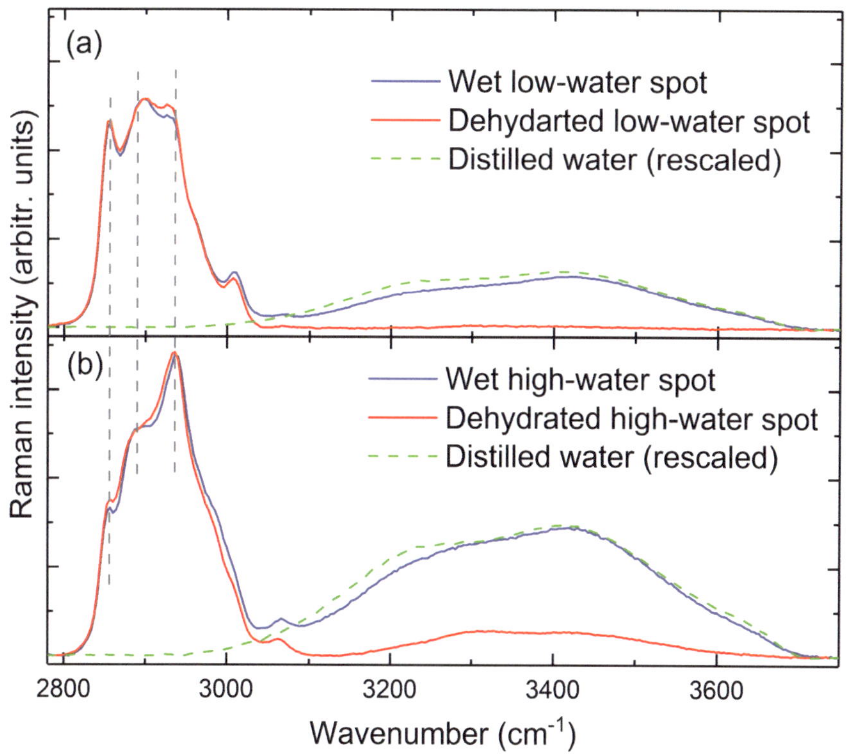

Figure 2.
Raman spectra of chicken skin at representative (a) low-water spots and (b) high-water spots before (blue curve) and after 24-hour air drying (red curve). Peaks marked with vertical dashed lines from low to high were located at 2854, 2895, and 2934 cm^{-1} respectively. The spectra were acquired with 85 mw, 866 nm laser light with 30 s exposure and 6 averages. All skin spectra were normalized according to the 2895 cm^{-1} signal for easier comparison, while the spectrum of distilled water was rescaled to match the corresponding OH signal of the skin.

emissions, the HW regions Raman signals were able to be acquired from human skins. Caspers et al. used a confocal Raman micro-spectrometer based on 720 nm excitation to obtain Raman spectra of skin showing clear water profile and detailed CH structures [31, 32]. Later, Choe et al. deconvoluted the CH and OH groups of Raman signal and studied the profile variation of bound water affected by protein and lipids interactions at different depths [33]; while Quatela et al. observed variation of spectral markers including OH and CH groups among different individuals [34]. It is worth noting that in Choe's work, two lasers operating at 671 nm and 785 nm were used. Such a dual-wavelength Raman setup is an alternative approach to acquiring the HW Raman signals when only a compact spectrometer with a fixed spectral range is available. For a compact Raman system designed for acquiring fingerprint region ~240-2000 cm^{-1} (corresponding to 800-930 nm) under 785 nm excitation, send in a second laser beam operating at 671 nm will effectively extend the spectral range to cover the region of ~2400-4100 cm^{-1} which include the main HW CH and OH groups.

In contrast to the study on the HW region Raman spectral variations with the depth of the skin tissue, Yang et al. investigated the lateral variations on the skin surface [25]. The team identified mainly two types of spots on the skin surface, i.e., high-water spots and low-water spots (**Figure 2**). Further analysis on other tissues, including muscle, fat, and tendons, the authors found that the skin tissues contain both fat and protein (keratin) and suggested the high-water spots on chicken skin are protein-rich while the low-water spots are lipids-rich. This suggestion is supported by the observation of the characteristic CH bonds (**Figure 3**) among these

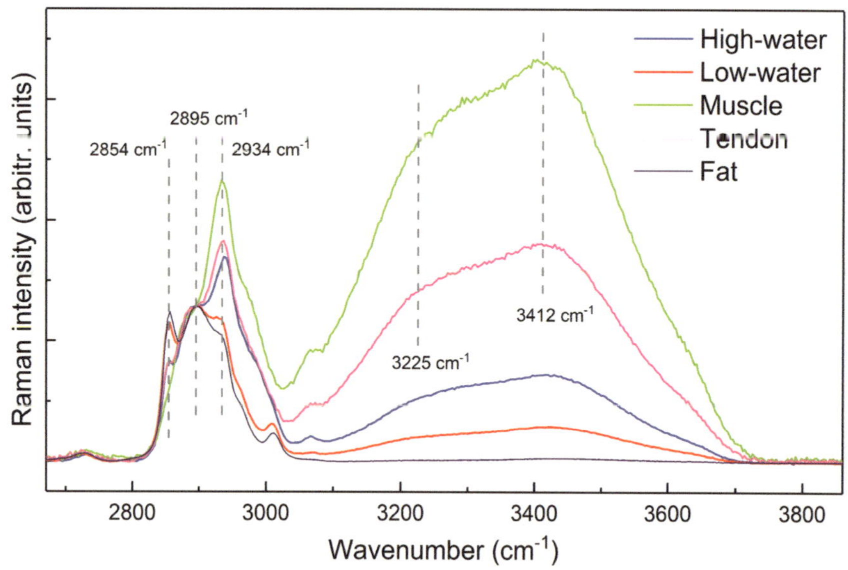

Figure 3.
Raman spectra of chicken tissues (a) before and (b) after dehydration (except fat). Spectra in (b) were vertically enlarged. Dehydrated fat was prepared with 6 hours of air drying at room temperature, while dehydrated muscle or tendon were prepared with 72 hours of air drying at room temperature or overnight oven drying at 45°C. All spectra were acquired under the condition of 85 mW laser at 866 nm illumination, with 30 s exposure, and averaged 6 times. All spectra in (a) and (b) were normalized according to the 2895 cm^{-1} signal.

tissues and the fact that the muscle tissue has the highest OH (3412 cm^{-1}) to CH (2895 cm^{-1}) signal intensity ratio, while the fat tissue has the lowest one among those tissues under investigation. Additional investigations on dehydrated tissues indicated that protein-rich tissues were more capable of retaining water and more resistant to dehydration. In other words, the protein component is positively correlated with skin hydration, in contrast to the fat component.

4. Water contents in mineral tissues

As mentioned earlier, the combination of the NIR laser lights near 850 nm and the InGaAs detector-based spectrometer allowed the observation of water contents in mineral tissues, including dental hard tissues and bones [22, 23]. Mineral tissues appeared more fluorescent than stratum corneum, and were challenging for Raman measurement using visible lights. Unal et al. characterized several HW peaks, including CH group and OH group, and suggest bound water in bone could be interacting with both collagen and mineral matrix [22]. He et al. discovered similar HW Raman signals from dental hard tissues, including dentin and enamel [23]. The authors demonstrated the spectral profile not only varied between enamel and dentin tissues but also varied among different locations within enamel tissues. As shown in **Figure 4**, representative spectra from dentin and enamel were stacked for comparison (rescaled Raman spectrum of distilled water was also provided for reference). The spectra were taken from the sagittal surface. The lower wavenumber region containing the fingerprint 960 cm^{-1} signal was acquired under 1064 nm laser excitation, while the HW region was acquired under 866 nm laser excitation. The spectra were normalized according to the 960 cm^{-1} signal intensity, and the same factor was applied to the HW region for consistency. The differences between the spectra indicated that both enamel and dentin contain water that is not 'free' like in

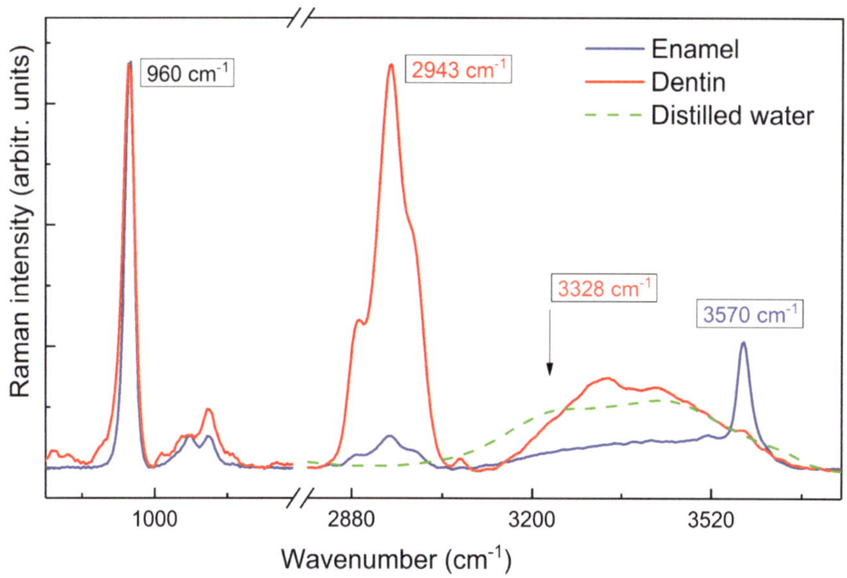

Figure 4.
Representative Raman spectra from enamel (blue curve) and dentin (red curve) portions from the sagittal surface of a tooth. Spectra below 1200 cm^{-1} were acquired under 20s 1064 nm excitation while the spectra above 3000 cm^{-1} were acquired under (30s × 16) 866 nm excitation. Water spectrum was scaled down for easier comparison.

distilled water, and dentin contains a greater number of organic components than enamel, evidenced by the much greater C-H stretch peak located at 2943 cm^{-1}. The author suggested part of the water in dentin interacts with the surrounding environment, likely the organic matrix, evidenced by the appearance of an additional peak at 3328 cm^{-1}, which could be partially contributed by the N-H bond (but not all considering its intensity) [35].

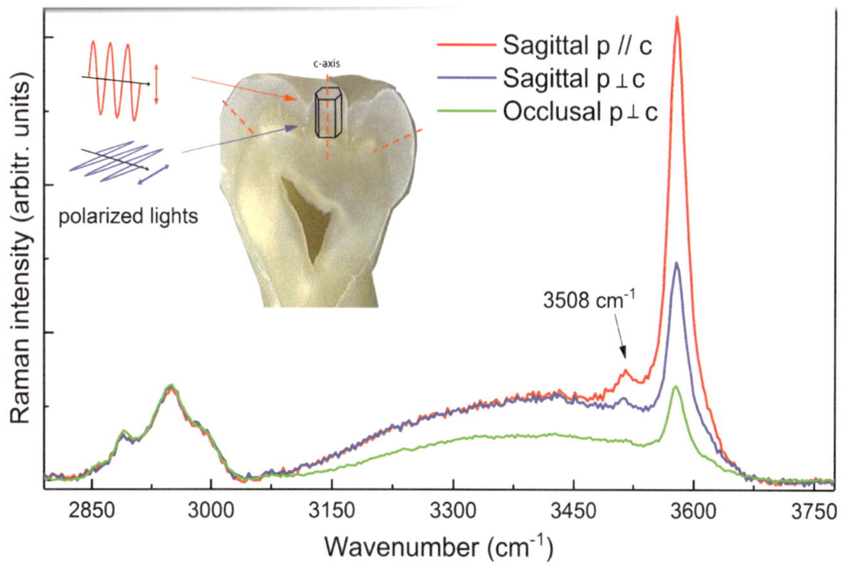

Figure 5.
Representative HW Raman spectra under linearly polarized laser excitation from internal and external surfaces of a tooth enamel. The inset shows the picture of the sagittal surface of a tooth, with dashed red line showing the approximate c-axis of HAP crystals of enamel layer. Experimental conditions are 30s × 16 integration time and 50 mW, 866 nm laser light illumination.

Another major difference between the enamel and dentin spectra is the appearance of a sharp 3570 cm^{-1} Raman signal in enamel, but the same signal is barely noticeable in dentin. Additional spectra acquired from different spots on enamel showed that the signal varied with locations (**Figure 5**). Further, the signal intensity from the same spot will also change with the polarization of the incident light. Because a very similar signal appears in synthetic hydroxyapatite crystals, the authors assign the peak at 3570 cm^{-1} to the OH radicals of hydroxyapatite crystals. This signal is polarization-dependent because most of the hydroxyapatite crystals align along the c-axis (see dashed red line in the inset tooth picture), which is roughly perpendicular to the external enamel surface. The difference in the signal between enamel and dentin tissues is mainly due to the sizes of hydroxyapatite crystals. The signal is stronger from internal enamel spots than external spots because the internal tissues keep better integrity.

The spectrum of enamel also showed a small difference between the internal and the external spots on enamel: a peak located at 3508 cm^{-1} that only obviously appearing in the spectra from internal enamel spots. This signal is absent in the spectra of single hydroxyapatite crystals or any hydroxyapatite powders reported before. Based on its similar location as the 3570 cm^{-1} signals this signal is suggested to be structure water bound to the mineral matrix of unaffected enamel tissue. In other words, this signal is likely the water that is bound to hydroxyapatite crystal, similar to how the OH radical forms during the crystal growth. The enamel may lose these bound waters when demineralization happens to the surface enamel.

5. HW Raman detection on pigmented biological materials

Pigmented tissues, from both animals and plants, are usually high fluorescent for Raman studies; thus, probing HW Raman signals from these tissues are difficult with visible lights. Muscle tissues showing in the previous section are one example

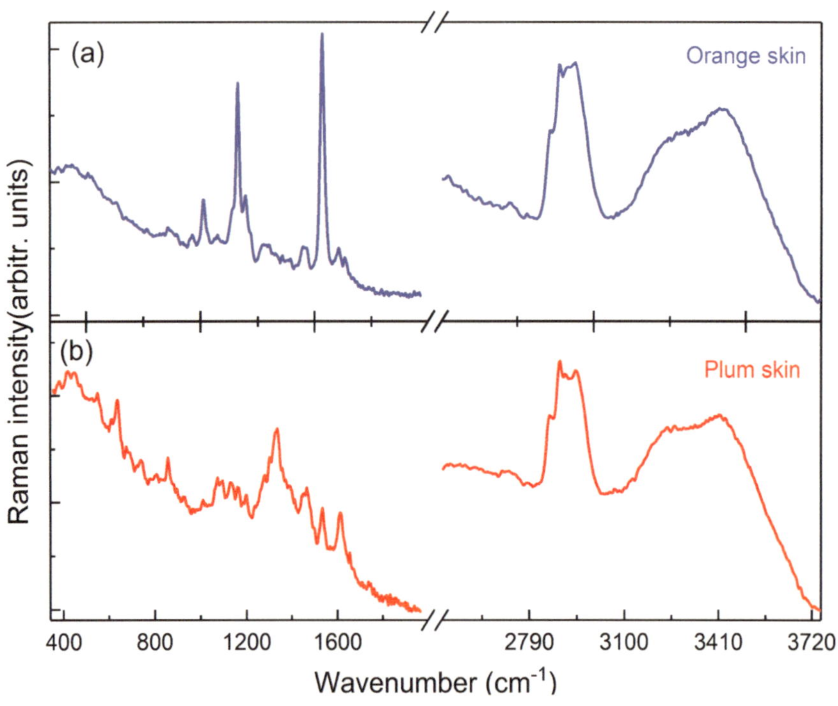

Figure 6.
Raman spectra acquired from skins of orange (a) and plum (b).

that requires NIR laser excitation (e.g., 866 nm). Santos et al. demonstrated acquiring HW Raman signals from pigmented skin lesions with 976 nm excitation [36]. Acquiring the HW Raman signals from orange and plum skins are two other examples of pigmented tissues, [24] as shown in **Figure 6**.

6. Options of InGaAs based Raman spectroscopy for HW signal detection

Based on the InGaAs spectrometer, there are basically three options to set up a Raman spectroscopy to acquire full range Raman spectrum. For a spectrometer that combined an InGaAs detector and a tunable spectrograph, one laser and one spectrometer will allow the acquisition of a full range Raman spectrum that includes the HW region. For a compact InGaAs spectrometer designed to work with a 1064 nm laser, it will typically have a fixed range starting ~1100 and ending ~1350 nm. In order to acquire a full range spectrum, one could add a second laser (e.g., 850 nm) to extend the range to cover the HW region. Alternatively, one could use a single laser (e.g., 850 nm) but use a second spectrometer (e.g., CCD detector based compact spectrometer) to record fingerprint region Raman signals. The advantage of the dual laser plus one spectrometer is it suppress the fluorescence to the maximum extent as 1064 nm is fluorescence free many biological tisses. The disadvantage is that the HW and fingerprint region must be taken consecutively, resulting in different sampling if the specimen is moving or changes with time. While the one laser plus two spectrometers will ensure HW and fingerprint regions are taken simultaneously on a same spot.

Author details

Shan Yang
Department of Chemistry, Physics and Atmospheric Sciences, Jackson State University, Jackson, MS, USA

*Address all correspondence to: Shan.Yang@jsums.edu

References

[1] Banwell, C. N. *Fundamentals of Molecular Spectroscopy*; McGraw-Hill, 1983.

[2] Tsuda, H.; Arends, J. Orientational Micro-Raman Spectroscopy on Hydroxyapatite Single Crystals and Human Enamel Crystallites. J Dent Res 1994, *73* (11), 1703-1710. https://doi.org/10.1177/00220345940730110501.

[3] Penney, C. M.; Goldman, L. M.; Lapp, M. Raman Scattering Cross Sections. Nature Physical Science 1972, *235* (58), 110-112. https://doi.org/10.1038/physci235110b0.

[4] Golcuk, K.; Mandair, G. S.; Callender, A. F.; Sahar, N.; Kohn, D. H.; Morris, M. D. Is Photobleaching Necessary for Raman Imaging of Bone Tissue Using a Green Laser? Biochimica et Biophysica Acta (BBA) - Biomembranes 2006, *1758* (7), 868-873. https://doi.org/10.1016/j.bbamem.2006.02.022.

[5] Movasaghi, Z.; Rehman, S.; Rehman, I. U. Raman Spectroscopy of Biological Tissues. Applied Spectroscopy Reviews 2007, *42* (5), 493-541. https://doi.org/10.1080/05704920701551530.

[6] Bumbrah, G. S.; Sharma, R. M. Raman Spectroscopy – Basic Principle, Instrumentation and Selected Applications for the Characterization of Drugs of Abuse. Egyptian Journal of Forensic Sciences 2016, *6* (3), 209-215. https://doi.org/10.1016/j.ejfs.2015.06.001.

[7] Vandenabeele, P.; Castro, K.; Hargreaves, M.; Moens, L.; Madariaga, J. M.; Edwards, H. G. M. Comparative Study of Mobile Raman Instrumentation for Art Analysis. Analytica Chimica Acta 2007, *588* (1), 108-116. https://doi.org/10.1016/j.aca.2007.01.082.

[8] Chase, B. A New Generation of Raman Instrumentation. *Appl. Spectrosc.,* AS 1994, *48* (7), 14A-19A.

[9] Schrader, B.; Dippel, B.; Erb, I.; Keller, S.; Löchte, T.; Schulz, H.; Tatsch, E.; Wessel, S. NIR Raman Spectroscopy in Medicine and Biology: Results and Aspects. Journal of Molecular Structure 1999, *480-481*, 21-32. https://doi.org/10.1016/S0022-2860(98)00650-4.

[10] Zhang, J.; Li, M.; Feng, Z.; Chen, J.; Li, C. UV Raman Spectroscopic Study on TiO_2. I. Phase Transformation at the Surface and in the Bulk. *J. Phys.* Chem. B 2006, *110* (2), 927-935. https://doi.org/10.1021/jp0552473.

[11] Strom, D.; Breuninger, S.; Fischer, H.; Richter, A.; Schmidt, U.; Hollricher, O. Criteria for High-Quality Raman Microscopy (accessed 2019 -07 -14).

[12] Holst, G. C.; Lomheim, T. S. CMOS/CCD Sensors and Camera Systems, Second Edition | (2011) Spie Press Book.

[13] Nikzad, S.; Hoenk, M. E.; Greer, F.; Jacquot, B.; Monacos, S.; Jones, T. J.; Blacksberg, J.; Hamden, E.; Schiminovich, D.; Martin, C.; Morrissey, P. Delta-Doped Electron-Multiplied CCD with Absolute Quantum Efficiency over 50% in the near to Far Ultraviolet Range for Single Photon Counting Applications. Appl. Opt. 2012, *51* (3), 365. https://doi.org/10.1364/AO.51.000365.

[14] Lesser, M. P. Improving CCD Quantum Efficiency. In *Instrumentation in Astronomy VIII*; International Society for Optics and Photonics, 1994; Vol. 2198, pp. 782-791. https://doi.org/10.1117/12.176777.

[15] Adar, F.; Atzeni, S.; Gilchrist, J. R.; Goldstone, L.; Noonan, J. Detector choice is vital to spectroscopy (accessed 2019 -07 -14).

[16] Hirschfeld, T.; Chase, B. FT-Raman Spectroscopy: Development and

Justification. Appl Spectrosc 1986, *40* (2), 133-137. https://doi.org/10.1366/0003702864509538.

[17] García-Flores, A. F.; Raniero, L.; Canevari, R. A.; Jalkanen, K. J.; Bitar, R. A.; Martinho, H. S.; Martin, A. A. High-Wavenumber FT-Raman Spectroscopy for in Vivo and Ex Vivo Measurements of Breast Cancer. Theor Chem Acc 2011, *130* (4-6), 1231-1238. https://doi.org/10.1007/s00214-011-0925-9.

[18] Church, J. S.; Corino, G. L.; Woodhead, A. L. The Analysis of Merino Wool Cuticle and Cortical Cells by Fourier Transform Raman Spectroscopy. Biopolymers 1997, *42* (1), 7-17. https://doi.org/10.1002/(SICI)1097-0282(199707)42:1<7::AID-BIP2>3.0.CO;2-S.

[19] das Chagas e Silva de Carvalho, L. F.; Sato, É. T.; Almeida, J. D.; da Silva Martinho, H. Diagnosis of Inflammatory Lesions by High-Wavenumber FT-Raman Spectroscopy. Theor Chem Acc 2011, *130* (4-6), 1221-1229. https://doi.org/10.1007/s00214-011-0972-2.

[20] Baraga, J. J.; Feld, M. S.; Rava, R. P. In Situ Optical Histochemistry of Human Artery Using near Infrared Fourier Transform Raman Spectroscopy. PNAS 1992, *89* (8), 3473-3477.

[21] SPECTROSCOPY: Indium gallium arsenide detectors open up the near-IR https://www.laserfocusworld.com/detectors-imaging/article/16555710/spectroscopy-indium-gallium-arsenide-detectors-open-up-the-nearir (accessed 2021 -04 -23).

[22] Unal, M.; Yang, S.; Akkus, O. Molecular Spectroscopic Identification of the Water Compartments in Bone. *Bone* 2014, *67*, 228-236. https://doi.org/10.1016/j.bone.2014.07.021.

[23] He, W.; Livingston, J. C.; Sobiesk, E. R.; Zhou, J.; Zhu, X.; Duan, Y.; Yang, S. A Pilot Study on High Wavenumber Raman Analysis of Human Dental Tissues. J Raman Spectrosc 2020, *51* (4), 630-634. https://doi.org/10.1002/jrs.5812.

[24] He, W.; Li, B.; Yang, S. High-Frequency Raman Analysis in Biological Tissues Using Dual-Wavelength Excitation Raman Spectroscopy. Appl Spectrosc 2020, *74* (2), 241-244. https://doi.org/10.1177/0003702819881762.

[25] Yang, S.; Sen, C.; Thompson, R.; Zhou, J.-G.; Akkus, O. An in Vitro Raman Study on Compositional Correlations of Lipids and Protein with Animal Tissue Hydration. Vibrational Spectroscopy 2020, *107*, 103022. https://doi.org/10.1016/j.vibspec.2020.103022.

[26] Nedić, M.; Wassermann, T. N.; Larsen, R. W.; Suhm, M. A. A Combined Raman- and Infrared Jet Study of Mixed Methanol–Water and Ethanol–Water Clusters. Phys. Chem. Chem. Phys. 2011, *13* (31), 14050. https://doi.org/10.1039/c1cp20182d.

[27] Starciuc, T.; Guinet, Y.; Hedoux, A.; Shalaev, E. Water Content Thresholds in Glycerol/Water System: Low- and High-Wavenumber Raman Spectroscopy Study. Journal of Molecular Liquids 2021, *321*, 114678. https://doi.org/10.1016/j.molliq.2020.114678.

[28] Santos, L. F.; Wolthuis, R.; Koljenović, S.; Almeida, R. M.; Puppels, G. J. Fiber-Optic Probes for in Vivo Raman Spectroscopy in the High-Wavenumber Region. Anal. Chem. 2005, *77* (20), 6747-6752. https://doi.org/10.1021/ac0505730.

[29] Carvalho, L. F. C. S.; Bonnier, F.; Tellez, C.; dos Santos, L.; O'Callaghan, K.; O'Sullivan, J.; Soares, L. E. S.; Flint, S.; Martin, A. A.; Lyng, F. M.; Byrne, H. J. Raman Spectroscopic Analysis of Oral Cells in the High Wavenumber Region. Experimental and Molecular Pathology 2017, *103* (3), 255-262. https://doi.org/10.1016/j.yexmp.2017.11.001.

[30] Barroso, E. M.; Smits, R. W. H.; Bakker Schut, T. C.; ten Hove, I.; Hardillo, J. A.; Wolvius, E. B.; Baatenburg de Jong, R. J.; Koljenović, S.; Puppels, G. J. Discrimination between Oral Cancer and Healthy Tissue Based on Water Content Determined by Raman Spectroscopy. *Anal.* Chem. 2015, *87* (4), 2419-2426. https://doi.org/10.1021/ac504362y.

[31] Caspers, P. J.; Lucassen, G. W.; Bruining, H. A.; Puppels, G. J. Automated Depth-Scanning Confocal Raman Microspectrometer for Rapidin Vivo Determination of Water Concentration Profiles in Human Skin. Journal of Raman Spectroscopy 2000, *31* (8-9), 813-818. https://doi.org/10.1002/1097-4555(200008/09)31:8/9<813::AID-JRS573>3.0.CO;2-7.

[32] Caspers, P. J.; Bruining, H. A.; Puppels, G. J.; Lucassen, G. W.; Carter, E. A. In Vivo Confocal Raman Microspectroscopy of the Skin: Noninvasive Determination of Molecular Concentration Profiles. Journal of Investigative Dermatology 2001, *116* (3), 434-442. https://doi.org/10.1046/j.1523-1747.2001.01258.x.

[33] Choe, C.; Lademann, J.; Darvin, M. E. A Depth-Dependent Profile of the Lipid Conformation and Lateral Packing Order of the Stratum Corneum in Vivo Measured Using Raman Microscopy. The Analyst 2016, *141* (6), 1981-1987. https://doi.org/10.1039/C5AN02373D.

[34] Quatela, A.; Miloudi, L.; Tfayli, A.; Baillet-Guffroy, A. In Vivo Raman Microspectroscopy: Intra- and Intersubject Variability of Stratum Corneum Spectral Markers. Skin Pharmacology and Physiology 2016, *29* (2), 102-109. https://doi.org/10.1159/000445079.

[35] Lauritsen, A. K.; Pereira, J. E. M.; Juranyi, F.; Bordallo, H. N.; Larsen, L.; Benetti, A. R. Probing Water Mobility in Human Dentine with Neutron Spectroscopy. J Dent Res 2018, *97* (9), 1017-1022. https://doi.org/10.1177/0022034518763051.

[36] P. Santos, I.; van Doorn, R.; Caspers, P. J.; Bakker Schut, T. C.; Barroso, E. M.; Nijsten, T. E. C.; Noordhoek Hegt, V.; Koljenović, S.; Puppels, G. J. Improving Clinical Diagnosis of Early-Stage Cutaneous Melanoma Based on Raman Spectroscopy. *Br J Cancer.* 2018, *119* (11), 1339-1346. https://doi.org/10.1038/s41416-018-0257-9.

Chapter 7

Resonance Raman Spectroscopy Investigation of the Interaction of Molecules Adsorbed on Solid Acid Surfaces

Lucia Kiyomi Noda

Abstract

Many solid acids with very strong acid sites, as some zeolites, transition metal exchanged montmorillonites, sulfated metallic oxides, are known to have the oxidizing ability, which can be related to the catalytic activity of these materials. The interaction of these solid acids with aromatic molecules can give rise to several oxidation products. Intermediate species of aromatic molecules formed by interaction with strong solid acids had been reported, as radical cations, proving the oxidizing ability of the solids. Besides radical cations, charge transfer complexes between the solid acids and aromatic molecules can be formed. These radical cations and charge transfer complexes usually show absorption bands in the visible region, opening the possibility of studying these species by Resonance Raman Spectroscopy (RRS). Benzene and substituted benzenes, phenothiazine, t-stilbene, adsorbed on solid acids, are examples of molecules that had been investigated by RRS. Exciting the spectrum with suitable radiation makes it possible to observe the RRS of the species of interest even when its concentration is low, because of the preferential enhancement of the vibrational modes of the chromophore. A review of RRS studies of molecules adsorbed on solid acids is presented. RRS proved valuable in characterizing intermediate species as radical cations or charge transfer complexes formed on the solid acids.

Keywords: resonance Raman spectroscopy, solid acids, radical cation, charge transfer complex

1. Introduction

Solid acids are the most important solid catalysts nowadays, considering not only the total amount used but also the final economic impact. Solid catalysts have a great advantage over liquid catalysts. They are, in fact, generally almost fully recovered from reaction products without any operation or with quite easy procedures.

Probably infrared spectroscopy is the most largely used spectroscopy technique to study the adsorption of molecules on solid surfaces. Other techniques need an ultra-high vacuum, while IR spectroscopy can be investigated over a wide range of temperatures and pressures. However, most of the materials used as adsorbents, for instance, semiconductor and insulator surfaces, have strong IR signals, so the spectrum is predominantly that of the adsorbent, which is a disadvantage [1].

Compared to infrared absorption spectroscopy, Raman scattering has an inherently low sensitivity. But this limitation was largely surpassed by the development of high-performance single-stage spectrometers combined with notch filters and CCD cameras. Moreover, the sensitivity of Raman spectroscopy can be increased by taking advantage of resonance effects, for example in resonance Raman spectroscopy, coherent anti-Stokes Raman spectroscopy (CARS), surface-enhanced Raman spectroscopy (SERS), or tip-enhanced Raman spectroscopy (TERS) [2].

The interaction of molecules adsorbed on solid acids can be weak or strong, but in solids with higher acidity, the interaction is often so strong that electron abstraction may occur, occurring in fact a redox reaction. It depends not only on the acidity of the solid acid but also on the ionization potential of the adsorbed molecule. Organic molecules with relatively low ionization potential can be ionized giving rise to radical cations. However, depending on some conditions, a charge transfer complex is observed. Radical cations of molecules with extended conjugation, as polyenes and some aromatic molecules, have electronic absorption shifted to the visible region of the spectrum, opening the possibility of exploring the Resonance Raman effect with the enhancement of the chromophore responsible for the electronic transition.

The theory behind the resonance Raman Effect will not be presented here, but for whoever is interested, the excellent article by Clark and Dines is recommended [3].

In the present review the interaction of organic molecules with high acidity solids and the species resulting from the interaction, observed through Resonance Raman spectroscopy (RRS), is presented. Three types of solid acids, zeolites, clays, and sulfated metallic oxides, are chosen because of the greater number of RRS studies with these solid acids.

2. Zeolites

Zeolites are crystalline aluminosilicates with general chemical composition $M_{x/n}^{x+}[Al_xSi_yO_{2(x+y)}]zH_2O$. Its lattice consists of a network of SiO_4^{4-} and AlO_4^{4-} tetrahedra with Si or Al atoms at the centers and oxygen atoms in each corner. The presence of aluminum atoms replacing silicon atoms in the zeolite network creates an excess negative charge. Charge-compensating cations (M^+) are introduced into the structure to compensate for the negative charge. These readily exchangeable cations are not covalently bound to the zeolite framework [4].

The Si/Al ratio may vary from Si/Al = 1 (faujasite X) to Si/Al = ∞ (silicalite). The number of cations in the framework and the zeolite properties as the thermal and chemical stability or the polarity of the internal surfaces are determined by the Al content.

Electron transfer processes at interfaces are central to many important chemical processes and are vital to energy conversion and storage technologies. Investigations of the mechanism of electron transfer reactions at interfaces are often difficult to perform because of the heterogeneity of the surfaces and because the reaction intermediates, often radicals, are generally highly reactive and difficult to isolate for extended periods of time to allow detailed characterization [5].

A common way to stabilize these intermediates is their isolation in frozen rare gas or halocarbon matrices.

Zeolites and other porous solids are an alternative to the study of electron-transfer processes, allowing the stabilization of radical cations for longer periods of time [6]. Garcia and Roth had made an extensive review about the generation and reaction of organic radical cation in zeolites [7].

Some acid zeolites have the ability to generate spontaneously organic radical cations upon adsorption of organic electron donors or the radical cations can be generated by the action of radiation or light. The restricted mobility within zeolite pores limits the tendency of free radicals or radical cations to dimerize and prevents access to reagents that typically would cause their decay in solution.

The spontaneous ionization depends on the ionization potential of the guest as well as on the ionizing capacity of the host, which is directly related to the Si/Al ratio and to the nature of the charge balancing cation. The highest yield was found from H^+. Ramamurthy et al. showed that an increasing Al content enhanced the electron acceptor ability [4]. They reported the generation of stable radical cations from α,ω-diphenyl polyenes (trans-stilbene, trans, trans-1,4-diphenyl butadiene, all-trans- 1,6-diphenylhexatriene, and all-trans-1,8-diphenyl-1,3,5,7- octatetraene upon inclusion in the channels of pentasil zeolites.

A systematic spectroscopic investigation of radical species of several organic molecules formed upon the interaction with zeolites was done by Moissette in the 2000s and 2010s. Most of the investigated molecules are aromatic, as trans-stilbene anthracene and other ones. UV–visible, ESR, and Raman spectra were obtained with excitation in and off-resonance, but in the present review, only the resonance Raman results of Moissette are presented. Only the Moissette results with t-stilbene and N, N, N′, N′ -tetramethyl-p-phenylenediamine (TMPD) and N, N, N′, N′ -tetramethylbenzidine (TMB) are included in this review, but Moissette investigated a larger number of molecules, as can be found in the literature.

2.1 Trans-stilbene

Many Moissette studies were done with trans-stilbene, a molecule considered as a prototype system for the photoisomerization reaction and as a model for photosensitized electron-donor structure [8]. Trans-Stilbene (E-1,2-diphenylethene, t-St) has a relatively low ionization potential value (I_g = 7.65 eV in the gas phase) and has the appropriate dimensions to enter the channels of medium-pore zeolites [9]. Moissette et al. investigated the formation of t-stilbene radicals in HZSM-5 and HFER zeolites, that differ in their pore size [10]. As HFER has a narrower pore size the radical cation is the only stable species while in the larger pore diameter HZSM-5, the electron transfers are faster and the radical cation (RC) evolves after several minutes/hours to a charge transfer complex (CTC). Three exciting laser lines, at 473, 515, and 633 nm, have been used to take advantage of the RR effect. The wavelengths have been chosen to correspond to the absorptions of the charge separated states and could give rise to resonance Raman enhancement of the specific vibrational modes of the radical cation or the charge transfer complex. The RC has a characteristic band at 475 nm, while the bands at 565 and 625 nm are characteristic of CTC 1 and the bands at 375 and 700 nm are characteristic of CTC 2. Exciting with 473 nm Raman bands at 1285 and 1605 cm^{-1} and a weak band at about 1565 cm^{-1} were observed. The spectrum excited with 515 nm presents the same main Raman bands characteristic of RC at 1290 and 1609 cm^{-1} as well as the contributions at 1560 and 1550 cm^{-1}. The resonance effect of the RC species is not as marked as at 473 nm given that the excitation is in the lower energy side of the absorption band. A broadening is observed especially at about 1600 cm^{-1} corresponding to the position of CTC contribution. Exciting with 633 nm broad and intense bands centered at 1596 cm^{-1} and 1192 cm^{-1} containing several contributions are observed. Several bands of lower intensity are also observed around 1320 cm^{-1}. These features were assigned to the resonance-enhanced bands of the CTC 1 species. The authors do not assign Raman bands to CTC 2 species, but the broadening of the Raman bands exciting with 633 nm may be a sign of the presence of another species.

When t-St was adsorbed in the medium size channel of nonacidic NaZSM-5 zeolite, the interaction between Na^+ cation and t-St occurred through one phenyl group coordinated to the Na^+ cation [11]. The similarity between Raman spectra of t-St in solution and occluded in NaZSM-5 led the authors to conclude that the interaction was weak, with no radical formation. Only with laser UV (266 nm) photoionization t-St$^{•+}$ was generated. Exciting with 4880 nm resonance Raman bands of the radical cation was observed. With 632 nm excitation, a different spectrum was observed, assigned to a primary t-St$^{•+}$-electron pair. Contrary to Moissette that did not observe spontaneous ionization of t-stilbene in NaZSM-5 zeolite, Ramamurthy [4] reported the formation of radical cation species upon adsorption of t-stilbene on the zeolite. Ramamurthy made the adsorption of t-stilbene diluted in a trimethylpentane solution. As Moissette did the adsorption experiments in the solid-state, this fact may have led to differences in the adsorption behavior.

2.2 N, N, N′, N′ -tetramethyl-p-phenylenediamine (TMPD) and N, N, N′, N′ -tetramethylbenzidine (TMB)

N, N, N′, N′ -tetramethyl-p-phenylenediamine (TMPD) and N, N, N′, N′-tetramethylbenzidine (TMB) were used as probes to characterize the nature of active sites in acid zeolite HZSM-5 [12]. These two amines are strong electron donors and are known to have low ionization potentials in the gas phase and to exhibit proton affinity. The rod-shaped N, N, N′, N′ - tetramethyl-p-phenylene-diamine (TMPD) and N, N, N′, N′ - tetramethylbenzidine (TMB) molecules can penetrate within the porous void of ZSM-5 zeolites.

After several days of the mixing of TMPD and HZSM-5 powders, an intense UV–visible spectrum with the vibronic structure of TMPD$^{•+}$ between 500 and 650 nm is observed. The vibrational progression of the TMPD$^{•+}$ band (with peaks separated by ~1500 cm^{-1}) is assigned to the stretching mode of the aromatic ring. RR and off-resonance Ramana spectra of the mixed TMPD and H6ZSM-5 powders were obtained with 514.4 and 632.0 nm excitation and 1064 nm, respectively. All the spectra are very similar, with wavenumbers and intensities characteristic of the TMPD$^{•+}$ species in solution. Raman bands assignable to neutral TMPD or protonated H2TMPD are absent, which provided evidence of complete ionization of TMPD upon sorption in the experimental conditions.

Exposure of HZSM-5 to TMB for several days led to an intense absorption around 465 and 900 nm with the vibronic structure of TMB$^{•+}$ between 400 and 500 nm. After several weeks a shoulder around 500 nm is induced in the spectra. This last feature was previously assigned to the TMB^{2+} dication [13]. Raman spectra of TMB adsorbed on HZSM-5 after several days of exposure were obtained with excitation at 514.5, 488.9, and 1064 nm. Exciting at 5145 nm within the electronic absorption of TMB^{2+} the resonance Raman spectrum of the dication is observed predominantly, compared with the bands of the TMB^{2+} in solution as charge-transfer bromide salt. The Raman spectrum obtained with 4880 nm excitation shows the simultaneous presence of TMB$^{•+}$ and TMB^{2+} in the porous void of the zeolite. However, it is not possible to give a quantitative estimation of the amount of dication. Nonetheless, the dication is probably a minor species as the FT-Raman spectrum displays only the radical cation. It should be noted that TMB^{2+} decomposes quickly in neutral organic solutions, but it can be stabilized in acid solution. The acidic property of HZSM-5 appears suitable to stabilize the dication. After 15 days of interaction between TMB and the zeolite the absence of a Raman band assignable to neutral TMB or protonated H2TMB provided evidence of complete ionization of TMB upon sorption under the experimental conditions.

2.3 Phenothiazine

Phenothiazines represent an important class of bioactive molecules. The photochemistry of phenothiazine (PTZ) (**Figure 1**) and its derivatives have been

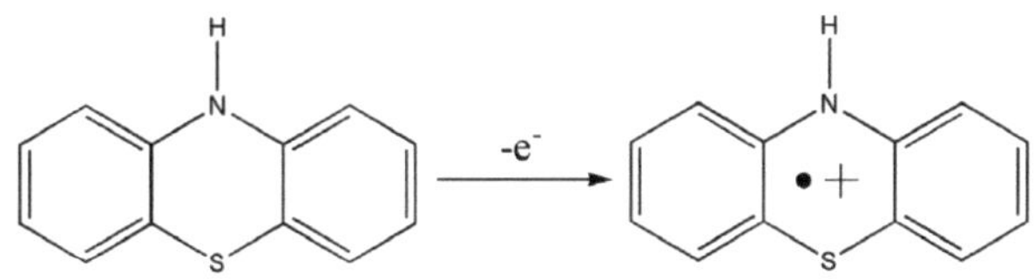

Figure 1.
Phenothiazine and its radical cation $PTZ^{\bullet+}$.

extensively studied because of its pharmacological interest [14]. Due to the low oxidation potential of PTZ, it is proposed that its role is to donate electrons or to transfer charge to the drug's receptor sites. This idea was corroborated by the discovery of many phenothiazine charge-transfer complexes. There has been an extensive study of the formation and reactivity of PTZ's stable radical cations [15], which are involved in an alternative to the biological activity mechanism [16].

As reported by Hester et al., the most intense electronic transition of PTZ radical cation ($PTZ^{\bullet+}$) in the visible region can be found at 513 nm [17]. A resonance Raman investigation of $PTZ^{\bullet+}$ in an aqueous solution was carried out by the authors, following the contour of the band. An enhancement was observed in several Raman bands, with the most intensified one being the one at 476 cm^{-1}. This band was tentatively assigned to a CNC angular deformation mode. A band at 644 cm^{-1}, assigned to the C-N stretching mode was also enhanced. Both bands have frequencies shifted to higher values compared to the ground state of the parent compounds which led them to suggest that the electron density in the radical cation of phenothiazine is localized on the N atom.

An RRS of $PTZ^{\bullet+}$ obtained from the interaction of the neutral molecules with the mordenite zeolite, which has strong oxidizing sites was performed [18]. Phenothiazine is adsorbed in larger amounts in the acid sites of mordenite due to its larger pores. The observation of a pink color soon after PTZ was adsorbed on the mordenite is indicative of the formation of the radical cation $PTZ^{\bullet+}$. **Figure 2** shows the strongest band in the PTZ visible spectrum, located at 516 nm, which was attributed by Hester et al. to the radical cation $PTZ^{\bullet+}$ [17]. The electronic spectrum calculated by the TDDFT method for $PTZ^{\bullet+}$ is also displayed (**Figure 3**).

The RR spectra were excited with laser lines close to the most intense electronic transition of $PTZ^{\bullet+}$ located at 516 nm.

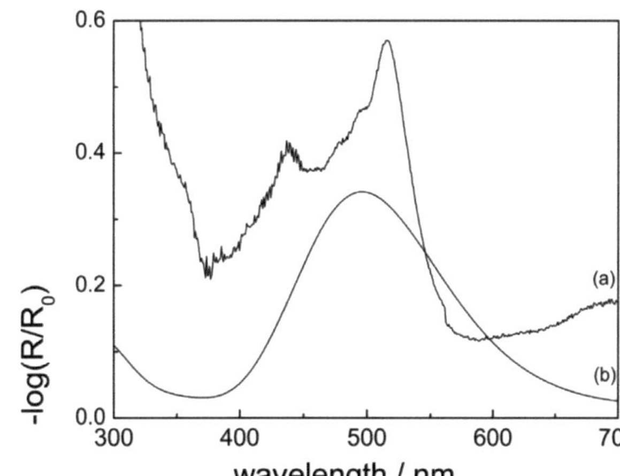

Figure 2.
(a) Diffuse reflectance spectrum of PTZ adsorbed on mordenite; (b) TDDFT (time-dependent density functional theory) calculated electronic spectrum of $PTZ^{\bullet+}$ [18].

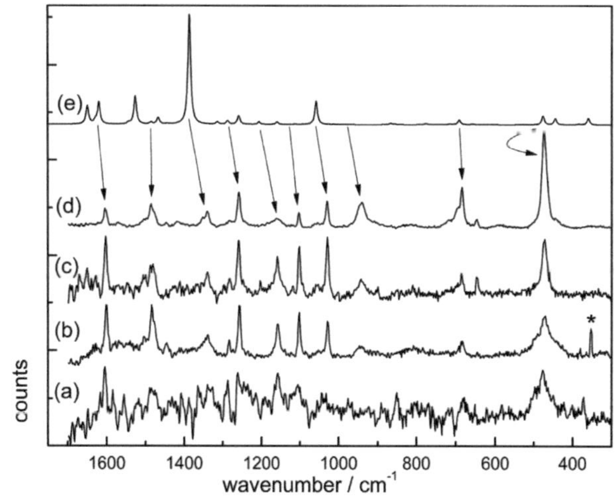

Figure 3.
RR spectra of PTZ adsorbed on mordenite zeolite at the exciting wavelengths: a) 457.9 nm; b) 488.0 nm; c) 496.5 nm; d) 514.5 nm; e) calculated. Arrows indicate the corresponding bands. Asterisk indicates a plasma line [18].

Exciting with 457.9 nm radiation gives rise to the weakest Raman spectrum. Raman excitation with radiations near 514.5 nm (488.0 and 496.5 nm).

It was noticed that excitation with 457.9 nm radiation gave the weaker Raman signal, while other wavelengths led to the increase of the Raman signal, with characteristic $PTZ^{•+}$ Raman bands showing up. Raman spectra excited near 514.5 nm (488.0 and 496.5 nm) also show enhancement of several vibrational modes, involving ring, CNC, and CSC vibrations, while with excitation at 514.5 nm there is a striking difference, giving the highest RR intensity and the best signal/noise ratio together with significative enhancement of the band at 476 cm^{-1}. This fact can be explained by the resonance with the electronic transition band at ca. 516 nm.

The most intensified Raman band at 476 cm^{-1} was assigned to a CSC bending mode, according to DFT (density functional theory) B3LYP/6- 31G(d,p) and TDDFT (time-dependent DFT) calculations, distinct from the previous assignment of Hester et al. Luchez et al. [19] also have obtained the $PTZ^{•+}$ radicals in mordenite, HZSM-5, and H-FER zeolites, following the reaction by UV–visible and Raman spectra. Two wavelengths, 1064 nm and 6328 nm, were used to excite the Raman spectra, but none of them reached the energy of the main electronic transition of $PTZ^{•+}$, so in this condition, it is not possible to obtain the RR spectrum of this species.

3. Clays

Clay minerals are aluminosilicates, most of them of layered structures with tetrahedral silicate and octahedral aluminate sheets. The tetrahedral cation, Si^{4+}, can be replaced by Al^{3+}` or Fe^{3+} cations, and the octahedral cations normally are Al^{3+}, Mg^{2+}, Fe^{2+}, and Fe^{3+} [20].

Smectite clays, (which have been most often used in organic reactions), have a 2:1 layer type, where one octahedral sheet is sandwiched by two tetrahedral sheets. If in the 2:1 layer-type aluminosilicate, the Al^{3+} is substituted by divalent cations, as Mg^{2+} (montmorillonite) in an octahedral sheet, results in a negative charge [21]. To balance the layer charge, cations are introduced between the layers. These cations are hydrated and exchangeable. In natural clays they are typically Na^+, K^+ Ca^{2+}, and Mg^{2+}; similar ions can be exchanged with these ions. Smectite clays have the unique properties of cation exchange capacity, intercalation, and swelling ability, which are important factors in their activity for various reactions involving organic molecules. Clays dried to low water content can behave as acids [22, 23]. Cation-exchanged montmorillonites act as strong Brönsted acids, where the reactive protons are derived from the dissociation of hydrated water molecules because of polarization by exchangeable cations. This situation may be represented by the following equilibrium reaction (Eq. 1):

$$\left[M(H_2O)_x\right]^{n+} \rightleftharpoons \left[M(OH)(H_2O)_{x-1}\right]^{(n-1)+} + H^+ \quad (1)$$

Equilibrium reaction between *[M(H₂O)x}n*$^+$ and the respective species following dissociation of hydrated water molecule and *H*$^+$.

Brönsted acidity results from the terminal hydroxy groups on the external surface and the hydrated cations in the space between the layers of the clay. Lewis acidity is due to central metal ions such as Al^{3+}, Mg^{2+}, and Fe^{3+} in the lattice and also from other metallic cations in the interlamellar space. Often the central metal ions are fully bonded to adjacent oxygen atoms in an octahedral site; so, these intact octahedra within the crystal arrays are unable of originating much Lewis acidity. Nevertheless, in the case of zeolites with small particle size, for instance, MMT K10 (typically 5–10 μm), what happens is that the sandwich layers are squashed to a remarkable extent, generating many broken edges of stacked layers, which contribute to Brønsted as well as Lewis acidity

The Lewis acid sites interaction between the metal cation and adsorbed organic molecules, especially with electron-donating character, results in electron transfer from the organic molecule to metal cation [24] with the consequent formation of radical cations and the reduction of metal ions to lower valency.

Mortland and Pinnavaia reported the interaction of some aromatic organic molecules with transition metal ions (Cu^{2+}, Fe^{3+}) exchanged montmorillonite. They observed the formation of colored adsorption complexes, which was explained due to the electron transfer from the aromatic molecule to the metal ion [25, 26].

The formation of aromatic radical cations was also observed by Pinnavaia et al. upon the adsorption of aromatic molecules on Cu^{2+}, Fe^{3+} and VO^{2+} cation exchanged layered silicates (hectorite). Electron spin resonance spectra were useful to identify the radical cations while UV absorption spectra showed the presence of two species. Besides the radical cation, a charge transfer complex between the metal cation and the aromatic molecule was also observed. Infrared spectra of the adsorbed samples also showed that after some days oligomers and polymers of benzene, toluene, and anisole were formed [27].

3.1 p-dimethoxy benzene

Y. Soma, M. Soma, and I. Harada investigated the interaction of several aromatic molecules with clays, mainly transition metal exchanged montmorillonites using Raman spectroscopy in order to characterize the chromophore responsible for the absorption in the UV–visible region. P-dimethoxy benzene was the first one chosen by the authors [28] because the RR spectra of the radical cation of this molecule in solution had already been investigated by Ernstbrunner et al. [29]. The Raman spectra of p-dimethoxy benzene adsorbed on Cu^{2+}- and Ru^{3+}- montmorillonite were similar to the respective spectrum of p-dimethoxy benzene radical cation reported by Ernstbrunner et al. for the species in solution. RR enhancement in the blue region (maximum with excitation at 457.9 nm) was observed in agreement with the absorption band maxima at 435 and 450 nm. The most intensified band was the ring stretching at 1622 cm^{-1}. It is also noticeable that the ring breathing band at 820 cm^{-1} which is the strongest band in the neutral molecule, is very weak in the cation. The enhancement of the ring stretching mode and the weakness of the ring breathing mode are also observed in the adsorbed samples on Cu^{2+}- and Ru^{3+}-montmorillonite.

3.2 Anisole

When anisole is adsorbed on Cu(II)-montmorillonite, the interaction between the adsorbate and the clay gives rise to several species, depending on the presence of water [30]. In a dry atmosphere, the sample has a green-blue color. In the UV visible spectrum, a broadband from ~550 nm to ~850 nm and other bands with the maximum at ca. 470 nm are observed. The RR spectra with excitation wavelengths of 6100 nm and 4570 nm are shown in **Figure 4(b)** and **(c)**, respectively, along with the Raman spectrum of liquid anisole. The RR spectrum excited with 6100 nm has bands at 995, 1208, 1325, 1528, and 1618 cm^{-1}, that resembles the spectrum of 4,4′-dimethoxybiphenyl but the inter ring CC stretching at 1325 cm^{-1} is upshifted, indicating the formation of a radical cation species of this molecule, similar to observed in the spectrum of 4,4′-dimethoxybiphenyl on Cu^{2+}- and Ru^{3+}- montmorillonite, reported in the same article [30]. Exciting with 457.9 nm a more complex spectrum is observed, as can be seen in **Figure 4(c)**. The authors made a deconvolution of the spectrum, which allowed the distinction of two spectra, the one in black is similar to the spectrum observed with 610.0 nm excitation, due to 4,4′-dimethoxybiphenyl radical cation, and the other in white has a resemblance with the spectrum of liquid anisole, but with shifts in some bands, as the ring breathing mode (925 cm^{-1} compared to 995 cm^{-1} in the liquid anisole spectrum). Another difference is the intensification of the ring stretching band at around 1600 cm^{-1}, which is a signal of radical cation formation.

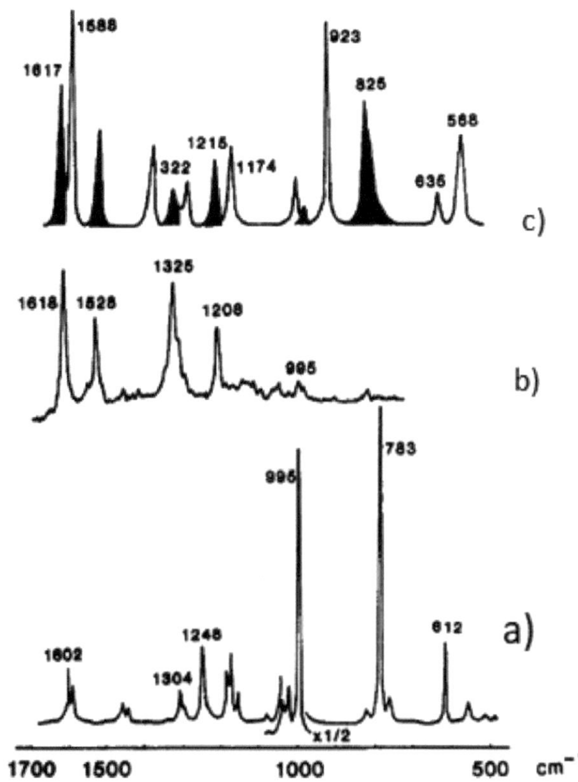

Figure 4.
(a) Raman spectrum of liquid anisole with 524.5 nm exciting wavelength and RR spectra of anisole adsorbed on Cu(II)- montmorillonite with (b) 610.0 nm and c) 457.9 nm exciting wavelengths. Adapted from Soma et al. [30].

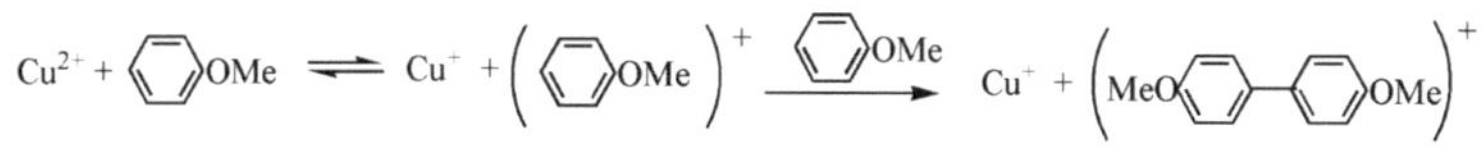

Figure 5.
Reactions occurring with anisole in Cu (II)-montmorillonite. Adapted from [30].

The authors proposed the following reactions to take place with anisole in the interlayer of the Cu(II)-montmorillonite (**Figure 5**).

The first reaction, a redox reaction forming Cu^{+} and the radical cation of anisole is favored in a dry atmosphere. In the second one anisole, radical cation reacts with neutral anisole, forming 4,4′-dimethoxybiphenyl radical cation.

3.3 Benzene

Soma et al. studied benzene adsorbed on Cu^{2+}-montmorillonite by RR spectroscopy [31]. Two species were observed: type II (red form), formed upon adsorption in an exhaustively dried atmosphere, and type I (yellow form) when exposed to humid air. Type I has a band with an absorption maximum at 380 nm while type II has an absorption maximum at 520 nm, besides a strong broad absorption that begins at 800 nm and extends into the near-infrared region.

The RR spectra of benzene adsorbed on Cu^{2+}_ montmorillonite with excitation at 4579 and 5145 nm are very different from the Raman spectra of liquid benzene, resembling the spectra of p-phenylene molecules, e.g. p-terphenyl or p-quaterphenyl, indicating that oligomerization occurred on the clay surface. The presence of the inter ring C-C stretching at 1324 cm^{-1} (type II) or 1240 cm^{-1} (type I) is strong evidence of oligomerization. Type I and Type II are assigned to poly p-phenylene and poly p-phenylene radical cation, respectively, based on the results of RR spectra of poly p-phenylene (PPP) and AsF_5-PPP complex reported by Tzinis [32]. He observed a shift to a higher frequency of the inter ring C-C stretching in the complex compared to the bulk PPP. Based on the results of RR spectra and absorption spectra, it was concluded that benzene adsorbed on Cu^{2+}-montmorillonite is polymerized to PPP cation (type II) and reversibly reduced to PPP in humid air (type I) as follows (**Figure 6**).

$$kCu^{2+} + n(C_6H_6)_{phys} \longrightarrow kCu^{+} + (C_6H_4)_n^{k+} \underset{-H_2O}{\overset{+H_2O}{\rightleftharpoons}} kCu^{2+} + (C_6H_4)_n$$

Figure 6.
Reactions of benzene with Cu(II)-montmorillonite. Adapted from [31].

$(C_6H_6)_{phys}$ means physically adsorbed benzene, $(C_6H_4)_n$ and $(C_6H_4)_n^{k+}$, PPP and its cation, respectively.

Benzene adsorbed on Fe^{3+}- and Ru^{3+}- montmorillonite was also investigated by RR spectroscopy and showed a similar behavior; two species being formed, type I in humid air and type II in the dry atmosphere [33]. The authors noted, however, a difference between benzene /Fe^{3+}- and benzene/Cu^{2+}- montmorillonite: for the first one the transformation of type II to type I in humid air did not occur easily mainly after repeating the reaction, in contrast to what was observed with Cu^{2+}- montmorillonite complex.

3.4 Benzidine

Aqueous suspensions of benzidine and montmorillonite clay (BZ-MMT system) give rise to samples with blue, yellow, purple, and orange-brown colors, depending on the pH of the suspension. In order to assign the species formed in these samples, in situ RR spectroscopy investigation was performed [34].

From pH 3 to 9, the main species responsible for the blue color is the radical cation of BZ (BZ(+center dot)) which has λ(max) at 600 nm. At pH lower than 3, the major species that gives rise to the yellow color is the dication, BZ(2+) with λ(max) at 440 nm. Following the dehydration of the blue BZ-MMT slurry, a yellow color appears, which was assigned to BZ(+center dot) and BZ(2+) species, according to its RR spectrum. At pH higher than 9, the orange-brown colored sample was characterized as poly(benzidine) (PBZ). When the blue BZ-MMT aqueous suspension was left standing for 3 weeks a purple color appeared and it was found to be from a mixture of BZ(+center dot) and BZ(2+) species together with PBZ. Synthetic Syn1 and pillared MMT clays in aqueous suspensions at different pHs were also tested to investigate the possibility of BZ oxidation. In these acidic clays, only thc BZ(+center dot) was identified, BZ(2+) species was not observed probably due to the majority presence of the less reactive BZ(+center dot). At pHs above 9, the benzidine polymerization through the reaction between unprotonated BZ(+center dot) radical cations is catalyzed by the clay.

4. Sulfated metal oxides

The preparation and the highly acid properties of sulfated metal oxide were reported for the first time by Hino and Arata [35] and Hino et al. [36], who investigated the catalytic properties of sulfated TiO_2 and ZrO_2. These metal oxides were able to catalyze the isomerization of n-butane at room temperature, which only very strong acid catalysts are able to do. Thereafter they have been considered by some authors as superacids. In sulfated metal oxides, there are Lewis acid sites and Brönsted acid sites. The super acidity of these materials is attributed to the Brönsted and Lewis acid sites that have increased acidity due to the inductive effect of sulfate, which is coordinately bound to the Ti^{4+}. The sulfate group act as an electron-withdrawing group, making Ti^{4+} more electron-deficient, that is, a strong Lewis acid site, which in turn also withdraws electron density from the -OH group, creating or increasing the Bronsted acid site, as can be seen in **Figure 7**.

The unusual catalytic properties of these materials are attributed to the presence of very strong Lewis or Brönsted acid sites or both types of sites or to their oxidative ability. The oxidative ability was postulated by Ghenciu and Farcasiu [38], who observed that benzene was oxidized when adsorbed on sulfated ZrO_2 and heated at 373 K. Several oxidation products, such as phenyl esters and phenols were formed, which were evidence of the presence of oxidizing sites [38].

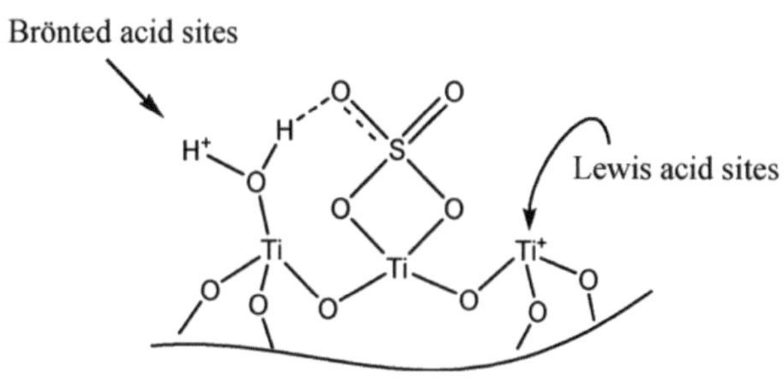

Figure 7.
Ilustrative structure of sulfated titania, representing the sulfate linked to the metal on the chelate form and indicating the Lewis and Brönsted acid sites. [37].

The strong ionizing properties of sulfated ZrO_2 and its relation to the catalytic activity were investigated by Chen et al. [39]. Benzene was used as a probe to study to evaluate the strong ionizing properties of sulfated ZrO_2 due to its high ionizing potential value (9.24 eV). In the ESR spectrum, a hyperfine structured peak was observed and attributed to a biphenyl cation. They proposed the following explanation: a two-step process took place as a consequence of the interaction of benzene with sulfated ZrO_2. A CT complex between sulfated ZrO_2 and benzene was formed in the first step. Then a complete electron transfer from benzene to the ZrO_2 occurred, originating a benzene radical cation, this, in turn, reacted with a benzene molecule, forming the biphenyl radical cation.

As with sulfated ZrO_2, sulfated TiO_2 (a solid acid) also may have the ability to interact with aromatic compounds and form CT complexes and may oxidize aromatic molecules with the formation of radical cations.

4.1 Benzene

A RR spectroscopy study using a commercial sulfated TiO_2, following its interaction with benzene was reported by our group [40]. Besides the remarkable enhancement of ν(CC) ring modes, the appearance of non-totally symmetric vibrations was observed in the Raman spectra. The presence of benzene was confirmed by the characteristic benzene ring breathing mode. The preferential enhancement of the ν(CC) ring modes and the appearance of non-totally symmetric vibrations were explained by the RR enhancement due to benzene to Ti(IV) CT transition [41]. Negligible amounts of radicals were detected in the ESR spectrum, because of the moderate strength of the acid sites of the sulfated TiO_2 used in the study. The RR spectra of benzene on sulfated TiO_2 with 457.9 and 476.5 nm exciting wavelengths are shown in **Figure 8**. The bands of anatase TiO_2 (marked with asterisks) acts as an internal standard to measure the Raman signal intensification. It is noteworthy the striking enhancement of the band at ca. 1600 cm^{-1}, due to benzene ring CC stretching mode exciting with 476.5 nm. This wavelength corresponds to an absorption band in this energy range, assigned to a charge transfer transition. This sulfated TiO_2 was not able to oxidize benzene, a molecule with relatively high ionization potential, but a strong interaction occurred, as can be seen in the RR spectra.

The experimental Raman spectrum of the benzene radical cation is unknown. However, the Raman spectra of the phenylene oligomeric radical cations such as biphenyl [42] and p-terphenyl that are formed after adsorption on zeolites [43]

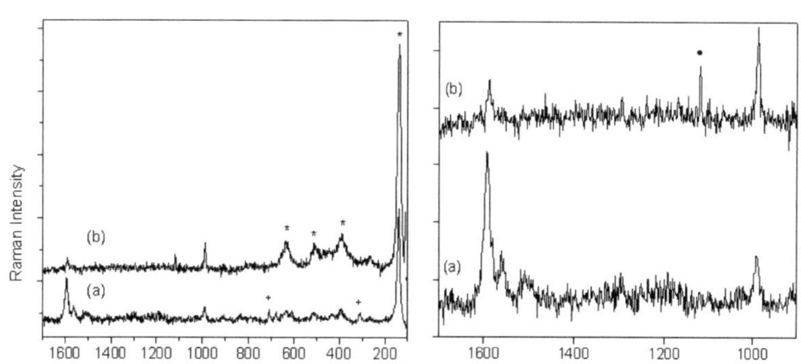

Figure 8.
RR spectra of benzene on sulfated TiO_2 with two exciting wavelengths: (a) 457.9 nm and (b) 476.5 nm. On the left: The 100–1700 cm^{-1} spectral region and on the right: The 900–1700 cm^{-1} region. $^*TiO_2$ bands, +plasma lines of the 457.9 argon-ion laser, Hg lamp line.

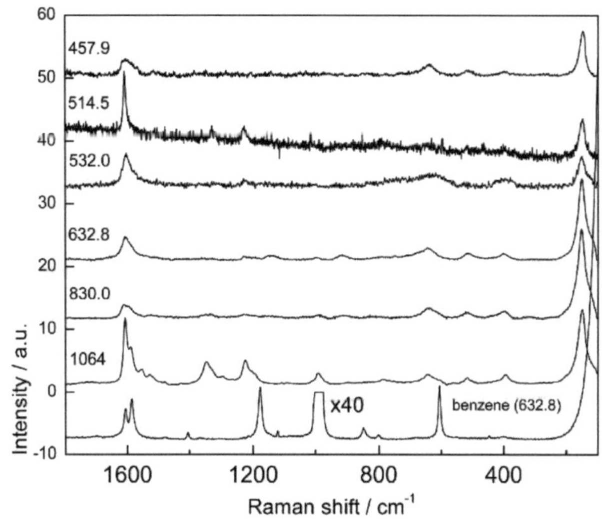

Figure 9.
RR spectra of benzene/sulfated TiO_2^* *sample and liquid benzene at the indicated exciting wavelengths [45].*
** = sulfated* TiO_2 *prepared by sol–gel method.*

have been reported in the literature. These species present the RR effect as reported in the literature for some radical cations.

We have performed an investigation of benzene adsorbed on a very highly acid sulfated TiO_2, synthesized by sol–gel method [44] at room temperature. A tentative assignment of the species formed from the interaction was done with Raman and electron spin resonance (ESR) spectroscopy [45]. From the absorption spectra in the visible and near-infrared (NIR), it was possible to identify these species as phenylene oligomer radical cations. The UV–visible-near IR spectrum of benzene/sulfated TiO_2 sample shows two maxima, one between 400 and 500 nm and the other in the NIR region that begins at ca. 1100 nm and increases towards the infrared. An explanation to this spectrum is based on the RR spectroscopy study of the interaction of benzene with Fe(III)-doped montmorillonite, reported by Soma et al. [31]. They observed two kinds of spectra. The dried sample had an absorption band at 510 nm and was named Type I. When moisture was allowed to contact the dried sample, the band at 510 nm shifted to 380 nm and the spectrum was named Type II. Both types have an absorption that begins at 660 nm and extends to the NIR region. Type I and Type II are assigned to PPP and PPP radical cation, respectively. The Raman spectra of benzene with sulfated TiO_2, excited with several excitation wavelengths are seen in **Figure 9**. It is evident that the spectra do not resemble the spectrum of liquid benzene, mainly because of the absence of the ring breathing mode at 998 cm^{-1}, the most intense band of neat benzene. It is noteworthy the intensification of some bands, notably the one at ca. 1600 cm^{-1}, with the maximum of intensification using exciting wavelengths of 514.5 nm and 1064 nm. These two wavelengths correspond to the two absorptions in the visible-near IR regions of the benzene/sulfated TiO_2 sample. The behavior is similar to that observed by Soma et al. RR spectra were observed and the species that gave rise to these spectra are phenylene oligomers or phenylene oligomer radical cations.

5. Conclusions

Several examples of the use of RR spectroscopy as a tool to identify species formed from the interaction of solid acids with several aromatic molecules were presented. The solid acids have strong acid sites, which can oxidize molecules adsorbed on them or originate charge-transfer complexes. The solid acids were zeolites, clays, and sulfated metal oxides, because of their strong acid sites. Selected works from the literature that report spontaneous ionization of molecules adsorbed on the solid acids and the follow-up of the intermediate species by RR spectroscopy were reported. The intermediate species are radical cations or charge transfer complexes that have absorptions in the visible and in some cases near the IR region. The RR spectra were able to characterize the intermediate species and also to provide information of the chromophore responsible for the electronic transition, because of the preferential RR enhancement of its vibrational modes.

Acknowledgements

The author is grateful to Molecular Spectroscopy Laboratory (Institute of Chemistry, São Paulo University, São Paulo) for the Raman spectra on Jobin-Yvon U-1000 spectrometer and to FAPESP (process 2017/06194-2) for the financial support.

Author details

Lucia Kiyomi Noda
Federal University of São Paulo, Diadema, Brazil

*Address all correspondence to: lucia.noda@unifesp.br

References

[1] Hendra PJ, Passingham G, Warnes GM, Burch R, Rawlence DJ. Fourier transform Raman spectroscopy in the study of species adsorbed on catalyst surfaces. Chemical Physics Letters. 1989;**164**(2,3):178-184. DOI: 10.1016/0009-2614(89)85012-2

[2] Hess H. New advances in using Raman spectroscopy for the characterization of catalysts and catalytic reactions. Chemical Society Reviews. 2021;**50**:3519-3564. DOI: 10.1039/d0cs01059f

[3] Clark RJH, Dines TJ. Resonance raman spectroscopy, and its application to inorganic chemistry. Angewandte Chemie, International Edition. 1986;**25**(2):131-158. DOI: 10.1002/anie.198601311. ISSN 0570-0833

[4] Ramamurthy V, Caspar JV, Corbin DR. Modification of photochemical reactivity by zeolites – generation, entrapment and spectroscopic characterization of radical cations of alpha-omega-diphenyl polyenes within the channels of pentasil zeolites. Journal of the American Chemical Society. 1991;**113**:594-600. DOI: 10.1021/ja00002a030

[5] Moissette A, Lobo RF, Vezin H, Al-Majnouni KA, Brémard C. Long Lived Charge Separated States Induced by trans-Stilbene Incorporation in the Pores of Brønsted Acidic HZSM-5 Zeolites: Effect of Gallium on the Spontaneous Ionization Process. Journal of Physical Chemistry C. 2010;**114**:10280-10290. DOI: 10.1021/jp103838b

[6] Werst DW, Trifunac AD. Observation of radical cations of by swiftness or by stealth. Accounts of Chemical Research. 1998;**31**:651-657. DOI: 10.1021/ar970234w

[7] Garcia H, Roth HD. Generation and reactions of organic radical cations in zeolites. Chemical Reviews. 2002;**102**(11):3947-4008. DOI: 10.1021/cr980026x

[8] Hureau M, Moissette A, Smirnov KS, Jobic H. Combined spectroscopic and modeling study of trans-stilbene molecule in cation-exchanged zsm-5 zeolites. Journal of Physical Chemistry C. 2012;**116**:15510-15518. DOI: 10.1021/jp305631q

[9] Vezin H, Moissette A, Hureau M, Brémard C. Trans-stilbene incorporation in acidic medium-pore ZSM-5 zeolite: A pulsed EPR study. ChemPhysChem. 2006;**7**:2474-2477. DOI: 10.1002/cphc.200600279

[10] Moissette A, Hureau M, Moreau M, Cornard JP. Pore selectivity and electron transfers in HZSM-5 single crystals: A Raman microspectroscopy mapping and confocal fluorescence imaging combined study. Physical Chemistry Chemical Physics. 2020;**22**:12745-12756. DOI: 10.1039/d0cp02018d

[11] Moissette A, Brémard C, Hureau M, Vezin H. Slow interfacial electron hole transfer of a trans-stilbene radical cation photoinduced in a channel of nonacidic aluminum rich ZSM-5 zeolite. Journal of Physical Chemistry C. 2007;**111**:2310-2317. DOI: 10.1021/jp066216q

[12] Moissette A, Gener I, Brémard C. Probing the spontaneous ionization of aromatic amines by adsorption in activated acidic zeolite HZSM-5. Journal of Raman Specroscopy. 2002;**33**:381-389. DOI: 10.1002/jrs.836

[13] Guichard V, Bourkba A, Poizat O, Buntinx G. Vibrational studies of reactive intermediates of aromatic amines. 2. Free-radical cation and dication resonance Raman spectroscopy of N, N, N', N'-tetramethylbenzidine and N, N, N', N'-tetramethylbenzidine. The Journal of Physical Chemistry.

1989;**93**:4429-4435. DOI: 10.1021/j100348a012

[14] Jaszczyszyn A, Gasiorowski K, Swiatek P, Malinka W, Ciezlik-Boczula K, Petrus JB, et al. Chemical structure of phenothiazines and their biological activity. Pharmacological Reports. 2012;**64**(1):16-23. DOI: 10.1016/S1734-1140(12)70726-0

[15] Turro NJ, Khudyakov IV, Van Willigen H. Photoionization of phenothiazine: EPR detection of reactions of the polarized solvated electron. Journal of the American Chemical Society. 1995;**117**(49): 12273-12280. DOI: 10.1021/ja00154a029

[16] Philips DL, Pan D. Raman and density functional study of the S_0 state of phenothiazine and the radical cation of phenothiazine. The Journal of Physical Chemistry. A. 1999;**103**(24):4737-4743. DOI: 10.1021/jp990399h

[17] Hester RE, Williams KPJ. Free radical studies by resonance Raman spectroscopy: Phenothiazine, 10-methylphenothiazine and phenoxazine radical cations. Journal of the Chemical Society, Perkin Transactions 2. 1981;**77**:852-859. DOI: 10.1039/P29810000852

[18] Noda LK, Gonçalves NS. Assignment of the electronic transition of phenothiazine radical cation in the visible region - a resonance Raman spectroscopy and theoretical calculation investigation. Journal of Molecular Structure. 2019;**1191**:253-258. DOI: 10.1016/j.molstruc.2019.04.053

[19] Luchez F, Carre´ S, Moissette A, Poizat O. Sorption and spontaneous ionization of phenothiazine within channel type zeolites: Effect of the confinement on the electron transfers. RSC Advances. 2011;**1**:341-350. DOI: 10.1039/c1ra00220a

[20] Soma Y, Soma M. Chemical reactions of organic compounds on clay surfaces. Environmental Health Perspectives. 1989;**83**:205-214. DOI: 10.2307/3430656

[21] Brindley GW, Brown G. Crystal Structures of Clay Minerals and Their X-ray Identification. London: Mineralogical Society of Great Britain; 1980. p. 518

[22] Laszlo P. Chemical reactions on clays. Science. 1987;**235**:1473-1477. DOI: 10.1126/science.235.4795.1473

[23] Pinnavaia TJ. Intercalated clay catalysts. Science. 1983;**220**:365-371. DOI: 10.1126/science.220.4595.365

[24] Theng BKG. Clay-activated organic reactions. In: Developments in Sedimentology (H. Van Olphen and F. Vaniale, Eds.)vol. 35. Elsevier: Amsterdam; 1982. pp. 197-238

[25] Mortland MM, Pinnavaia TJ. Formation of copper(II) arene complexes on the interlamellar surfaces of montmorillonite. Nature. 1971;**229**:75-77. DOI: 10.1038/physci229075a0

[26] Pinnavaia TJ, Mortland MM. Interlamellar metal complexes on layer silicate. I. Copper(II)-arene complexes on montmorillonite. The Journal of Physical Chemistry. 1971;**75**:3957-3962. DOI: 10.1021/j100695a007

[27] Pinnavaia TJ, Hall PL, Cady SS, Mortland MM. Aromatic radical cation formation on the intracrystal surfaces of transition metal layer lattice silicates. The Journal of Physical Chemistry. 1974;**78**:994-999. DOI: 10.1021/j100603a010

[28] Soma Y, Soma M, Harada I. Raman spectroscopic evidence of formation of p-dimethoxybenzene cation on Cu- and Ru- montmorillonites. Chemical Physics Letters. 1983;**94**(5):475-478. DOI: 10.1016/0009-2614(83)85035-0

[29] Ernstbrunner E, Girling RB, Grossman WEL, Hester RE. Free radical studies by resonance Raman spectroscopy. 1. 1,4-dimethoxybenzene radical cation. Journal of the Chemical Society, Perkin Transactions 2. 1978;**2**:177-184. DOI: 10.1039/p29780000177

[30] Soma Y, Soma M, Harada I. Reactions of aromatic molecules in the interlayer of transition-metal ion exchanged montmorillonite studied by Resonance Raman Spectroscopy. 2. Monosubstituted benzenes and 4,4′-disbstituted biphenyls. The Journal of Physical Chemistry. 1985;**89**(5):738-742. DOI: 10.1021/j150658a021

[31] Soma Y, Soma M. Resonance Raman spectra of benzene adsorbed on Cu^{2+}-montmorillonite. Formation of poly(p-phenylene) cations in the interlayer of the clay mineral. Chemical Physics Letters. 1983;**99**(2):153-156. DOI: 10.1016/0009-2614(83)80549-1

[32] Tzinis C-H, Baughman RH, Risen WM Jr. Raman Spectra of Highly Conducting Poly-p-phenylene Complexes. Office of Naval Research; 1980. Providence, Rhode Island: Technical Report No. TR-80-01

[33] Soma Y, Soma M, Harada I. The reaction of aromatic molecules in the interlayer of transition-metal ion-exchanged montmorillonite studied by resonance Raman spectroscopy. 1. Benzene and *p* –Phenylenes. The Journal of Physical Chemistry. 1984;**88**: 3034-3038. DOI: 10.1021/j150658a021

[34] Nascimento GM, Barbosa PSM, Constantino VRL, Temperini MLA. Benzidine oxidation on cationic clay surfaces in aqueous suspension monitored by in situ resonance Raman spectroscopy. Colloids and Surfaces A: Physicochemical and Engineering Aspects. 2006;**289**:39-46. DOI: 10.1016/j.colsurfa.2006.04.005

[35] Hino M, Arata K. Reactions of butane and isobutane catalysed by titanium oxide treated with sulphate ion. Solid superacid catalyst. Journal of the Chemical Society, Chemical Communications. 1979;**24**:1148-1149. DOI: 10.1039/C39790001148

[36] Hino M, Kobayashi S, Arata K. Solid catalyst treated with anion. 2. Reactions of butane and isobutane catalyzed by zirconium oxide treated with sulfate ion. Solid superacid catalyst. Journal of the American Chemical Society. 1979;**101**:6439-6441. DOI: 10.1021/ja00515a051

[37] Almeida RM, Noda LK, Gonçalves NS, Meneghetti SMP, Meneghetti MR. Transesterification reaction of vegetable oils, using superacid sulfated TiO2–base catalysts. Applied Catalysis A. 2008;**347**:100-105. DOI: 10.1016/j.apcata.2008.06.006

[38] Ghenciu A, Farcasiu D. Oxidizing ability as the defining factor of reactivity of sulfated zirconia. Chemical Communications. 1996;**2**:169-170. DOI: 10.1039/CC9960000169

[39] Chen FR, Coudurier G, Joly JF, Vedrine JC. Superacid and catalytic properties of sulfated zirconia. Journal of Catalysis. 1993;**143**(2):616-626. DOI: 10.1006/jcat.1993.1304

[40] Noda LK, Rosales R, Gonçalves NS, Sala O. Evidences for charge-transfer complex formation in the benzene adsorption on sulfated TiO2 – a resonance Raman spectroscopy investigation. Journal of Raman Specroscopy. 2008;**39**(3):415-420. DOI: 10.1002/jrs.1843

[41] Gonçalves NS, Noda LK. Spectroscopic study of the charge-transfer complexes TiCl4/styrene and TiCl4/polystyrene. Journal of Molecular Structure. 2017;**1146**:750-754. DOI: 10.1016/j.molstruc.2017.06.029

[42] Buntinx G, Poizat O. Triplet (T1) state and radical cation resonance

Raman investigation of biphenyl derivatives. The Journal of Chemical Physics. 1989;**91**:2153-2162. DOI: 10.1063/1.457023

[43] Belhadj F, Moissette A, Bernard C, Hureau M, Derriche Z. Effects of spatial constraints and Brönsted acid site locations on para-terphenyl ionization and charge transfer in zeolites. ChemPhysChem. 2011;**12**:1378-1388. DOI: 10.1002/cphc.201000825

[44] Noda LK, Almeida RM, Probst LFD, Gonçalves NS. Characterization of sulfated TiO2 prepared by the sol–gel method and its catalytic activity in the n-hexane isomerization reaction. Journal of Molecular Catalysis A. 2005;**225**(1):39-46. DOI: 10.1016/j.molcata.2004.08.025

[45] Gonçalves NS, Rettori D, Silva GMG, Noda LK. Spectroscopic study of radical cation species formed on sulfated TiO_2 upon benzene adsorption. Vibrational Spectroscopy. 2018;**99**:80-85. DOI: 10.1016/j.vibspec.2018.08.012

Chapter 8

Raman Spectroscopy for Characterization of Hydrotalcite-like Materials Used in Catalytic Reactions

Luciano Honorato Chagas, Sandra Shirley Ximeno Chiaro, Alexandre Amaral Leitão and Renata Diniz

Abstract

This chapter covers a brief review of the definition, structural characteristics and main applications of hydrotalcite, an interesting multifunctional material which finds applicability in different areas. Particularly, some catalytic reactions using hydrotalcite or mixed oxides derived from these materials are addressed (Ethanol Steam Reforming, Photochemical conversions, Hydrodesulfurization). The use of Raman Spectroscopy associated with other techniques, such as powder X-ray diffraction (XRD), Extended X-ray Absorption Fine-Structure (EXAFS), Temperature Programmed Reduction of hydrogen (H_2-TPR), Fourier-Transform Infrared (FTIR) and Density Functional Theory (DFT) simulations, to characterize this type of material is addressed through examples described in the current literature. In this sense, multidisciplinary efforts must be made in order to increase the understanding of the properties of these materials and the catalytic behavior in the most varied reactions.

Keywords: hydrotalcite, heterogeneous catalysis, Raman spectroscopy, nanomaterials, photocatalysis

1. Introduction

Hydrotalcite is a hydroxycarbonate of magnesium and aluminum which occur in nature as a layered double hydroxide (LDH). The LDHs are represented by the general formula $[M^{2+}{}_{(1-x)}M^{3+}{}_{x}(OH)_2]x^{+}(A_{x/n})^{n-}\cdot nH_2O$, where M^{2+} and M^{3+} are divalent and trivalent cations, respectively, and An^{-} is a charge compensation anion. Their structure consists of brucite-type layers (**Figure 1**), with the substitution of divalent with trivalent cations resulting in a positively charged layer, compensated by interlayer anions [1, 2]. In addition, water molecules are present in interlayer spaces collaborating for the stabilization of the crystalline arrangement through varying hydrogen bonds.

The mineral hydrotalcite has the molecular formula $Mg_6Al_2(OH)_{16}CO_3.4H_2O$, however, the synthetic hydrotalcite-like materials can have a wide compositional cation variability, different cationic ratios and varying anions in the interlayer region.

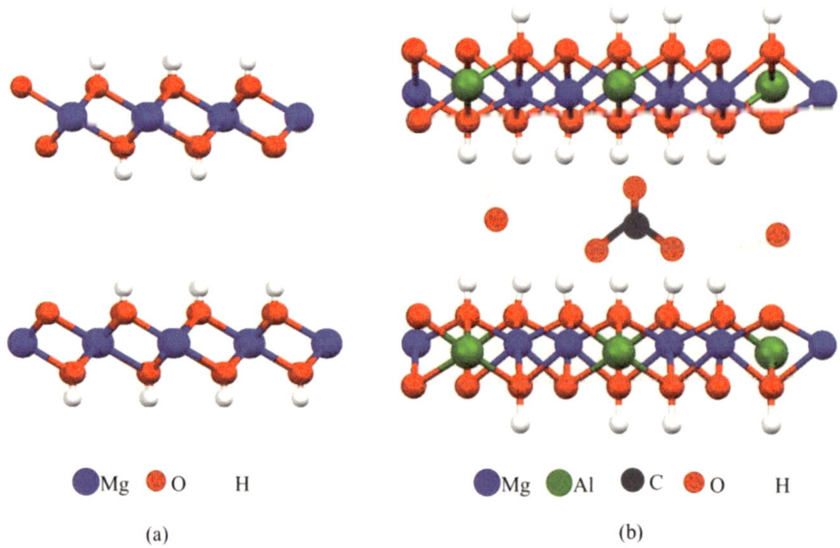

Figure 1.
Schematic representations of (a) Brucite and (b) Hydrotalcite.

Undoubtedly the LDH most used in several applications is the MgAl-LDH system containing carbonate as interlayer ions. However, several other cations and anions are part of countless possible compositions. Co-precipitation is the main preparation method of LDHs, which consists of slowly adding an aqueous solution containing the metal ions in a proper ratio over an aqueous solution containing the hydroxyl ions (usually NaOH) and the anion to be intercalated under vigorous stirring during a certain time. The obtained precipitate is filtered, washed with deionized water, and dried. This is the least cost method for producing LDH. Mostly the production requires strict control of temperature, stirring, and pH to avoid the formation of impurities such as simple hydroxides. In the industrial case, the separation of impurities can raise the price of the final product.

These materials can be prepared by several methodologies other than co-precipitation. An alternative is the hydrothermal method, which permits obtaining high particle size and high product purity [3, 4]. This method is similar to the co-precipitation, meantime, after stirring the suspension is transferred to a Teflon autoclave for hydrothermal treatment. The temperature and aging time influences directly in the crystallinity and morphology uniformity of the material [3–8]. Urea hydrolysis is a satisfactory example of hydrothermal synthesis in which urea is used as a source of carbonates and hydroxyls anions, providing better crystallinity due to very slow precipitation [9–11].

The ion exchange method is widely used; mainly for the intercalation of drugs [12, 13]. This method consists of inserting the precursor LDH in an aqueous solution containing the anion to be intercalated. The exchange of anions in the interlamellar space occurs after pH adjustment and constant stirring, generating a new LDH after the experimental procedures. The solid precipitate is then filtered, washed with deionized water, and dried in an oven. Increased interlayer distance may occur depending on the size of the intercalated ion, as exemplified in **Figure 2**.

The LDHs reconstruction method is known as the memory effect, an intrinsic property of this type of material, which is characterized by the regeneration of the lamellar structure after thermal decomposition (**Figure 3**).

The thermal decomposition of this type of material is a complex sequence of steps that involve dehydration, dehydroxylation (loss of hydroxyls), and loss of

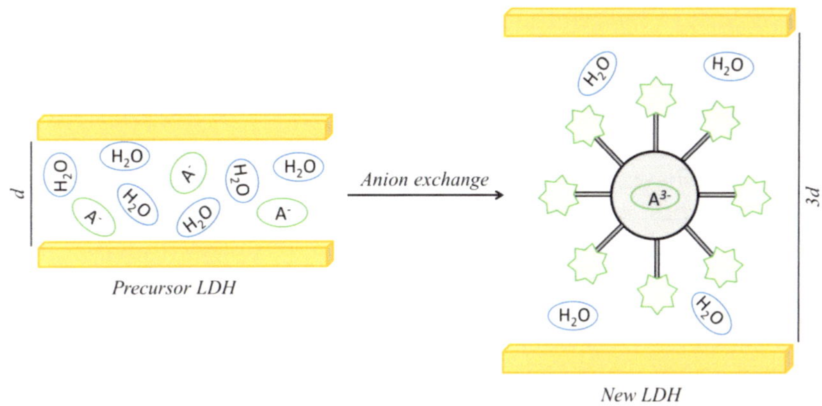

Figure 2.
Schematic representation of ion exchange method.

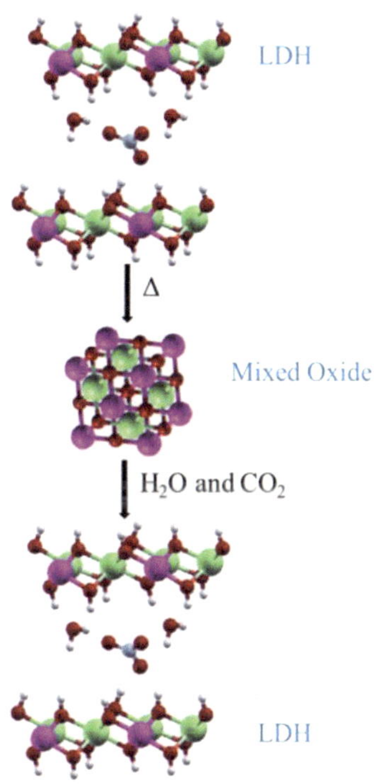

Figure 3.
Schematic representation of memory effect.

carbonate in the starting material [14]. The initial lamellar structure forms mixed metal oxides as the final products of thermal decomposition. When subjected to hydration, these oxides are capable of regenerating the initial lamellar structure. This method can also be used for ion exchange, since the solution used for rehydration may contain anions different from those contained in the initial LDH.

In addition to the aforementioned methodologies, the following deserve mention: sol–gel synthesis [15, 16], salt oxide method [17, 18], sonochemical assisted synthesis [19–21], and manual grinding method [22, 23]. The methodologies mentioned here do not exhaust the possibilities of synthesis of LDH. It should be noted that each method has its advantages and disadvantages and can be improved and applied according to the specifics of the desired product.

2. Applications of hydrotalcite-like materials

The great current interest in LDHs is due to the wide range of possible applications in different areas. This is reflected in a large number of recent reviews and papers published in high-impact journals from a variety of research fields. In the biomedical area, for example, Shirin *et al.* [24] described recent advances in the structure, properties, synthesis, functionalization, and drug delivery applications of LDHs. It was emphasized that compared to other nanomaterials, LHDs are better candidates for release a drug/gene in a controlled manner and deliver them efficiently in the target sites. This is due to its structure and high surface volume ratio. Moreover, the stability in a definite pH range, biocompatibility, high loading, and high anion exchange capacities improve the bioavailability and allow slow release of intercalated drug decreasing the frequency of drug administration.

Jin *et al.* [25] reported that the physicochemical properties of the LDH nanoparticles favor the high biocompatibility and low toxicity for use in the human biological system. The synthetic MgAl-LDH Talcid®, from Bayer, is a traditional example of this field. This worldwide commercialized stomach antacid and anti-pepsin has a structure analogous to that of the hydrotalcite mineral, being able to keep the stomach pH stable between 3 and 5. The intercalation of LDHs with anionic biomolecules, forming the so-called hydrotalcite nanohybrids, is also addressed in the work of Jin *et al.* [25]. DNA, small interfering ribonucleic acid (siRNA), anti-cancer drugs, and contrast agents are among the anionic bioactive molecules that can be intercalated in the hydrotalcite structure. Because it is naturally sensitive to the biological acid medium, LDH allows for controlled drug/gene delivery. In addition, its physicochemical properties such as particle size and morphology can be controlled by varying the synthesis conditions, enabling the minimization of toxicity.

Still, in relation to the use of LDHs in the biomedical area, Meirelles and Raffin [26] published a technological and scientific prospection related to patents and articles involving composites employed in therapeutic devices. Despite the growing interest in the area, the authors considered the number of patents low (on average less than 10 documents per year in the last decade), and attributed this to the lack of regulation on nanomaterials used in the development of medications. Additionally, for further advances in nanomedicine applications are necessary: to improve the synthesis methodologies to enable uniform particle size distribution; understand how the number/density of surface modifiers affects biological performance; increase the molecular selectivity; study long-term side effects; to develop diverse imaging modalities for studies at molecular level providing comprehensive biological information; and to develop functional LDHs loading multiple antigens and biological adjuvants [27].

Another extremely relevant sector where LDHs are widely used is the removal of contaminants from water. Access to clean water and an efficient basic sanitation system are essential factors for socioeconomic development and the reduction of millions of deaths annually worldwide. Certainly, this situation is further aggravated by the pandemic caused by the new corona virus. Moreover, the need for decontamination of rivers and fountains is increasing, due to pollution caused by industrial and human waste. In this sense, LDHs have been extensively researched for use in water purification [28–30]. Its anion exchange property associated with high surface area, compositional versatility, and higher adsorption capacity are characteristics that differentiate them from other mineral adsorbents such as aluminum or iron oxy-hydroxides [28]. Nevertheless, scientists in this field agree that more knowledge is needed to be able to apply these materials on a large scale [28–31]. Techniques of preparation, functionalization and thermal activation must be improved aiming to increase the understanding of their behavior on an atomic scale. In this way, it will be possible to correlate structure, chemical composition, morphology, and surface properties to maximize the adsorption capacity and, consequently, achieve better performances in removing pollutants from water.

Polymer nanocomposites (PNCs) containing LDHs are also considered as an important alternative for water purification systems. These materials have the interesting ability to combine the characteristics of the polymeric matrix and the LDH, forming nanocomposites with multifunctional properties. Pandey *et al.* summarized eight methods used to water decontamination: adsorption, coagulation and flocculation, membrane separations, ion exchange, oxidation, advanced oxidation process, biodegradation, and microbial treatment [32]. The authors stated that is required a combination of processes to insure adequate quality of water, and PNCs can be used in all these processes, permitting efficient decontamination of metal ions, dyes, and microbes. Excellent arsenic absorption and regeneration ability were reported for PNCs, and factors like synthesis, calcination and LDH composition were pointed as crucial for achieve a better performance [33]. Wang *et al.* synthesized a functionalized hydrotalcite/graphene oxide hybrid nanosheets and used as nanofiltration membrane for water desalination. The exfoliated hydrotalcite and graphene oxide were incorporated into polyamidde membrane, generating a material with singular characteristics, achieving enhanced water flux and superior water softening performance [34]. However, according to Mohapi *et al.* [31], there are challenging conditions for these materials to be manufactured and used efficiently, such as: the selection of appropriate nanomaterials that possess specific interfacial interaction, the compatibility of nanomaterials with polymer matrix, and the homogeneous dispersion of LDH particles in the polymeric matrix. Given the above, the use of LDHs or PNCs for water contaminants elimination requires a multidisciplinary knowledge of characterization techniques, whether they are X-ray diffraction, microscopic or spectroscopic.

The great interest in the application of LDHs and their composites with various substances is not limited only to the mentioned areas. A series of possible preparations, characterizations, and industrial applications were exemplified in some interesting reviews [35, 36]. The review of Yan *et al.* report the use of these materials in the selective catalytic reduction of NO_x with NH_3, which is one important task for non-power industry (steel, cement, waste incineration, etc.), capable of enabling energy conservation and emission reduction [37]. The uniform interlayer galleries of the LDHs allow its application as membranes for gas/liquid separations [38]. Their excellent anion exchange capacity stimulates the use as host-guest materials applied in the pesticide-related field [39]. The improved thermal, mechanical and rheological properties of PNCs containing LDHs significantly enhance the performance on flame retardancy and physical properties of the paper and epoxy resin [40, 41]. Low cost plastic films can be produced and used in agricultural area. The work of Xie *et al.* showed that PNCs composed by low density polyethylene and intercalated LDH with lauryl phosphoric acid ester potassium can be applied for this purpose [42]. Charttejee *et al.* reported the synergistic effect present in bionanocomposites made from LDH and different biopolymers [43]. The importance of this theme is related to environmental protection, since biopolymers are environment friendly, fully degradable and sustainable materials. In the area of fertilizers, for example, Borges *et al.* [44] stressed the importance of new methods or products to achieve improvements in the management of nutrients and to reduce environmental impacts. The use of LDHs for corrosion protection of aluminum alloys has also been identified as a new alternative to replace chromate-based coatings due to the harmful action of chromium species on human health and the environment [45]. Electrochemical capacitors, or supercapacitors, based on LDHs also have been studied as novel and sustainable energy storage technology [46]. Furthermore, the structural characteristics of hydrotalcite-like materials are identified as suitable for use as building materials in the construction industry in addition to factors such as low cost and high availability in mineral reserves [47]. Thus, LDHs find applications in these and in several other fields of research. In the next section, emphasis will be placed on its use in catalytic reactions.

3. Some catalytic reactions using hydrotalcite or mixed oxides derived from these materials

One of the areas in which LDHs find wide application is in catalysis. Particularly, in heterogeneous catalysis and photocatalysis, the LDHs are very used as a catalyst or, mainly, as a precursor of mixed oxides, which, in turn, can be used in various industrial catalytic processes [48–54]. Below will be highlighted some interesting catalytic reactions using hydrotalcite-like materials as catalyst or catalyst precursor. Examples of how Raman spectroscopy can be used to characterize these materials will also be discussed.

3.1 Ethanol steam reforming

The use of hydrogen as an alternative to fossil fuels is an area of great interest to industries, governments, and researchers worldwide. In addition to its pollution-free characteristic, the high energy value and rich resources further enhance research on this subject. The catalytic steam reforming is the technology applied industrially for the hydrogen production. Light hydrocarbons are used in this process, especially methane. In this case, the reaction occurs in two steps: firstly, methane reacts with water vapor generating CO and H_2; after, CO undergoes water gas shift (WGS) reaction generating CO_2 and more H_2 Eqs. (1) and (2).

$$CH_4 + H_2O \rightarrow CO + 3H_2 \qquad \Delta H^{o}_{298K} = 206\,kj\,mol^{-1} \tag{1}$$

$$CO + H_2O \rightarrow CO_2 + H_2 \qquad \Delta H^{o}_{298K} = -41\,kj\,mol^{-1} \tag{2}$$

An alternative for the hydrogen production is the use of alcohols as source. Particularly, in relation to other alcohols, ethanol has some advantages from a socio-environmental point of view, such as ease of obtaining from renewable sources, large volume of production due to existing industrial facilities, ease of handling, and being non-toxic. The Eq. (3) summarizes the global ethanol steam reforming (ESR) reaction, but several other reactions such as WGS Eq. (2) and Boudouard reaction Eq. (4) occur concurrently [55].

$$C_2H_5OH + 3H_2O \rightarrow 2CO_2 + 6H_2 \qquad \Delta H^{o}_{298K} = 174\,kj\,mol^{-1} \tag{3}$$

$$2CO \rightarrow CO_2 + C \qquad \Delta H^{o}_{298K} = -172.5\,kj\,mol^{-1} \tag{4}$$

Several systems are tested in the ESR reaction and the most relevant results are obtained for catalysts containing noble metals such as Pt, Pd, Ru or Rh [56, 57]. On the other hand, Ni, Co or Cu based catalysts can provide a better cost–benefit ratio. However, the greatest impediment to the industrial use of this process is to control the secondary reactions that occur and lead to the deactivation of the catalyst. The resolution for this involves the development of active, selective, and stable catalysts. Thus, mixed oxides formed from LDHs have been used for this purpose.

Passos *et al.* [58] combined Quick-EXAFS and Raman spectroscopies in *operando* conditions to monitoring activation, reaction, and deactivation of NiCu catalysts obtained from hydrotalcite-like precursors. The catalyst activation was performed by heating the LDH precursor under two steps. Firstly, the LDH precursor was calcined under air atmosphere until 210°C. After purge with He, the material was further heated up to 500°C under 5% of H_2/He in order to form the metallic nanoparticles. The EXAFS and XANES analyzes showed that the activation method used is more efficient than the conventional one, as it completely reduces the copper and nickel particles producing metallic particles at lower temperatures.

Raman and mass spectroscopies were used to monitoring the evolution of ethanol conversion and products obtained during the ESR reaction. Full conversion was achieved during the initial 30 min, however several byproducts were observed revealing the occurrence of parallel reactions. After 50 min a decrease in ethanol conversion is accompanied by a decrease in selectivities to CO_2 and H_2. The catalyst deactivation was monitored by increase of D and G bands observed respectively at 1336 and 1586 cm^{-1} in Raman spectra. These bands are respectively characteristics of large aromatic ring systems and ordered graphitic carbon species. Concomitantly, mass spectroscopy showed that these coke deposits originate from decomposition reactions of ethylene, acetaldehyde, and methane, in addition to the polymerization of ethylene and the Boudouard reaction. Besides filamentous and graphitic, amorphous coke species also were detected by Raman analysis, reaching 30% of coke deposits after 180 min on stream. The amorphous coke species were identified through vibrations at 1278 and 1500 cm^{-1}, by deconvolution of Raman signals. These species come from acetaldehyde and ethylene reactions, and encapsulate the metallic sites accelerating the deactivation process. Oxidative regeneration was performed and ESR reaction was restarted. Then, metallic particles were recovered (100% of Cu^0 and 85% of Ni^0) due to the H_2 formed in the ESR reaction, leading to a second reaction cycle with performance equivalent to the first one.

Sikander *et al.* published a detailed review addressing the hydrogen production using hydrotalcite based catalysts [59]. Besides ESR reaction, emphasis was given to reactions such as: methane steam reforming, methanol steam reforming, dry reforming of hydrocarbons, methane partial oxidation, and sorption enhanced reaction process. The authors highlighted that the main drawback in conventional hydrogen production systems is the high carbon deposition on catalytic surface. In this sense, it is necessary to produce in situ catalytic regeneration conditions. Undoubtedly, the work of Passos *et al.* [58] represents a contribution in this regard. Indeed, the structure and high surface area of the LDH based catalysts are physicochemical characteristics that make these materials able to overcome these challenges. New compositions and the association of LDHs with other materials, forming nanocomposites, are pointed out as the future of catalysts for hydrogen generation.

3.2 Photochemical conversions

During a photocatalytic process, a semiconductor surface is excited by ultraviolet–visible radiation. After absorbing energy equivalent to or greater than its band gap, an electron (e^-) is promoted from the valence band (VB) to the conduction band (CB), where holes (h^+) are produced. Then, photocatalytic reactions occur through charge conductors, derived from this electronic promotion between bands, leading to the reduction of molecules adsorbed by the excited electrons present in the CB or to the oxidation of molecules by the positively charged holes in the VB (**Figure 4**). Thus, the generation of photoactivated electron hole pairs allows conducting a widespread range of important chemical reactions [60–64] in an economical and environmentally sustainable manner as alternative to substitute the traditional processes. Because of low costs, recyclability capacity, wide light absorption range, and adjustable band gap, the hydrotalcite-like materials are intensively studied as photocatalysts.

As mentioned in Section 2, LDHs can be used for water purification. Major sources of environmental contamination are found in industrial waste, mainly from the dyeing industries, leading to problems such as low biodegradability, changes in color, smell and pH, in addition to low oxygen availability. The biological treatment is the most used for decontamination of water containing dyes, however it is considered slow and several poisonous molecules cannot be biologically treated. Other techniques are considered expensive and not all the usual techniques are capable of efficiently eliminating all toxic elements. In this regard, photocatalysis emerges as an alternative, enabling the development of more efficient and less environment harmful systems.

De Carvalho *et al.* [65] carried out the application of a niobium oxide catalyst supported on mixed oxide derived from LDH in the photodegradation of the methylene blue dye. The synthesis of ZnAl-LDH was performed by co-precipitation. After thermal decomposition, the precursor generated a mixed oxide, which was submitted to wetness impregnation for incorporation of niobium oxide. In the tests, after only three hours of sun exposure, the applied catalyst led to almost 100% degradation of the dye without the need for any additives. After degradation, the catalyst was recovered and reapplied in another three reaction cycles without significant loss of catalytic activity. This study showed the importance of using photocatalysis in advanced oxidation processes as a method for destroying water polluting molecules.

In another work, De Carvalho *et al.* [66] tested this system in oxidative and photochemical conversion of anilines to azoxybenzenes. Beyond ZnAl, MgAl and MgZnAl-LDH were also synthesized and tested as supports for niobium oxide, yet MgZnAl catalyst was more successful, leading to azoxybenzenes yields up to 92%. The XRD patterns showed wide profiles associated to mixed oxides with low crystallinity. In this case, the presence of niobium species on the support surface was verified through Raman spectroscopy coupled to an optical microscope with a CCD detector. The Raman mapping was measured in the region characteristic of

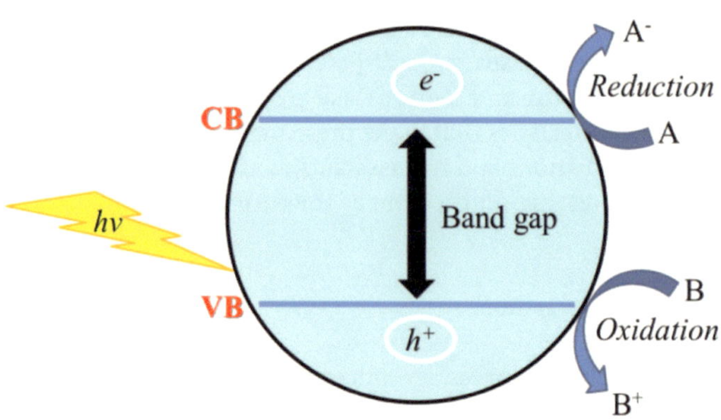

Figure 4.
Schematic representation of a reaction catalyzed by a semiconductor.

niobium oxides (between 960 and 750 cm^{-1}). Integration of this area revealed that the relationship between zinc content and surface area is inversely proportional. This directly affects the dispersion of niobium oxide on the support surface, because the greater the amount of zinc in the support, the greater the number of NbOx clusters, that is, the lesser the dispersion. However, even with heterogeneity in the active phase distribution, the most effective catalyst was the one impregnated on the mixed oxide derived from MgZnAl-LDH. DFT calculations and acid-basic characterization tests showed that the balance between acidic and basic sites is responsible for the greater activity of this catalyst. Moreover, DFT calculations revealed that the charge transfer between nitrogen of aniline and niobium is the first step of the mechanism of photocatalytic synthesis of azoxybenzenes, suggesting chemisorption between the reactant and the catalyst surface. Anyway, this work is an example of how multidisciplinary efforts should be used to characterize materials and understand reaction mechanisms.

3.3 Hydrodesulfurization

In order to improve the air quality, governments in various countries have announced new regulations to reduce the level of sulfur, nitrogen, and other contaminants which are present in transportation fuels [67]. Therefore, refiners need to decrease the concentration of contaminants, particularly in gasoline and middle distillates [68]. Gasoline from fluid catalytic cracking (FCC gasoline), which represents 30–50% of the total gasoline pool, is by far the most important sulfur contributor in gasoline, up to 90% [69]. Although, the olefins, which are important contributors to the octane rating in commercial petrol, are also present in FCC fraction. As a result, FCC gasoline is the focus for sulfur reduction.

The conditions used in the catalytic hydrodesulfurization (HDS) process such as high pressure, high temperature, and high hydrogen consumption make the process expensive. Several alternative methods, such as adsorption or alkylation have been developed in recent years. However, the key technical problem for the HDS of FCC gasoline is to perform a deep sulfur removal and, at the same time, to reduce the loss of the olefins occurring in the HDS process, by minimizing the hydrogenation (HYD) [70].

To preserve the olefins responsible for the octane number, it is necessary to improve the selectivity of the conventional catalysts (sulfide CoMo/γ-Al_2O_3) without loss of octane number. In this connection, one of the key parameters which determine the activity of the CoMo HDS catalysts is the type of support. Aiming to reduce the loss of octane number, Zhao *et al.* [70] used sulfide CoMo catalysts supported on the MgAl, CuAl and ZnAl mixed oxides obtained from hydrotalcite compounds. The authors observed that the catalysts give lower levels of olefin hydrogenation than the traditional γ-Al_2O_3 supported catalyst. In this sense, Coelho *et al.* reported a series of papers devoted to the preparation, characterization and catalytic evaluation of CoMgMoAl catalysts derived from LDHs [71–73]. Initially, the co-precipitation method was used to prepare terephthalate-intercalated CoMgAl-LDH. Next, the anion exchange process was used to substitution of terephthalate by polyoxometalate and preservation of the LDH structure. Then, the calcination of this LDH generates CoMgMoAl mixed oxide, which is precursor of the sulfide active phase. The sulfide characterizations and catalytic tests showed that olefin hydrogenation is associated with un-promoted Mo sites while the improvement in activity and selectivity for HDS is due to the increase in the number of Mo sites promoted by Co.

Presently, we characterize an MgAl-LDH and evaluate the use of their derived mixed oxide as support for HDS catalyst in comparison with γ-Al_2O_3. The powder

X-ray diffraction (XRD), BET, Temperature Programmed Reduction of hydrogen (H_2-TPR), and Fourier Transform Spectroscopies (FTIR and FT-Raman) were used to characterize the materials. FTIR spectra were recorded on a BOMEN MB-102 spectrometer using pressed KBr pellets and 4 cm^{-1} of spectral resolution to verify the vibrational modes present in the samples. Good signal-to-noise ratio was obtained from the accumulation of 128 scans. Raman spectra were acquired on a LabRAM HR-UV 800/Jobin-Yvon equipment, with He–Ne (633 nm) laser and CCD detector. The resolution was 2 cm^{-1} in the range between 1200 and 100 cm^{-1}. Moreover, the hydrodesulfurization of thiophene and hydrogenation of cyclohexene were the reactions chosen to evaluate the activity and selectivity of CoMo sulfide catalysts.

The MgAl-LDH and Boehmite are commercial samples provided by Petrobras-Cenpes. These samples were calcined at 500°C for 3 h under air, to obtain the oxide supports (named as MgAl and γ-Al_2O_3, respectively). The supports were submitted to incipient wetness impregnation of solutions containing the Mo and Co salts using the appropriate amount of ammonium heptamolybdate and cobalt nitrate, to obtain catalysts with 10% of MoO_3 and 3% of CoO on the surface. After calcination at 450°C for 1 h, the oxide catalysts were denominated CoMo/MgAl and CoMo/γ-Al_2O_3.

The spectroscopic data obtained for the MgAl-LDH are shown in **Figure 5**. In the FTIR spectrum, the strong and wide absorption band centered at 3464 cm^{-1} is due to the contribution of the asymmetric stretching modes of the lamellar hydroxyl groups (ν_{OH}) and of the interlamellar water molecules [74]. The shoulder at 3085 cm^{-1} is characteristic of the ν_{OH} symmetrical stretching of water molecules interacting by hydrogen bonding with interlamellar carbonate ions. The poor absorption at 1633 cm^{-1} is attributed to the deformation mode of water molecules (δ_{OH}). The band at 1354 cm^{-1} is assigned to the asymmetric stretch mode, ν_3, of carbonate, and the small band at 1074 cm^{-1} is assigned to the symmetrical mode, ν_1, of carbonates connected to OH groups, as suggested by absorption around 3000 cm^{-1}. This absorption is expected due to the decrease in the symmetry of the carbonate groups (from D_{3h} to C_{2v}), caused by different types of interaction of these anions with interlamellar water molecules and hydroxyl groups present in brucite-like layers [75, 76]. The band at 775 cm^{-1} corresponds to the mode of deformation outside the plane of carbonate ions, and the mode of deformation in the plane is

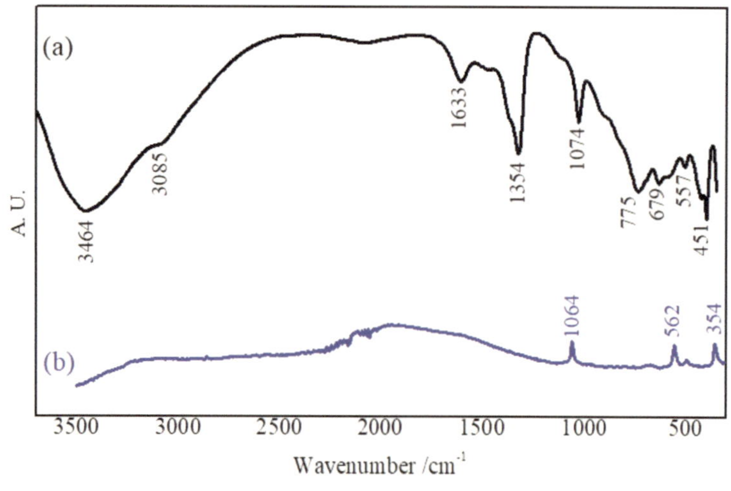

Figure 5.
FTIR (a), and FT-Raman (b) spectra of MgAl-LDH.

observed at 679 cm^{-1}. Still in the region of low wavenumber, the absorption around 557 cm^{-1} is attributed to the vibration of the carbonate-water units [76], however this absorption can also be attributed to the M-O-M, O-M-O, and M-OH lattice vibration modes (where M is Mg or Al) [77]. Finally, the 451 cm^{-1} band is attributed to the contribution of the Mg-O and Al-O stretching modes.

All data obtained through infrared analysis are in agreement with that observed in the FT-Raman spectrum. This spectrum features three typical LDH bands [76, 78]. The three weak absorptions at 1064 cm^{-1}, 562 cm^{-1}, and 354 cm^{-1} are attributed as symmetrical stretching of the carbonate ion, CO_3^{2-} units linked by hydrogen interaction to interlamellar water molecules, and Mg-O stretching, respectively. Therefore, the vibrational study carried out using infrared and Raman spectroscopies suggests that carbonate ions are present in the crystalline network of the sample and are involved in hydrogen bonds.

In the infrared spectrum of the Boehmite sample, it is verified the presence of a band at 3315 cm^{-1} commonly attributed to the asymmetric stretching modes ν_{OH} from water molecules and hydroxides (**Figure 6a**). In addition, a band at 3094 cm^{-1} is observed due to the symmetric stretching ν_{OH} of hydroxyl groups interacting through hydrogen bonds. Around 1639 cm^{-1}, absorption attributed to the δ_{OH} mode is noted. The $\delta_{Al\text{-}OH}$ mode is observed at 1074 cm^{-1}. The $\nu_{Al\text{-}O}$ vibrational modes, with maximum absorption at 625 cm^{-1}, appear as part of a wide and intense band in the region between 900 and 600 cm^{-1}, which in turn still contains the contribution of lattice vibrational modes [79].

In the Raman spectrum (**Figure 6b**), the 680 and 500 cm^{-1} bands are assigned to the asymmetric and symmetric stretch modes $\nu_{Al\text{-}OH}$, respectively. In addition, the intense band at 364 cm^{-1} stands out, due to the $\nu_{Al\text{-}O}$ stretching mode. Thus, these results corroborate with the infrared analysis, suggesting that the sample has a typical Boehmite spectrum.

All the peaks in the XRD patterns were indexed (**Figure 7**) [9, 80–84]. The cell parameters for the MgAl-LDH were refined using the Checkcell software [68] (**Table 1**). The input values were a = 3.0424 and c = 22.6641 Å of a rhombohedral *R-3 m* space group. The interlayer distance value calculated from the more intense reflection (d_{003}) is consistent with the values found in the literature [9, 81].

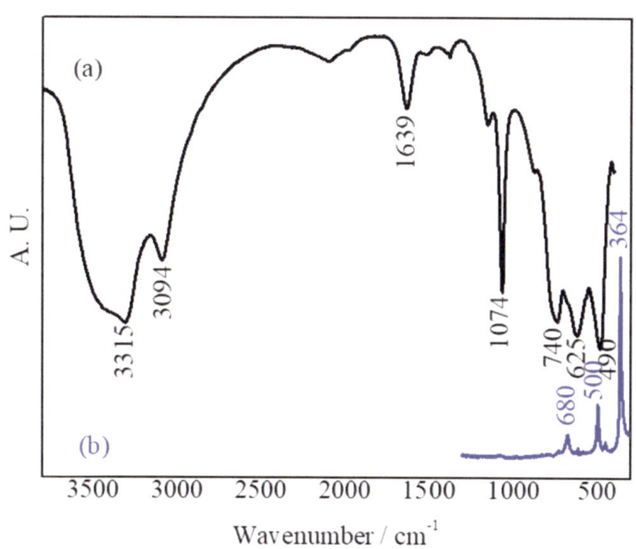

Figure 6.
FTIR (a), and FT-Raman (b) spectra of Boehmite.

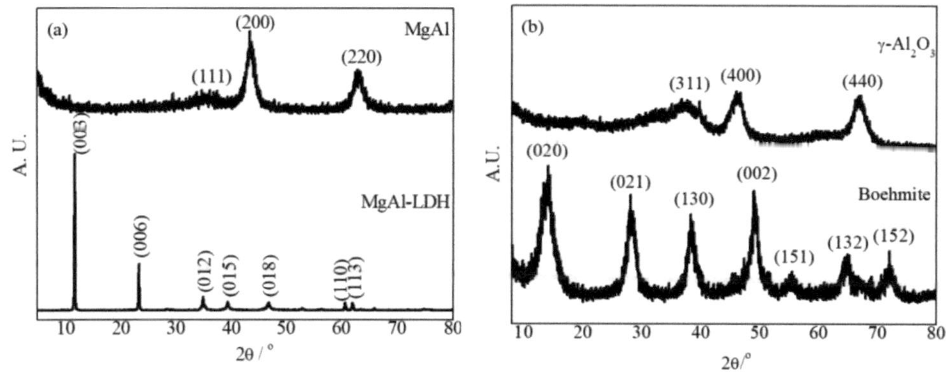

Figure 7.
X-ray diffraction patterns of (a) MgAl-LDH and MgAl, and (b) Boehmite and γ-Al_2O_3.

CS (nm)	*a* (Å)	*c* (Å)	d_{003} (Å)	Support	CS (nm)	*a* (Å)
56.8	3.026 (7)	22.607 (5)	7.49		11.4	4.149 (4)

Table 1.
Crystallite size (CS) and lattice parameters for MgAl-LDH and its derived mixed oxide.

Subtraction from this value of the brucite layer width (4.80 Å) provides an interlayer spacing around 2.70 Å, which is of the same order of magnitude of the size of carbonate anions in vertical orientation. This result agrees with Raman and FTIR data and suggests that these anions have reduced mobility in the crystal lattice as they are involved in strong electrostatic interactions with lamellar hydroxyls and water molecules located in the interlayer spaces [82, 83]. The other precursor exhibits a typical Boehmite profile; consisting of an orthorhombic unit cell with space group *Cmcm* [84–86].

The diffraction patterns obtained for the supports are also show in **Figure** 7. The XRD profile of MgAl mixed oxide is typical of rock-salt phase [87–89]. The cell parameter *a* (**Table 1**), lower than the pure MgO (4.211 Å), suggests an isomorphous Mg^{2+}/Al^{3+} substitution, giving rise to an oxide solid solution containing Mg and Al [87–89]. In turn, the calcination of Boehmite led to the formation of γ-Al_2O_3, with spinel-type structure (*Fd-3 m* space group).

The XRD patterns obtained for both CoMo catalysts show no major differences relative to the respective supports, despite the decrease of the peak intensities, indicating a decrease in crystallinity after impregnation (**Figure 8**). This aspect indicates good dispersion of the impregnated phases in accordance with surface areas, which decreased slightly in relation to the supports (from 191 to 175 m^2g^{-1} for CoMo/MgAl, and from 240 to 201 m^2g^{-1} for alumina based catalyst). However, in this case, the XRD technique does not allow the identification of the crystalline phases present on the supports. For this, Raman spectroscopy is extremely useful.

Typical Raman spectrum of the supported CoMo oxide catalysts is shown in **Figure 9**. The main bands around 995, 818, 665, 378, 337 and 291 cm^{-1} are characteristics of MoO_3 [90, 91]. The most intense bands at 995 cm^{-1} and 818 cm^{-1} correspond to symmetric ($\nu_{Mo=O}$) and asymmetric stretching ($\nu_{Mo\text{-}O\text{-}Mo}$) vibrational modes, respectively [92]. Additionally, a low intense band assigned to $CoMoO_4$ phase is observed at 950 cm^{-1} [93]. The $CoMoO_4$ oxide is known as a good precursor for HDS catalysts, because it could lead to the formation of the active CoMoS Type II phases [94]. Thus, the Raman spectrum suggests that $CoMoO_4$ and MoO_3 species coexist on the surface of the supported CoMo oxide catalysts.

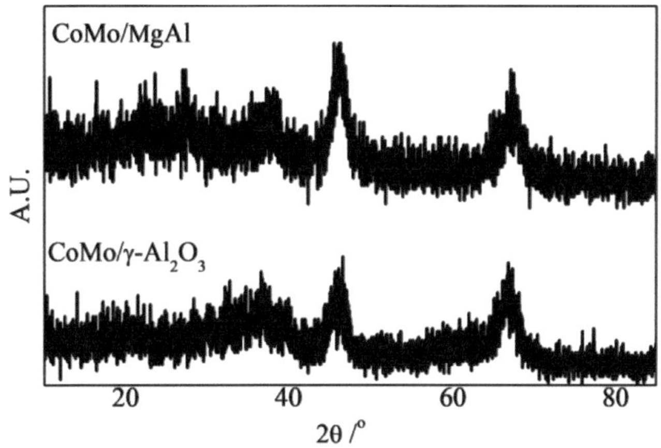

Figure 8.
X-ray diffraction patterns of the CoMo catalysts.

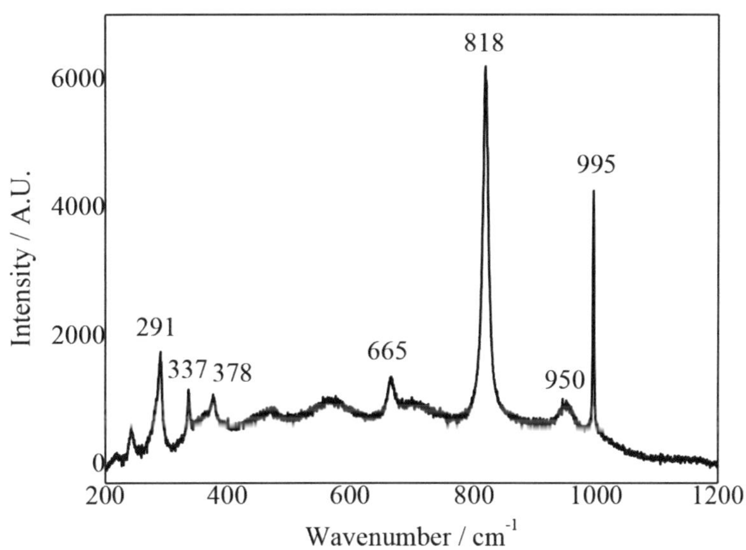

Figure 9.
Raman spectrum of CoMo/γ-Al_2O_3 catalyst.

H_2-TPR technique was used to verify the reduction behavior of impregnated species on the supports. **Figure 10** shows the H_2-TPR profiles of supported oxide catalysts. The CoMo/MgAl catalyst exhibit two peaks (375 and 556°C), while the CoMo/γ-Al_2O_3 catalyst display a peak at 460°C. In the literature, was reported that for catalysts containing only cobalt oxide dispersed on alumina (Co/γ-Al_2O_3) there is a peak around 340°C, assigned to the reduction of Co_3O_4, and peaks between 600 and 700°C, attributed to the reduction of Co^{2+} ions in different chemical environments [95]. When there is only molybdenum oxide on alumina (Mo/γ-Al_2O_3), the $Mo^{6+} \rightarrow Mo^{4+}$ reduction generally occurs at 500°C. This indicates that Mo^{6+} cations are easily reduced. Moreover, commonly above 800°C, are observed peaks related to different reduction steps ($MoO_3 \rightarrow MoO_2 \rightarrow Mo^0$) [96, 97].

H_2-TPR profiles obtained in the present work suggest an interaction between cobalt and molybdenum species, considering that the first peak is observed between 375 and 460°C (temperature higher than Co/γ-Al_2O_3 and lower than Mo/γ-Al_2O_3 reductions). Furthermore, for both samples the second reduction occurs above

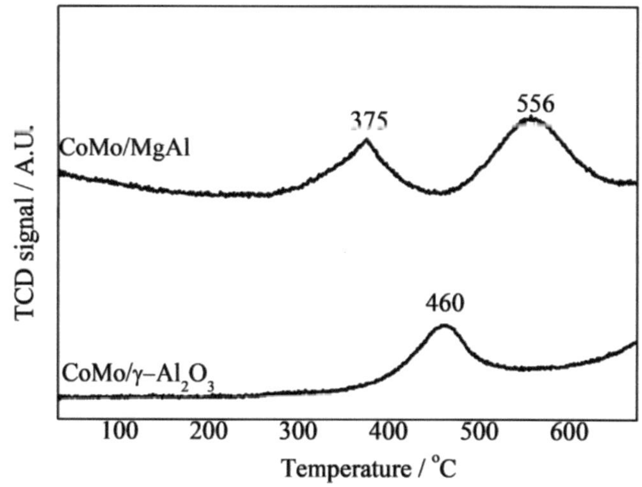

Figure 10.
H_2-TPR profiles of CoMo catalysts.

800°C. The species reducing at slightly higher temperatures may have a somewhat stronger interaction with support surface. Thus, they cannot be reduced and they are therefore probably lead to inactive phases in the HDS reaction. Similar profile is observed in the work of Liu *et al.* [95], suggesting the formation of $CoMoO_4$ in addition to MoO_3 on both supports, corroborating the Raman results showed earlier. It is also observed that the reduction occurs primarily for the sample supported on mixed oxide derived from LDH. This suggests that the interactions of impregnated species with this support are weaker than with γ-Al_2O_3. Additionally, the hydrogen consumption in H_2-TPR analyses is 1.8 and 2.2 mmol g^{-1} for CoMo/MgAl and CoMo/γ-Al_2O_3 respectively. This consumption is directly related to the amount of CoMo reducible species on the surface, which can be related to HDS activity.

For catalytic tests, the reactor was loaded with 300 mg of supported oxide catalyst and 900 mg of SiC (both 80–100 Tyler mesh) between quartz-wool plugs. The presulfiding of the supported oxide catalysts was carried out according to the following procedure: initially, the materials were dried at 150°C for 30 min under a 450 mL min^{-1} nitrogen flow. Then, the supported oxide catalysts were presulfided using a mixture of 1.66% CS_2 in n-heptane (v/v). The liquid was fed to the reactor at 20 mL h^{-1} under hydrogen flow (450 mL min^{-1}) and at atmospheric pressure. The sulfidation temperature was maintained at 280°C for 1 h, at 350°C for 30 min and, finally, at 400°C for 30 min. After sulfidation, the catalysts were tested at 280°C and 20 bar. The liquid feed consisting of 0.8% of thiophene and 17% of cyclohexene in n-heptane (v/v) was pumped to the reactor at 16.8 mL h^{-1} with a 450 mL min^{-1} hydrogen flow. The conversions were kept low in order to operate in differential regime.

The results show that the tested catalysts are active for both, HDS and HYD reactions, for which the main products were butenes and cyclohexane, respectively. Previous studies described calculations methods for the conversions of thiophene HDS and cyclohexene HYD, which were performed from the carbon balance for each of the reactants and the respective reaction products [73]. The results of catalytic performances are displayed in **Figure 11**. It is important to mention that the activity for HDS is practically the same for both catalysts, since thiophene conversions are around 14%. On the other hand, the catalyst supported on alumina has greater HYD activity than the catalyst supported on mixed oxide derived from LDH; in this case, the conversions of cyclohexene are 25 and 8.5%, respectively. Thus, the HDS/HYD ratios are 0.6 for sulfide CoMo/γ-Al_2O_3 and 1.7 for sulfide

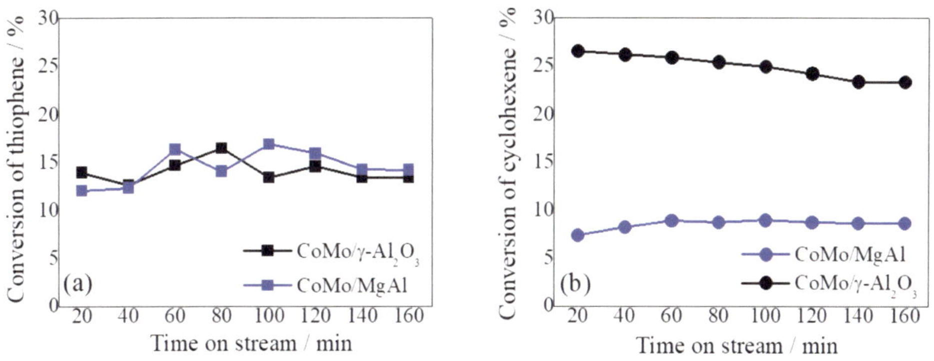

Figure 11.
Conversions of thiophene (a) and cyclohexene (b) for CoMo catalysts.

CoMo/MgAl. This result shows that the catalyst supported on mixed oxide derived from LDH is more selective for HDS reaction.

As in the work of Trejo *et al.* [68], the supported MoO_3 content is approximately 10%. This amount of molybdenum disperses widely in supports containing magnesium, forming $MgMoO_4$ which are easily sulfided. After sulfidation, the type II CoMoS phase is formed, characterized by promoting high activity for HDS due to the weak interaction with the support, as revealed by Raman spectrum. The opposite effect occurs when the material is supported on alumina, forming connections of the Mo–O–Al type, responsible for making the material sulfidation difficult. The H_2-TPR results corroborate these hypotheses, showing that the catalyst supported on MgAl oxide is reduced at lower temperatures than that supported on alumina, indicating less interaction between this support and the oxide precursor of the active phase.

In summary, supports based on mixed oxides derived from LDHs can be an alternative for use in HDS reactions. Raman spectroscopy is useful in the characterization of the support precursors and, associated with other characterization techniques, it is important in the identification of the active phases

4. Conclusion

An overview of hydrotalcite-like materials was presented and the most used preparation methods were described. The possibility of varied compositions and their unique structural characteristics make it possible to obtain materials with specific properties and applications in several areas of industrial interest. In catalytic systems particularly these materials are widely studied and several processes need further characterization. In this sense, Raman spectroscopy proves to be an extremely useful and versatile tool, as it can be used for the structural characterization of LDHs, derived mixed oxides, composites, and other materials used. Furthermore, Raman spectroscopy is a very sensitive technique that makes it possible to determine the chemical nature of reaction products, being able to monitor the entire catalytic process. Thereby, the association of Raman with other techniques will allow the evolution in the understanding of different materials and processes.

Acknowledgements

The authors acknowledge the Petrobras for Financial support. Moreover, we would like to express our gratitude toward Prof. Arnaldo C. Faro Jr. (from UFRJ)

for H_2-TPR and HDS catalytic tests facilities, and Prof. Renato B. Guimarães and Jackson A.L.C. Resende (from LDRx/UFF) for providing the XRD facilities.

Conflict of interest

The authors declare no conflict of interest.

Author details

Luciano Honorato Chagas[1*], Sandra Shirley Ximeno Chiaro[2], Alexandre Amaral Leitão[1] and Renata Diniz[3]

1 Universidade Federal de Juiz de Fora, Juiz de Fora, MG, Brazil

2 Petrobras-CENPES, Rio de Janeiro, RJ, Brazil

3 Universidade Federal de Minas Gerais, Belo Horizonte, MG, Brazil

*Address all correspondence to: hc.luciano@gmail.com

References

[1] Duan X, Evans D. G, Layered Double Hydroxides: Structure and Bonding. Berlin: Springer-Verlag; 2006. 234 p. DOI: 10.1007/b100426.

[2] Rives V. Layered Double Hydroxides: Present and Future. New York: Nova Science Publishers, Inc.: 2001. 439 p. ISBN: 978-1-61209-289-8.

[3] Sun Y, Zhou J, Cheng Y, Yu J, Cai W, Hydrothermal synthesis of modified hydrophobic Zn–Al-layered double hydroxides using structure-directing agents and their enhanced adsorption capacity for nitrophenol. Adsorpt. Sci. Technol. 2014;32:351-364. DOI: 10.1260/0263-6174.32.5.351.

[4] Dos Santos G E S, Dos Lins P V S, De Oliveira L M T M, Da Silva E O, Anastopoulos I, Erto A, Giannakoudakis D A, De Almeida A R F, Da Duarte J L S, Meili L, Layered double hydroxides/biochar composites as adsorbents for water remediation applications: recent trends and perspectives, J. Clean. Prod. 2020;124755. DOI: 10.1016/j.jclepro.2020.124755.

[5] Soliman H M A, Aly H F, Hydrothermal preparation and characterization of cobased layered double hydroxide and their catalytic activity, J. Adv. Nanomater. 2019; 4:1-10. DOI: 10.22606/jan.2019.41001.

[6] Guo X, Xu S, Zhao L, Lu W, Zhang F, Evans D G, Duan X, One-step Hydrothermal Crystallization of a Layered Double Hydroxide / Alumina Bilayer Film on Aluminum and Its Corrosion Resistance Properties. Langmuir. 2009;25:9894-9897. DOI: doi.org/10.1021/la901012w.

[7] Tao Q, Zhang Y, Zhang X, Yuan P, He H, Synthesis and Characterization of Layered Double Hydroxides With a High Aspect Ratio. 2006;179:708-715. DOI: 10.1016/j.jssc.2005.11.023.

[8] Ogawa M, Asai S, Hydrothermal synthesis of layered double hydroxide deoxycholate intercalation compounds, Chem. Mater. 2000;12:3253-3255. DOI: 10.1021/cm000455n.

[9] Chagas L H, De Carvalho G S G, Do Carmo W R, San Gil R A S, Chiaro S S X, Leitão A A, Diniz R, De Sena L A, Achete C A, MgCoAl and NiCoAl LDHs synthesized by the hydrothermal urea hydrolysis method: Structural characterization and thermal decomposition. Mat. Res. Bull. 2015;64:207-215. DOI: 10.1016/j.materresbull.2014.12.062.

[10] Mishra G, Dash B, Pandey S, Layered double hydroxides: A brief review from fundamentals to application as evolving biomaterials. Appl. Clay Sci. 2018;153:172-186. DOI: 10.1016/j.clay.2017.12.021.

[11] Barahuie F, Hussein M Z, Gani S A, Fakurazi S, Zainal Z, Synthesis of protocatechuic acid–zinc/aluminium–layered double hydroxide nanocomposite as an anticancer nanodelivery system. J. Solid State Chem. 2015;221:21-31. DOI: 10.1016/j.jssc.2014.09.001.

[12] Olya N, Ghasemi E, Ramezanzadeh B, Mahdavian M, Synthesis, characterization and protective functioning of surface decorated Zn-Al layered double hydroxide with SiO_2 nano-particles. Surf. Coat. Technol. 2020;387:125512. DOI: 10.1016/j.surfcoat.2020.125512.

[13] Crepaldi E L, Pavan P C, Valim J B, Anion exchange in layered double hydroxides by surfactant salt formation. J. Mater. Chem. 2000;10:1337-1343. DOI: 10.1039/A909436I.

[14] Costa D G, Rocha A B, Souza W F, Chiaro S S X, Leitão A A, Ab Initio Study of Reaction Pathways Related to

Initial Steps of Thermal Decomposition of the Layered Double Hydroxide Compounds. J. Phys. Chem. C. 2012;116:13679–13687. DOI: 10.1021/jp303529y.

[15] Danks A E, Hall S R, Schnepp Z, The evolution of 'sol–gel' chemistry as a technique for materials synthesis. Mater. Horiz. 2016;3:91-112. DOI: 10.1039/C5MH00260E.

[16] Mallakpour S, Hatami M, Hussain C M, Recent innovations in functionalized layered double hydroxides: fabrication, characterization, and industrial applications. Adv. Colloid Interf. Sci. 2020;283:102216. DOI: 10.1016/j.cis.2020.102216.

[17] Boehm H P, Steinle J, Vieweger C, $[Zn_2Cr(OH)_6]_X \cdot 2H_2O$, new layer compounds capable of anion exchange and intracrystalline swelling. Angew. Chemie Int. Ed. English 1977;16:265-266. DOI:.org/10.1002/ anie.197702651.

[18] Richetta M, Medaglia P G, Mattoccia A, Varone A, Pizzoferrato R, Layered double hydroxides: tailoring interlamellar nanospace for a vast field of applications. J. Mater. Sci. Eng. 2017; 06.DOI: 10.4172/2169-0022.1000360.

[19] Do Carmo W R, Haddad J F, Chagas L H, Beltrão M S S, De Carvalho G S G, De Oliveira L C A, Souza T E, Leitão A A, Diniz R. Effect of precursor synthesis on the physicochemical properties of Zn–Mg–Al mixed oxides. Appl. Clay Sci. 2015;116-117:31-38. DOI: 10.1016/j.clay.2015.08.010.

[20] Sanati S, Rezvani Z, Ultrasound-assisted synthesis of NiFe-layered double hydroxides as efficient electrode materials in supercapacitors. Ultrason. Sonochem. 2018; 48;199-206. DOI: 10.1016/j.ultsonch.2018.05.035.

[21] Mallakpour S, Dinari M, Behranvand V, Ultrasonic-assisted synthesis and characterization of layered double hydroxides intercalated with bioactive N,N′-(pyromellitoyl)-bis-l-α-amino acids. RSC Adv. 2013;3:23303-23308. DOI: 10.1039/c3ra43645d.

[22] Ay A N, Zümreoglu-Karan B, Mafra L, A simple mechanochemical route to layered double hydroxides: synthesis of hydrotalcite-like Mg-Al-NO_3-LDH by manual grinding in a mortar, Zeitschrift Für Anorg. Und Allg. Chem. 2009;635:1470-1475. DOI: 10.1002/zaac.200801287.

[23] Khusnutdinov V R, Isupov V P, Mechanochemical synthesis of nanocomposites based on Fe_3O_4 and layered double hydroxides. Mater. Today Proc. 2019;12:48-51. DOI: 10.1016/j.matpr.2019.03.061.

[24] Shirin V K A, Sankar R, Johnson A P, Gangadharappa H V, Pramod K, Advanced drug delivery applications of layered double hydroxide. J. Control. Release 2021;330:398-426. DOI: 10.1016/j.jconrel.2020.12.041.

[25] Jin W, Lee D, Jeon Y, Park D-H, Biocompatible Hydrotalcite Nanohybrids for Medical Functions. Minerals 2020;10:172. DOI: 10.3390/min10020172.

[26] Meirelles L M A, Raffin F N, Clay and Polymer-Based Composites Applied to Drug Release: A Scientific and Technological Prospection. J. Pharm. Pharm. Sci.2017; 20:115-134. DOI: 10.18433/J3R617.

[27] Cao Z, Li B, Sun L, Zhi L L, Xu P, Gu Z, 2D Layered Double Hydroxide Nanoparticles: Recent Progress toward Preclinical/Clinical Nanomedicine. Small Methods 2019;1900343. DOI: 10.1002/smtd.201900343.

[28] Dias A C, Fontes M P F, Arsenic (V) removal from water using hydrotalcites as adsorbents: A critical review. Appl.

Clay Sci. 2020;191:105615. DOI: 10.1016/j.clay.2020.105615.

[29] Chubar N, Gilmour R, Gerda V, Mičušíc M, Omastova M, Heister K, Man P, Fraissard J, Zaitsev V, Layered double hydroxides as the next generation inorganic anion exchangers: Synthetic methods versus applicability. Adv. Colloid Interface Sci. 2017;245:62-80. DOI: 10.1016/j.cis.2017.04.013.

[30] Theiss F L. Couperthwaite S J, Ayoko G A, Frost R L, A review of the removal of anions and oxyanions of the halogen elements from aqueous solution by layered double hydroxides. J. Colloid Interface Sci. 2014;417:356-368. DOI: 10.1016/j.jcis.2013.11.040.

[31] Mohapi M, Sefadi J S, Mochane M J, Magagula S I, Lebelo K, Effect of LDHs and Other Clays on Polymer Composite in Adsorptive Removal of Contaminants: A Review. Crystals 2020;10:0957. DOI: 10.3390/cryst10110957.

[32] Pandey N, Shukla S K, Singh N B, Water purification by polymer nanocomposites: an overview. Nanocomposites 2017;3:47. DOI: 10.1080/20550324.2017.1329983.

[33] Wang J, Zhang T, Li M, Yang Y, Lu P, Ning P, Wang Q, Arsenic removal from water/wastewater using layered double hydroxide derived adsorbents, a critical review. RSC Adv. 2018;8:22694. DOI: 10.1039/c8ra03647k.

[34] Wang X, Wang H, Wang Y, Gao J, Liu J, Zhang Y, Hydrotalcite/graphene oxide hybrid nanosheets functionalized nanofiltration membrane for desalination. Desalination 2019;451:209. DOI: 10.1016/j.desal.2017.05.012.

[35] Tichit D, Layrac G, Gérardin C, Synthesis of layered double hydroxides through continuous flow processes: A review. Chem. Eng. J. 2019;369:302. DOI: 10.1016/j.cej.2019.03.057.

[36] Taviot-Guého C, Prévot V, Forano C, Renaudin G, Mousty C, Leroux F, Tailoring Hybrid Layered Double Hydroxides for the Development of Innovative Applications. Adv. Funct. Mater. 2017;1703868. DOI: 10.1002/adfm.201703868.

[37] Yan Q, Hou X, Liu G, Li Y, Zhu T, Xin Y, Wang Q, Recent advances in layered double hydroxides (LDHs) derived catalysts for selective catalytic reduction of NO_x with NH_3. J. Hazard. Mater. 2020;400:123260. DOI: 10.1016/j.jhazmat.2020.123260.

[38] Lu P, Liu Y, Zhou T, Wang Q, Li Y, Recent advances in layered double hydroxides (LDHs) as two-dimensional membrane materials for gas and liquid separations. J. Membrane Sci. 2018;567:89. DOI: 10.1016/j.memsci.2018.09.041.

[39] Hashim N, Sharif S N M, Hussein M G, Isa I M, Kamari A, Mohamed A, Ali N M, Bakar S A, Mamat M, Layered hydroxide anion exchanger and their applications related to pesticides: a brief review. Mater. Res. Innov. 2017;21:129. DOI. 10.1080/14328917.2016.1192717.

[40] Wang S, Yang X, Wang F, Song Z, Dong H, Cui L, Effect of modified hydrotalcite on flame retardancy and physical properties of paper. BioResources 2019;14:3991. DOI: 10.15376/biores.14.2.3991-4005.

[41] Zhang Z, Qin J, Zhang W, Pan Y-T, Wang D-Y, Yang R, Synthesis of a novel dual layered double hydroxide hybrid nanomaterial and its application in epoxy nanocomposites. Chem. Eng. J. 2020;381:122777. DOI: 10.1016/j.cej.2019.122777.

[42] Xie J, Wang H, Wang Z, Zhao Q, Yang Y, Waterhouse G I N, Hao L, Xiao Z, Xu J, Innovative Linear Low Density Polyethylene Nanocomposite Films Reinforced with Organophilic Layered Double Hydroxides:

Fabrication, Morphology and Enhanced Multifunctional Properties. Sci. Rep. 2018;8:52. DOI:10.1038/s41598-017-18811-y.

[43] Chatterjee A, Bharadiya P, Hansora D, Layered double hydroxide based bionanocomposites. Apppl. Clay Sci. 2019;177:19. DOI: 10.1016/j.clay.2019.04.022.

[44] Borges R., Wypych F, Petit E, Forano C, Prevot V, Potential sustainable slow-release fertilizers obtained by mechanochemical activation of MgAl and MgFe layered double hydroxides and K_2HPO_4. Nanomaterials 2019;9:183. DOI: 10.3390/nano9020183.

[45] Bouali A C, Serdechnova M, Blawert C, Tedim J, Ferreira M G S, Zheludkevich M L, Layered double hydroxides (LDHs) as functional materials for the corrosion protection of aluminum alloys: A review. Appl. Mat. Today 2020;21:100857. DOI: 10.1016/j.apmt.2020.10085.

[46] Yan A L, Wang X C, Cheng J P, Research progress of NiMn layered double hydroxides for supercapacitors: A review. Nanomaterials 2018;8:747. DOI: 10.3390/nano8100747.

[47] Lauermannová A-M, Paterová I, Patera J, Skrbek K, Jankovský O, Bartuněk V, Hydrotalcites in Construction Materials. Appl. Sci. 2020;10:7989. DOI: 10.3390/app10227989.

[48] Debecker D P, Gaigneaux E M, Busca G, Exploring, tuning, and exploiting the basicity of hydrotalcites for applications in heterogeneous catalysis. Chem. Eur. J. 2009;15:3920-3935. DOI: 10.1002/chem.200900060.

[49] Climent M J, Corma A, Fornes V, Guil-Lopez R, Iborra S, Aldol condensations on solid catalysts: A cooperative effect between weak acid and base sites. Adv. Synth. Catal. 2002;344:1090-1096. DOI: 10.1002/1615-4169(200212)344:10<1090::AID-ADSC1090>3.0.CO;2-X.

[50] Choudary B M, Kantam M L, Reddy C R V, Rao K K, Figueras F, The first example of Michael addition catalysed by modified Mg-Al hydrotalcite. J. Mol. Catal. A Chem. 1999;146:279-284. DOI: 10.1016/S1381-1169(99)00099-0.

[51] Pillai U R, Sahle-Demessie E, Sn-exchanged hydrotalcites as catalysts for clean and selective Baeyer-Villiger oxidation of ketones using hydrogen peroxide. J. Mol. Catal. A Chem. 2003;191:93-100. DOI: 10.1016/S1381-1169(02)00347-3.

[52] Climent M J, Corma A, Iborra S, Heterogeneous catalysts for the one-pot synthesis of chemicals and fine chemicals. Chem. Rev. 2011;111:1072-1133. DOI: 10.1021/cr1002084.

[53] Climent M J, Corma A, Iborra S, Mifsud M, Velty A, New one-pot multistep process with multifunctional catalysts: Decreasing the E factor in the synthesis of fine chemicals. Green Chem. 2010;12:99-107. DOI: doi.org/10.1039/B919660A.

[54] Kaneda M, Mizugaki T, Design of high-performance heterogeneous catalysts using hydrotalcite for selective organic transformations. Green Chem. 2019;21:1361-1389. DOI: 10.1039/c8gc03391a.

[55] Homsi D, Rached J A, Aouad S, Gennequin C, Dahdah E, Estephane J, Tidahy H L, Aboukaïs A, Abi-Aad E, Steam reforming of ethanol for hydrogen production over Cu/Co-Mg-Al-based catalysts prepared by hydrotalcite route. Environ. Sci. Pollut. Res. 2017;24:9907-9913. DOI: 10.1007/s11356-016-7480-9.

[56] Ni M, Leung D Y C, Leung M K H, A review on reforming bio-ethanol for

hydrogen production. Int. J. Hydrogen Energy 2007;32:3238-3247. DOI: 10.1016/j.ijhydene.2007.04.038.

[57] Haryanto A, Fernando S, Murali N, Adhikari S, Current Status of Hydrogen Production Techniques by Steam Reforming of Ethanol: A Review. Energy Fuels 2005;19:2098-2106. DOI: doi.org/10.1021/ef0500538.

[58] Passos A R, Pulcinelli S H, Santilli C V, Briois V, Operando monitoring of metal sites and coke evolution during non-oxidative and oxidative ethanol steam reforming over Ni and NiCu ex-hydrotalcite catalysts. Catal. Today 2019;336:122-130. DOI: 10.1016/j.cattod.2018.12.054.

[59] Sikander U, Sufian S, Salam M A, A review of hydrotalcite based catalysts for hydrogen production systems. Int. J. Hydrogen Energy 2017;42:19851-19868. DOI: 10.1016/j.ijhydene.2017.06.089.

[60] Zhang G, Zhang X, Meng Y, Pan G, Ni Z, Xia S, Layered double hydroxides-based photocatalysts and visible-light driven photodegradation of organic pollutants: A review. Chem. Eng. J. 2020;392:123684. DOI: 10.1016/j.cej.2019.123684.

[61] Yang Z-Z, Wei J-J, Zeng G-M, Zhang H-Q, Tan X-F, Ma C, Li X-C, Li Z-H, Zhang C, A review on strategies to LDH-based materials to improve adsorption capacity and photoreduction efficiency for CO_2. Coord. Chem. Rev. 2019;386:154-182. DOI: 10.1016/j.ccr.2019.01.018.

[62] Shaw M H, Twilton J, MacMillan D W C, Photoredox Catalysis in Organic Chemistry. J. Org. Chem. 2016;81:6898–6926. DOI: 10.1021/acs.joc.6b01449.

[63] Mohapatra L, Parida K, A review on the recent progress, challenges and perspective of layered double hydroxides as promising photocatalysts. J. Mater. Chem. A, 2016;4:10744-10766. DOI: 10.1039/c6ta01668e.

[64] Jing G, Sun Z, Ye P, Wei S, Liang Y, Clays for heterogeneous photocatalytic decolorization of wastewaters contaminated with synthetic dyes: a review. Water Practice & Technology 2017;12:432-443. DOI: 10.2166/wpt.2017.046.

[65] De Carvalho G S G, Siqueira M M, Nascimento M P, Oliveira M A L, Amarante G W, Nb_2O_5 supported in mixed oxides catalyzed mineralization process of methylene blue. Helyion 2020;6:e04128. DOI: 10.1016/j.heliyon.2020.e04128.

[66] De Carvalho G S G, Chagas L H, Fonseca C G, De Castro P P, Sant'Ana A C, Leitão A A, Amarante G W, Nb_2O_5 supported on mixed oxides catalyzed oxidative and photochemical conversion of anilines to azoxybenzenes, New J. Chem. 2019;43:5863-5871. DOI: 10.1039/c9nj00625g.

[67] Kaufmann T G, Kaldor A, Stuntz, G F, Kerby M C, Ansell L L, Catalysis science and technology for cleaner transportation fuels. Catal. Today 2000;62:77-90. DOI: 10.1016/S0920-5861(00)00410-7.

[68] Trejo F, Rana M, Ancheyta J, CoMo/$MgO–Al_2O_3$ supported catalysts: An alternative approach to prepare HDS catalysts. Catal. Today 2008;130:327-336. DOI: 10.1016/j.cattod.2007.10.105.

[69] Brunet S, Mey D, Pérot G, Bouchy C, Diehl F, On the hydrodesulfurization of FCC gasoline: a review. Appl. Catal. A Gen. 2005;278:143-172. DOI: 10.1016/j.apcata.2004.10.012.

[70] Zhao R, Yin C, Zhao H, Liu C, Synthesis, characterization, and application of hydrotalcites in hydrodesulfurization of FCC gasoline. Fuel Process. Technol. 2003;81:201-209. DOI: 10.1016/S0378-3820(03)00012-2.

[71] Coelho T L, Arias S, Rodrigues V O, Chiaro S S X, Oliviero L, Maugé F, Faro Jr A C, Characterisation and performance of hydrotalcite derived CoMo sulphide catalysts for selective HDS in the presence of olefin. Catal. Sci. Technol. 2018;8:6204-6216. 10.1039/c8cy01855c.

[72] Coelho T L, Micha R, Arias S, Licea Y E, Palacio L A, Faro Jr A C, Influence of the Mg^{2+} or Mn^{2+} contents on the structure of NiMnAl and CoMgAl hydrotalcite materials with high aluminum contents. Catal. Today 2015;250:87-94. DOI: 10.1016/j.cattod.2014.07.015.

[73] Coelho T L, Licea Y E, Palacio L A, Faro Jr A C, Heptamolybdate-intercalated CoMgAl hydrotalcites as precursors for HDS-selective hydrotreating catalysts. Catal. Today 2015;250:38-46. DOI: 10.1016/j.cattod.2014.06.016.

[74] Herrero M, Benito P, Labajos F M, Rives V, Stabilization of Co^{2+} in layered double hydroxides (LDHs) by microwave-assisted ageing. J. Solid State Chem. 2007;180:873-884. DOI: 10.1016/j.jssc.2006.12.011.

[75] Cavani F, Trifirò F, Vaccari A, Hydrotalcite-type anionic clays: Preparation, properties and applications. Catal. Today 1991;11:173-301. DOI: 10.1016/0920-5861(91)80068-K.

[76] Frost R L, Reddy B J, Thermo-Raman spectroscopic study of the natural layered double hydroxide manasseite. Spectroch. Acta Part A 2006;65:553-559. DOI: 10.1016/j.saa.2005.12.007.

[77] Zhang H, Wen X, Wang Y, Synthesis and characterization of sulfate and dodecylbenzenesulfonate intercalated zinc–iron layered double hydroxides by one-step coprecipitation route. J. Solid State Chem. 2007;180:1636-1647. DOI: 10.1016/j.jssc.2007.03.016.

[78] Kloprogge J T, Hickey L, Frost R L, The effect of varying synthesis conditions on zinc chromium hydrotalcite: a spectroscopic study. Mater. Chem. Phys. 2005;89:99-109. DOI: 10.1016/j.matchemphys.2004.08.035.

[79] Pradhan J K, Bhattacharya I N, Das S C, Das R P, Panda R K, Characterisation of fine polycrystals of metastable η-alumina obtained through a wet chemical precursor synthesis. Mater. Sci. Eng. 2000; B77:185-192. DOI: 10.1016/S0921-5107(00)00486-4.

[80] Laugier J, Bochu B, Checkcell–A Software Performing Automatic Cell/Space Group Determination, Laboratoire des Matériaux et du Génie Physique de I'Ecole Supérieure de Physique de Grenoble (INPG), France, 2000.

[81] Constantino V R L, Pinnavaia T J, Basic Properties of $Mg^{2+}{}_{1-x}Al^{3+}{}_{x}$ Layered Double Hydroxides Intercalated by Carbonate, Hydroxide, Chloride, and Sulfate Anions Inorg. Chem. 1995;34:883-892. DOI: 10.1021/ic00108a020.

[82] Crivello M, Pérez C, Fernández J, Eimer G, Herrero E, Casuscelli S, Rodríguez-Castellón E, Synthesis and characterization of Cr/Cu/Mg mixed oxides obtained from hydrotalcite-type compounds and their application in the dehydrogenation of isoamylic alcohol. Appl. Catal. A Gen. 2007;317:11-19. DOI: 10.1016/j.apcata.2006.08.035.

[83] Costa D G, Rocha A B, Diniz R, Souza W F, Chiaro S S X, Leitão A A, Structural Model Proposition and Thermodynamic and Vibrational Analysis of Hydrotalcite-Like Compounds by DFT Calculations. J. Phys. Chem. C 2010;114:14133-14140. DOI: 10.1021/jp1033646.

[84] Farkas L, Gadó P, Werner P E, The structure refinement of Boehmite

(γ-AIOOH) and the study of its structural variability based on Guinier-Hagg powder data. Mat. Res. Bull. 1977;12:1213-1219. DOI: 10.1016/0025-5408(77)90176-3.

[85] Chagas L H, De Farias S B P, Leitão A A, Diniz R, Chiaro S S X, Speziali N L, De Abreu H A, Mussel W N, Structural comparison between samples of hydrotalcite-like materials obtained from different synthesis route. Quim. Nova 2012;35:1112. DOI: 10.1590/S0100-40422012000600008.

[86] Brühne S, Gottlieb S, Assmus W, Alig E, Schmidt M U, Atomic Structure Analysis of Nanocrystalline Boehmite AlO(OH). Cryst. Growth Des. 2008;8:489. DOI: 10.1021/cg0704044.

[87] Velu S, Suzuki K, Osaki K, Ohashi F, Tomura S, Synthesis of new Sn incorporated layered double hydroxides and their evolution to mixed oxides. Mater. Res. Bull. 1999;34:1707-1717. DOI: 10.1016/S0025-5408(99)00168-3.

[88] Miyata S, Physico-chemical properties of synthetic hydrotalcites in relation to composition. Clays Clay Miner. 1980;28:50. DOI: 10.1346/CCMN.1980.0280107.

[89] Mackenzie K J D, Meinhold R H, Sherriff B L, Xu Z, ^{27}AI and ^{25}Mg Solid-state Magic-angle Spinning Nuclear Magnetic Resonance Study of Hydrotalcite and its Thermal Decomposition Sequence. J. Mater. Chem. 1993;3:1263. DOI: 10.1039/JM9930301263.

[90] Desikan A N, Huang L, Oyama S T, Structure and Dispersion of Molybdenum Oxide Supported on Alumina and Titania, J. Chem. Soc. Faraday Trans. 1992;88:3357-3365. DOI: 10.1039/FT9928803357.

[91] Brown F R, Makovsky L E, Rhee K H, Raman Spectra of Supported Molybdena Catalysts 1. Oxide Catalysts. J. Catal. 1977;50:162-171. DOI: 10.1016/0021-9517(77)90018-5.

[92] Radhakrishnan R, Reed C, Oyama S T, Seman M, Kondo J N, Domen K, Ohminami Y, Asakura K, Variability in the Structure of Supported MoO_3 Catalysts : Studies Using Raman and X-ray Absorption Spectroscopy with ab Initio Calculations. J. Phys. Chem. B 2001;105:8519-8530. DOI: 10.1021/jp0117361.

[93] Medema J, Van Stam C, De Beer V H J, Konings A J A, Koningsberger D C, Raman spectroscopic study of CoMoγ-Al_2O_3 catalysts. J. Catal. 1978;53:386-400. DOI: 10.1016/0021-9517(78)90110-0.

[94] Li X, Chai Y, Liu B, Liu H, Li J, Zhao R, Liu C, Hydrodesulfurization of 4,6-Dimethyldibenzothiophene over CoMo Catalysts Supported on γ-Alumina with Different Morphology. Ind. Eng. Chem. Res. 2014;53:9665-9673. DOI: 10.1021/ie5007504.

[95] Liu X M, Xue H X, Li X, Yan Z F, Synthesis and hydrodesulfurization performance of hierarchical mesopores alumina. Catal. Today 2010;158:446-451. DOI: 10.1016/j.cattod.2010.06.032.

[96] El Kady F Y A, Abd El Wahed M G, Shaban S, Abo El Naga A O, Hydrotreating of heavy gas oil using CoMo/γ-Al_2O_3 catalyst prepared by equilibrium deposition filtration. Fuel 2010;89:3193-3206. DOI: 10.1016/j.fuel.2010.06.024.

[97] Nava R, Morales J, Alonso G, Ornelas C, Pawelec B, Fierro J L G, Influence of the preparation method on the activity of phosphate-containing CoMo/HMS catalysts in deep hydrodesulphurization. Appl. Catal. A 2007;321:58-70. DOI: 10.1016/j.apcata.2007.01.038.

Chapter 9

Raman Spectroscopy in the Analysis of Textile Structures

Dorota Puchowicz and Malgorzata Cieslak

Abstract

Raman spectroscopy as a non-destructive technique is very often used to analyze a historic or forensic material. It is also a very valuable method of testing textile materials, especially modified and functionalized. In the case of textiles, the advantages of this technique is the compatibility *inter alia* with FTIR, which is helpful in natural fibers identification or to distinguish between isomers and conformers of synthetic fibers. The work shows the possibility of special application of the Raman spectroscopy to the characterization of textile materials after modification and functionalization with nanoparticles. A functionalized textile structure with a metallic surface can provide a good basis for analytical studies using surface enhanced Raman spectroscopy as it was presented on the example of wool, cotton and aramid fibers.

Keywords: natural fibers, synthetic fibers, textile surface, functionalized textiles, SERS

1. Introduction

Raman spectroscopy is complementary method to FTIR Spectroscopy as both methods are based on detection of molecule vibrations. These spectroscopic techniques have different mechanism of vibrations detection and different physical phenomena are studied. Vibrations modifying the dipole moment of a molecule are active in IR spectroscopy, whereas vibrations modifying the polarizability of a molecule (i.e. stretching of C-C or C=C groups) are detected by Raman technique [1]. That is why Raman spectrum brings information about polymer backbone structure, conformation, orientation, crystallinity, density, etc. All this causes that Raman spectroscopy is a very useful tool in the analysis of polymers. As the intensity of Raman scattering does not depend directly on the volume of the tested sample application of the optical microscope together with the Raman spectrometer allows analysis of very small samples, including nanoparticles. Samples once studied in Raman spectrometer could be given for additional analysis by other techniques, if it is necessary. Therefore Raman spectroscopy is a valuable tool in analysis of ancient materials or forensic samples and also can be successfully applied in the textile materials study. Modern Raman spectrometers equipped with advanced software and volume mapping system allow the characterization of textile materials containing nanoparticles, their distribution on the surface of the material and inside the fiber analyzing its cross-section as well as in the case of colored materials, distribution of dyes, metal-dye interactions on the surface and inside the fiber structure [2–9]. The functionalization with metal nanoparticles and metal oxides plays now a special role, as it offers the possibility of giving textile products such properties as: bioactivity, UV protection, catalytic or conductive properties [2–15]. Knowledge of the structure of the fibers, their physical, chemical properties as well as surface characteristics [2–5, 16–20] is extremely important in carrying out any modification. Application of nanoparticles is closely related to the use of advanced research methods necessary for the analysis of modified surfaces. Research on the fibrous structures modification using nanoparticles is the another area of analysis in which Raman spectroscopy can be successfully applied. Furthermore presence of noble metal nanoparticles on the fiber surface affects its Raman spectrum causing the effect of strengthening the Raman signal from the fiber itself and the contained substances (e.g. dye, modifier) [21–23]. Thus functionalized textile structure can be a good substrate for analytical studies using surface enhanced Raman spectroscopy.

2. Raman spectra of textile fibers

Raman spectroscopy, in general, is limited by fluorescence which often hides the Raman effect. However thanks to development of electronics and computing brought the possibility of Raman scattering enhancement, fluorescence minimalization, application different light sources in Raman spectrometers [24] significantly broadened analytical applications of this technique [25]. All Raman spectra presented in this chapter are performed on Raman Renishaw In via dispersive Spectrometer with Leica microscope. CCD detector of high quantum efficiency and extremely low intrinsic noise enabled to obtain Raman signal from textile substrate and 785 nm light excitation allowed overcome the fluorescence. Raman spectroscopy provides information about the structure of the fiber, e.g. the degree of crystallinity. Raman *inter alia* plays an important role in the analysis of hygroscopic fibers because, unlike FTIR, water does not affect the spectrum and is very useful in analysis of organic systems with C-C or C=C bonding.

2.1 Natural fibers

Cotton is the most commonly used natural plant fiber and the second most popular fiber, next to polyester [26]. The main component of the cotton fiber is cellulose over 88%, other component pectin, wax, proteins and other organic materials do not exceed 2% [27]. In Raman spectrum (**Figure 1**) the most intense bands are the stretching vibrations of symmetric and asymmetric COC glycosidic ring breathing, skeletal stretching; at 1099 cm^{-1} asymmetric and 1125 cm^{-1} symmetric. Another vibrations characteristic for cotton are at: 331 and 381 cm^{-1} of CCC ring deformations bending, at 438 and 460 cm^{-1} of CCC, CCO ring deformation and skeletal bending, at 520 cm^{-1} COC glycoside linkage and CCC ring deformation bending, at 901,1001 cm^{-1} HCC, HCO skeletal rotating. Stretching vibrations of CC and CO of glycosidic ring are at 1156 cm^{-1}. Cluster of bands in the region between 1200 and 1500 cm^{-1} concerns the CH_2 twisting at 1297 cm^{-1}, CH_2 wagging 1341 cm^{-1}, CH_2 bending at 1384 cm^{-1}, CH_2 bending scissors at 1484 cm^{-1} [28, 29]. In this cluster should be also bending vibrations of alcohol COH that are overlapped. In the spectrum there are also present stretching bands of CH vibrations at 2903 cm^{-1}.

Wool and silk are animal protein fibers [30]. Wool consists mainly of keratin which is sulfur possessing protein. Silk consists mainly of fibroin protein. In analysis of wool and silk, by vibrational spectroscopy techniques, Raman spectroscopy is the method that allows for their unambiguous identification. IR Spectra of wool and silk are very similar (**Figure 2**). No significant differences, even in fingerprint region could be observed, even differences in relative peak intensities or wavenumber shifts are very subtle. Only the band shapes in the area over 2900 cm^{-1}

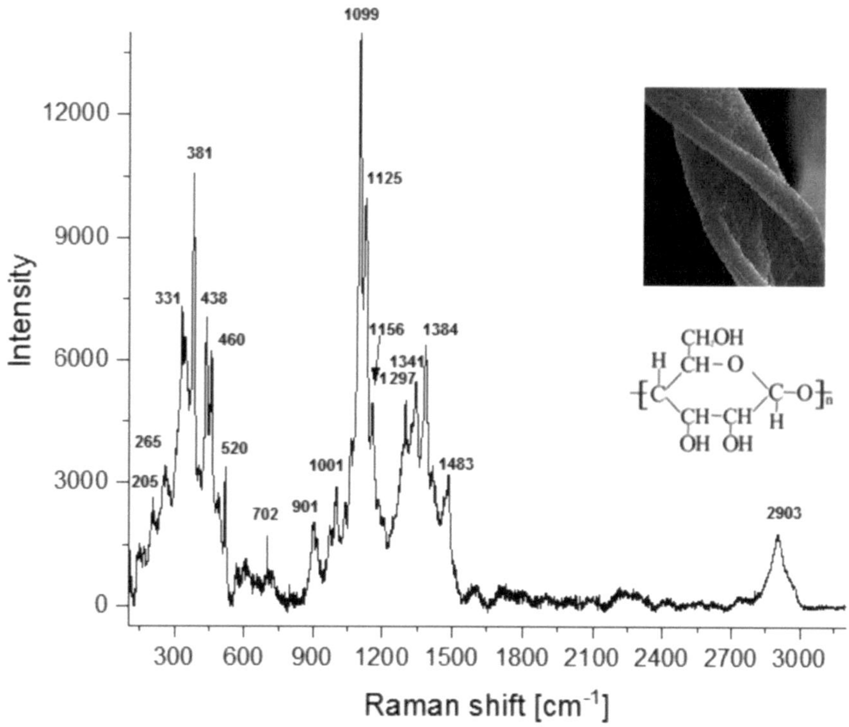

Figure 1.
Raman spectrum of cotton fiber.

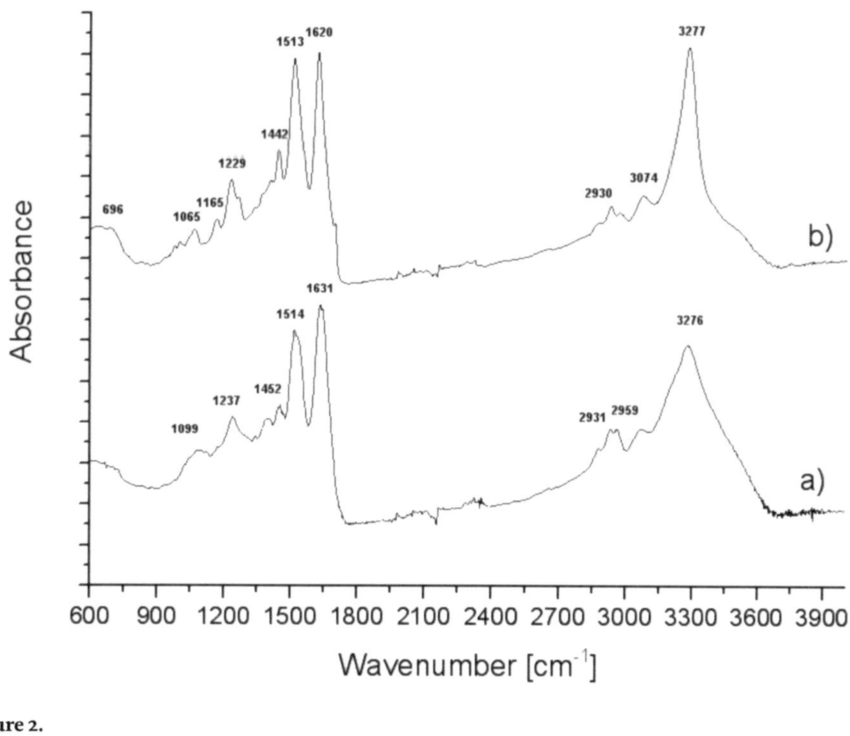

Figure 2.
IR spectrum of: a) wool, b) silk.

distinguishes these fibers. However these slight differences are visible when the wool and silk spectra are aligned as it is shown in **Figure 2a** and **b**. Therefore, the use for the fiber identification the complementary method is recommended.

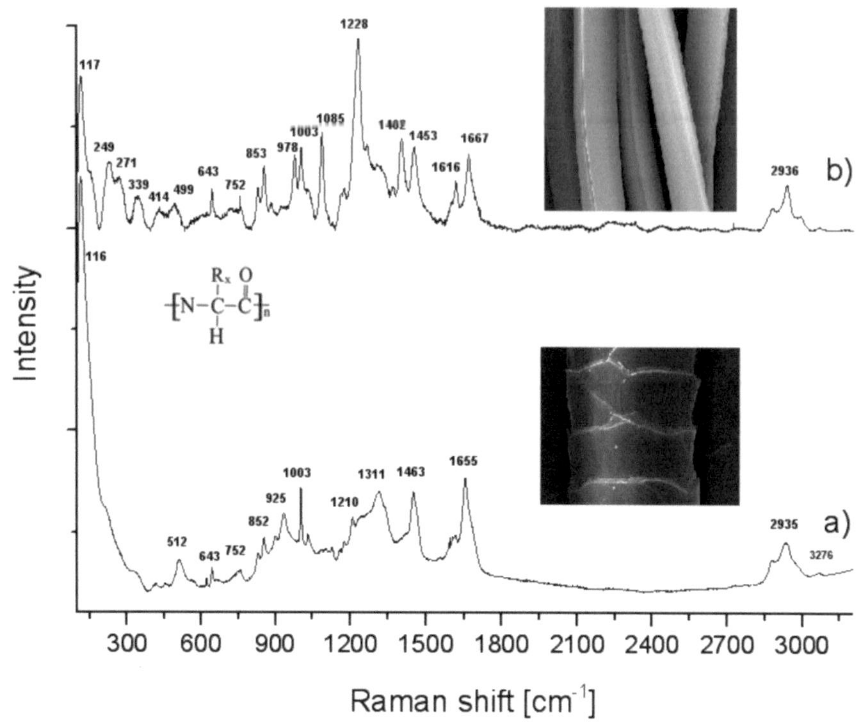

Figure 3.
Raman spectrum of: a) wool, b) silk.

The Raman measurement on wool and silk will help to distinguish between the two protein fibers, as the spectra obtained are clearly different (**Figure 3a** and **b**). This is possible thanks to the sensitivity of this technique to differences in the organic structures, in this case the sequence of amino acids [30]. Silk fibroin is the protein mainly consists of alanine (44,1%) and glycine (26,5%). There are also present in smaller amounts serine (11,8%), tyrosine (4,9%), aspartic acid (4,7%) and arginine (2,6%). The content of other amino acids does not exceed 1%, and, at the same time there is no cysteine, that is one the basic components of wool keratin. Wool keratin in the protein that mainly consists of glutamic acid (11,9,%), serine (10,4%), cysteine (10,3%), glycine (8,4%), leucine (7,7%), arginine (6,9%), proline (6,6%) and other amino acids [31]. The different sequence of amino acids directly translates into different Raman characteristics and the fingerprint of wool and silk differ substantially. The most important observed feature in Raman spectrum of wool is the presence of band at 512 cm^{-1} characteristic to S-S disulphide bridge coming from cysteine [31, 32].

While in case of silk there is sequence of amino acids skeletal vibrations in the region of 100–500 cm^{-1} and the dominating band of amide CH_2 bending at 1228 cm^{-1} [33]. The more detailed description of wool and silk Raman characteristic bands is listed in **Table 1**.

2.2 Synthetic fibers

Today most textile products are made of synthetic fibers (approx. 62%) [26], therefore they are the subject of research, including functionalization. Over the 50% of all fibers on the world market are polyester fibers (PET) Another popular synthetic fibers are nylon, acrylic, polypropylene, aromatic polyamides: meta- and para-aramid. Aramid fibers due to their properties, such flame retardance, mechanical or thermal resistance are the class of heat resistant, flame retardant and

Wool [21]		Silk [23]	
Raman shift [cm^{-1}]	**Description**	**Raman shift [cm^{-1}]**	**Description**
512	SS stretching	117,249, 271, 339, 414, 499	CCC def CCN, combinations of backbone deformations with pendant chains
643	CS stretching	643	NCH bending Amid IV
752	CCC, stretching	752	CH_2 CH_3 skeletal alanine
852	CH_2, HCS bending, $CCOO^-$ stretching	853	CC skeletal tyrosine
		978	CH_2
925	C-C skeletal bending	1003	CH_2 bending
1003	NCH bending	1085	CC skeletal
1210	NH deformation, C-N elonging	1228	CH_2 bending Amid III
1311,1463	CH_2 bending, COO^- bending	1402	CH_2 bend
1514, 1631	CH, NH bending	1453	CH_2 bending scissors
1665	C=O, COO^- stretching (Amide I)	1616	C=C stretching (Tyrosine)
2931, 2959	CH stretching	1667	Amide I
		2935	CH stretching

Table 1.
Raman characteristic bands of wool and silk.

strong synthetic fibers. They are important in military applications i.e. for firefighters clothing or bullet resistant vests [34]. In the case of synthetic polymers, it is also important, that the intensity of Raman scattering is stronger than fluorescence. Raman spectra of Polyester (PET), nylon fibers (Polyamide 6 (PA 6) and Polyamide (6.6 PA 6.6), polypropylene (PP), polyacrylic fiber (PAN), meta-aramid (mAr) and para-aramid (pAr) are presented below and discussed.

In Raman spectrum of polyethylene terephthalate (PET) fiber [35–37] there are two dominating very intense bands: at 1615 cm^{-1} that is C-C aromatic ring and at 1728 cm^{-1} – carbonyl C=O stretching. Another characteristic bands of PET are: 278 cm^{-1} deformation skeletal C-C, 702 cm^{-1} ring C-C stretch, 859 cm^{-1} C-C, COC bending, 998 cm^{-1}, 1096 cm^{-1} C-O and C-C stretch, 1181 cm^{-1} C-C ring stretch, 1289 cm^{-1} CO-O stretch, 1416, 1463 cm^{-1} CH_2 bending. Band at 142 cm^{-1} belongs to TiO_2 which is often used in the processing of the fiber, as a matting agent.

Raman spectrum of PET gives information about the polymer form. Textile polymers are not 100% crystalline, they also have amorphous areas, therefore the degree of crystallinity is determined for them. In the crystalline state the ethylene glycol units of PET have a *trans* structure while the amorphous state PET has a *gauche* structure of the ethylene glycol units [35]. An intense peak at 1,096 cm^{-1} indicates that analyzed PET fiber is in crystalline form. However, the carbonyl band at 1728 cm^{-1} is considered a better marker of crystallinity [35]. The highly crystalline samples give a narrow carbonyl peak; whereas for the amorphous the band width is demonstrably broader. In presented spectrum (**Figure 4**) peak at 1728 cm^{-1} is narrow, so the crystalline form is confirmed.

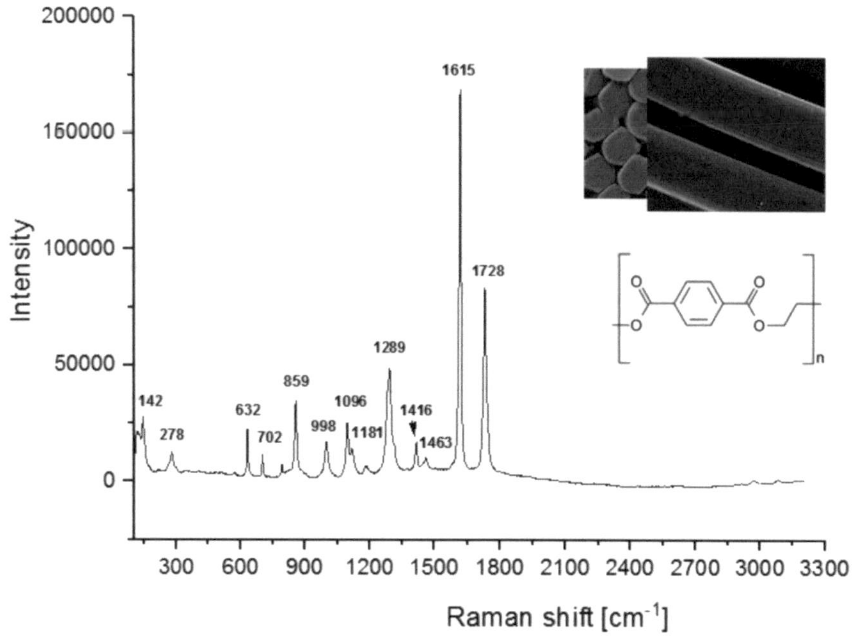

Figure 4.
Raman spectrum of PET fiber.

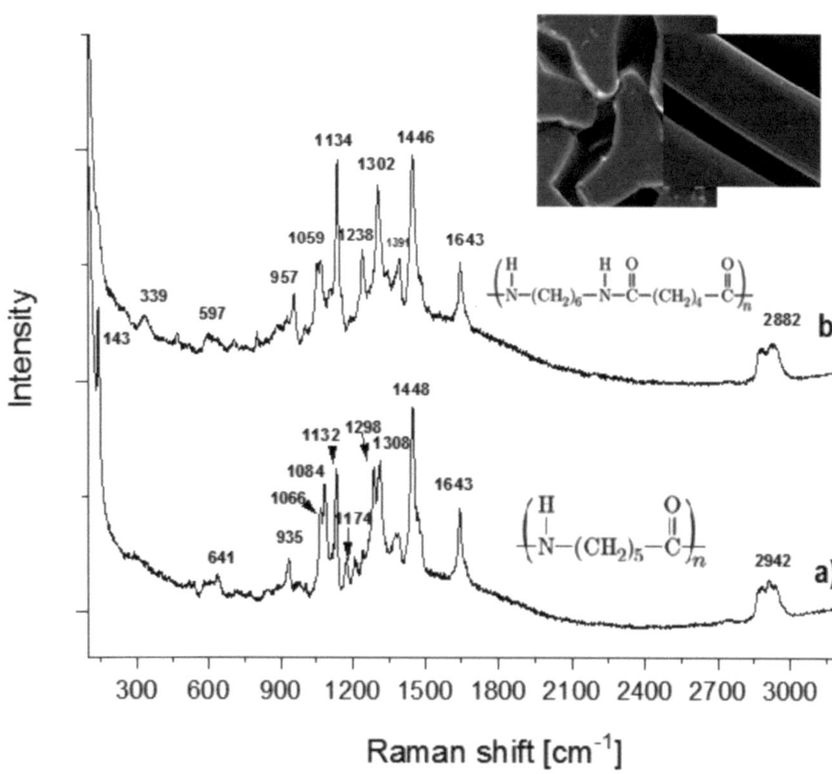

Figure 5.
Raman spectrum of polyamide fibers: a) PA 6, b) PA6.6.

Polyamides are polymers in which an amide group -NH-CO- joins the monomer units. Two most important commercially textile polyamides are Polyamide 6 (PA 6) and Polyamide 6.6 (PA 6.6). In both polyamides the monomeric units contains six carbon atoms. They can be distinguished most easily by examining the melting point. However, these analysis requires destruction of the sample. Raman spectroscopy is in this case very good method for polyamide type identification, used in practice i.e. in the carpet recycling [38]. Raman Spectrum of PA 6 (**Figure 5a**) shows CC deformation at 643 cm^{-1}, CCO stretching at 935 cm^{-1}, CC skeletal stretching at 1066, 1084 and 1132 cm^{-1}, CN stretching and NH bending of amide III at 1298 cm^{-1}, CH_2 twisting at 1308 cm^{-1}, CH_2 bending at 1448 cm^{-1}, 1643 cm^{-1} C=O stretching and CH stretching at 2942 cm^{-1} [39].

There is also present band at 143 cm^{-1} of TiO_2 used in fiber processing. While in the Raman spectrum polyamide 6.6 (**Figure 5b**) CC deformation at 643 cm^{-1}, CCO stretching at 957 cm^{-1}, CC skeletal stretching at 1059 and 1134 cm^{-1}, NH deformation at 1238 cm^{-1}, twisting at 1302 cm^{-1}, CH_2 bending at 1446 cm^{-1}, 1643 cm^{-1} C=O stretching and CH stretching at 2882 cm^{-1} are present [39]. The Raman spectra of both polyamides differ from each other the presence of three skeletal CC bands (1066, 1084 and 1132 cm^{-1}) in the case of PA 6 and two CC skeletal bands (1059 and 1134 cm^{-1}) in the case of PA 6.6. Moreover PA 6,6 spectrum does not have an amide III band (C–N stretching and N–H bending) at 1298 cm^{-1}, which is present in PA 6 spectrum.

Polypropylene fibers are the most commercially important polyolefin fibers, whose polymer chain consists of olefin units. Raman spectrum of PP (**Figure 6**) contains: CH_2-CH-CH_3 torsions in the polymer backbone at 107 and 175 cm^{-1} [40], CH_2 wagging and CH bending at 251 and 399 cm^{-1}, CH_2 wagging and CH bending at 320 and 455 cm^{-1}, CH_2 wagging, CH_2 bending and CCH_3 stretching at 528 cm^{-1}, CC backbone stretching, CH_2 wagging, CCH_3 stretching at 809 cm^{-1}, CC backbone stretching, CH_2 wagging, CCH_3 stretching, CH_3 bending at 847 cm^{-1}, CC backbone stretching, CH_3 rocking at 947 cm^{-1}, CH_3 rocking, CH_2 wagging, CH bending at 999 cm^{-1}, CC backbone stretching, CCH_3 stretching, CH bending at 1038 cm^{-1}, CC backbone stretching, CCH_3 stretching, CH bending and CH_3 bending at 1153 cm^{-1},

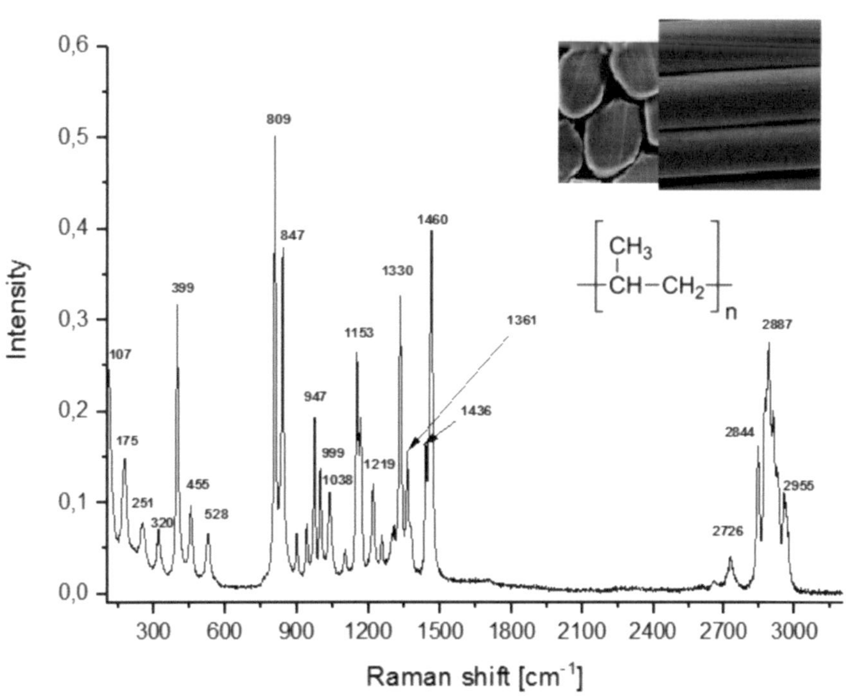

Figure 6.
Raman spectrum of PP fiber.

CC backbone stretching, CH bending and CH_2 twisting at 1219 cm^{-1}, CH bending, CH_3 symmetric bending at 1361 cm^{-1}, CH_3 asymmetric bending at 1436 cm^{-1}, CH_3 asymmetric bending and CH_2 bending at 1460 cm^{-1}, symmetric CH_2 stretching at 2726 cm^{-1}, symmetric CH_3 stretching at 2844 cm^{-1} and 2887 cm^{-1}, asymmetric CH_3 stretching and 2955 cm^{-1} [41].

The skeleton C-C vibrations of PP are sensitive to conformation effects, thus the vibrations at 809 cm^{-1} and 842 cm^{-1} are connected to crystallinity of PP. The band at 809 cm^{-1} corresponded to vibrations of the crystalline moieties, whereas the band at 842 cm^{-1} to non-crystalline part [42].

Polyacrylic fibers used in the manufacturing of textiles are composed of at least 85% by weight of polyacrylonitrile and 4–10% of non-ionic comonomer and 0,5–1% of an ionic co-monomer [43]. This is due that fibers produced from 100% polyacrylonitrile have poor elasticity and they are difficult to dye [44]. Polyacrylonitrile (PAN) is produced by the polymerization of cyanoethene and nitrile CN group is the characteristic element of this polymer. In IR spectroscopy is the very useful method in the subclasses of acrylic distinguishing the additional comonomers and additives [39]. In Raman spectrum mostly polyacrylic polymer is visible. Presence of another monomers is not evident in Raman spectrum, they show only minor variations in band shapes [39].

In the Raman spectrum of PAN fiber (**Figure** 7) the dominant band is the nitrile CN stretching band at 2245 cm^{-1}. Another characteristic bands of PAN concern the CH_2 bending at 1455 cm^{-1}, CH bending at 1314 cm^{-1}, CN twisting at 1225 cm^{-1}, CC skeletal stretching at 1118 and 1102 cm^{-1}, CH_2 twisting and CCN stretching at 829 cm^{-1}, CN wagging and bending 638 cm^{-1}, CN bending at 516 cm^{-1}, CC backbone deformation at 397 cm^{-1} and 283 cm^{-1}. Band at 142 cm^{-1} belongs to TiO_2 used in fiber processing. Weak carbonyl band at 1737 cm^{-1} belongs to vinyl acetate monomer [39].

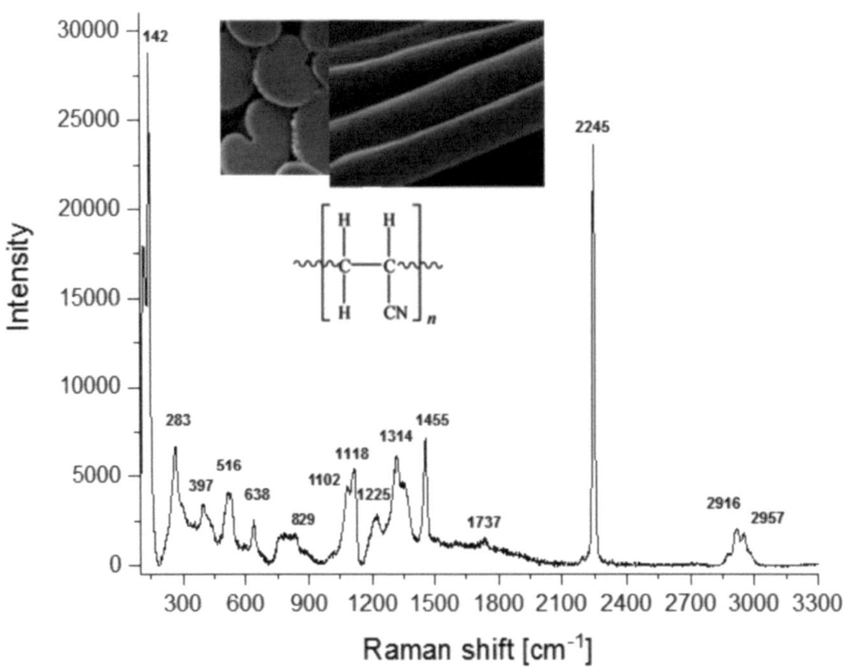

Figure 7.
Raman spectrum of PAN fiber.

Aramid fibers are made from aromatic polyamide polymers. Aramid chains possess amide groups which are directly connected to two aromatic rings: meta aramid (mAr) contains m-disubstituted benzene rings, para aramid (pAr) contains p-disubstituted benzene rings. Raman spectra of aramid fibers (**Figure 8**) are presented in the range 100–2000 cm^{-1} for better readability, as no bands were recorded in the 2000–3000 cm^{-1} area. Lack of the bands in the 2900–3200 cm^{-1} region characteristic of CH and NH stretching vibrations was interpreted to be due an orientational effects [45]. In the spectra of meta aramid (**Figure 8a**) characteristic Raman bands occur at 115 cm^{-1} (CC in-plane bending), at 192 cm cm^{-1} (ring out of plate CCC bending vibration), at 278 cm^{-1} (ring CCC asymmetric deformation, CN in-plane bending), at 659 cm^{-1} (ring puckering deformation, ring bending and asymmetric torsion, CH out-of-plane deformation), at 1003 cm^{-1} (trigonal ring breathing vibration CH in plane bending, ring and ring CH deformation), at 1250 cm^{-1} (NH bending and CN stretching), 1339 cm^{-1} (CH in-plane deformation), at 1420 and 1442 cm^{-1} (ring puckering vibration, aromatic CH bending), 1488 cm^{-1} (CH in-plane and NH in-plane bending), at 1547 cm^{-1} (NH in-plane bending), at 1606 cm^{-1} (CC aromatic ring stretching), at 1655 cm^{-1} (amide stretch C=O). In the spectrum of pAr (**Figure 8b**) characteristic Raman bands occur at 154 cm^{-1}, at 633 cm^{-1} (CC ring in plane deformation), at 698 cm^{-1} (CH out-of-plane deformation; CO bending), at 739 cm^{-1} (CO in-plane bending; ring asymmetric CH deformation; CN stretching), at 789 cm^{-1} (CH out-of-plane deformation, CCC ring puckering deformation), at 845 cm^{-1} (CH out-of-plane deformation; ring CC stretching, ring bending and asymmetric torsion), at 919 cm^{-1} (ring out-of-plane bending, CH in-plane bending, CH in-plane ring bending mode, CC stretching), 1186 cm^{-1} (ring CH deformation), at 1280 cm^{-1} (NH bending and CN stretching, CC stretching), at 1332 cm^{-1} (ring CH bending, ring CC stretching), at 1414 cm^{-1} (symmetric ring puckering/aromatic CH in-plane), at 1520 cm^{-1} (ring CH bending), at 1576 cm^{-1} (amide II vibration, bend (NH) and stretch (CN), ring CC stretching; NH bending), at 1615 cm^{-1} (aromatic ring CC stretching), 1655 cm^{-1} amide I (CO stretching) [45].

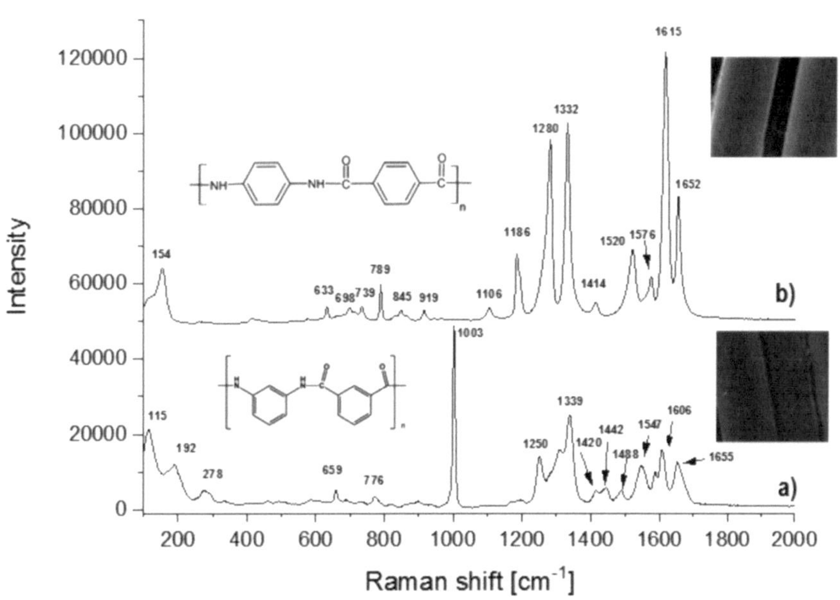

Figure 8.
Raman spectrum of aramid fibers: a) mAr, b) pAr.

Thus mAr could be identified by three characteristic Raman bands concerning CCC ring bending vibrations (at 115, 192 and 278 cm^{-1}), and the presence intense band of ring breathing vibration (at 1003 cm^{-1}). Spectrum of pAr exhibits characteristic Raman bands associated with p-substituted benzenes and may be identified by strong band at 154 cm^{-1} (unassigned), ring deformation bands at 789 cm^{-1} and 1182 cm^{-1}, NH bending and CN stretching band at 1280 cm^{-1}, ring CH bending and CC stretching at 1332 cm^{-1} and the intense band of aromatic ring CC stretching at 1615 cm^{-1}. The Raman spectroscopy is a very good method for nondestructive and unambiguous identification of the aramid fibers as the spectra of mAr are sufficiently different from those of the pAr to enable a definitive distinguishing.

3. Textile modified structures

Growing market demand for functional textile materials has followed the development of research on the fibrous structures modification using nanoparticles [2–23]. Modern multi-functional textiles are based very often on fibers surface modified with nanoparticles. The most extensively studied nanomaterial for textile modification is nanosilver. Over 200 publications a year concerning the textile functionalization by nanosilver appeared between 2015 and 2020 [46]. The second most studied nanoscale material, just behind silver is TiO2 with more than 100 publications per year since 2011 [46].

3.1 Characterization of functional textiles with silver nanowires (AgNWs)

Application of silver nanowires AgNWs for textile functionalization allows for obtaining bifunctional textiles with bioacive and conductive properties [4–5]. AgNWs colloid of 0,5% concentration was synthetized [4, 44, 46, 47] and nanowires were applied to the surface of the fabrics by dipping and drying method [4–5].

Raman rectangle map of functionalized cotton is presented on **Figure 9**. Characterization of cotton modified by AgNWs was described in publications [4, 12].

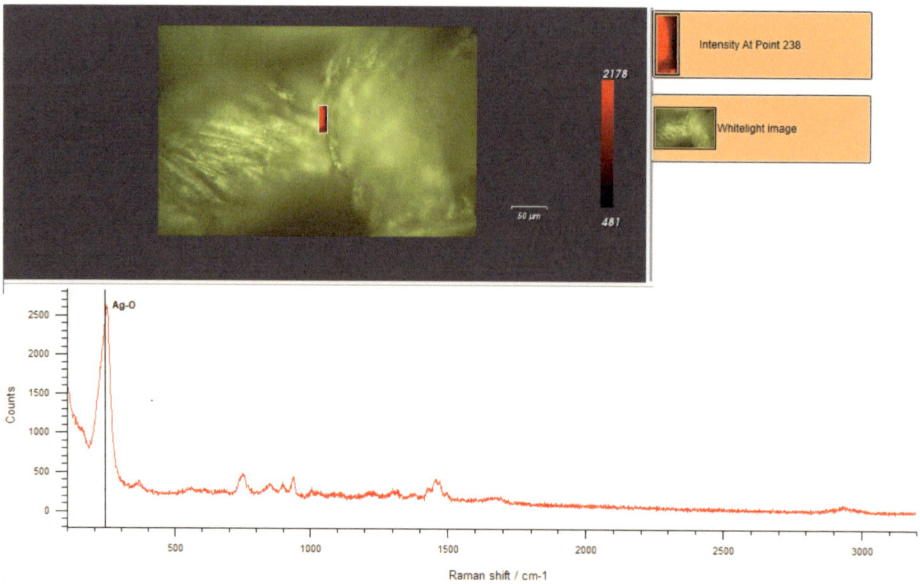

Figure 9.
Raman map of cotton fabric functionalized by AgNWs.

Raman technique thanks to Raman mapping enables the characterization of the selected surface area on a micro scale. The analyzed area is visualized in terms of the intensity of the band characteristic for a given material or modifier. Two or even three dimensional maps (2D and 3D maps) show the distribution of the modifier on the analyzed area. Band characteristic for AgNWs is in the region 240–250 cm^{-1} and it is the result of Ag-O coordination band that is the effect of interactions between silver and oxygen adsorbed on the surface [4, 22, 48]. In the case of modified cotton AgNWs, no cotton bands are visible on the surface. All analyzed surface is covered with a metallic AgNWs coating. The map was made according to the characteristic band of Ag-O coordination at 238 cm^{-1}.

Characterization of functionalized aramid fibers is presented on **Figures 10** and **11**. For the functionalized mAr, the most intense band of mAr is the ring breathing band at 1003 cm^{-1}. Raman maps were performed according to the 1003 cm^{-1} band and according to AgNWs band which is represented by Ag-O coordination band at 238 cm^{-1} (**Figure 10**). Both maps are also presented in 3D form.

Whereas for the functionalized pAr, the characteristic band of mAr was the NH bending and CN stretching band at 1279 cm^{-1}, characteristic band of AgNWs was Ag-O coordination band (**Figure 11**). In both figures (**Figures 10** and **11**), the blue color shows the area covered with nanowires, and the red color indicates that aramid predominates on the surface in the studied area.

3.2 Characterization functional textiles with TiO_2 on the surface

Titanium dioxide (TiO_2) applied to the modification of textile materials can give them such properties as i.e. photocatalytic, self-cleaning, bioactive, UV protective etc. [2, 9, 49–52]. Raman map of TiO_2 modified PP is presented on **Figure 12**.

Polypropylene as a one of the most used component of floorcoverings was also modified by TiO_2. Titanium dioxide was applied to limit the environmental tobacco

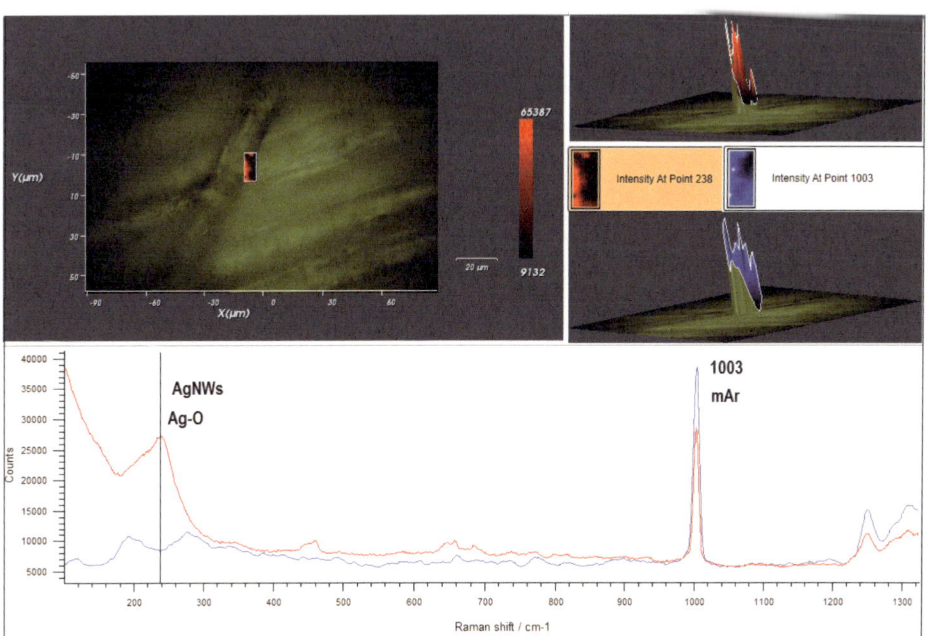

Figure 10.
Raman map of mAr modified by AgNWs.

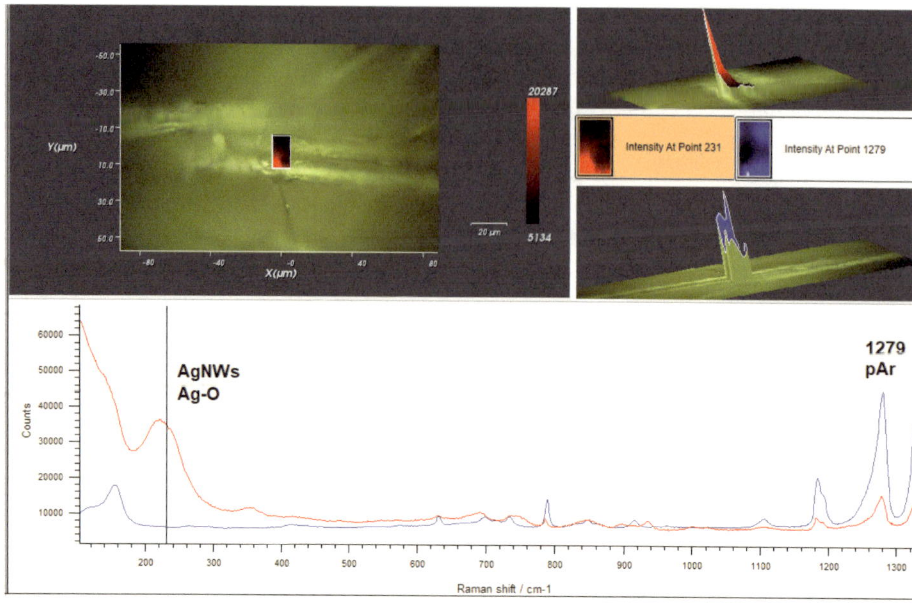

Figure 11.
Raman map of pAr modified by AgNWs.

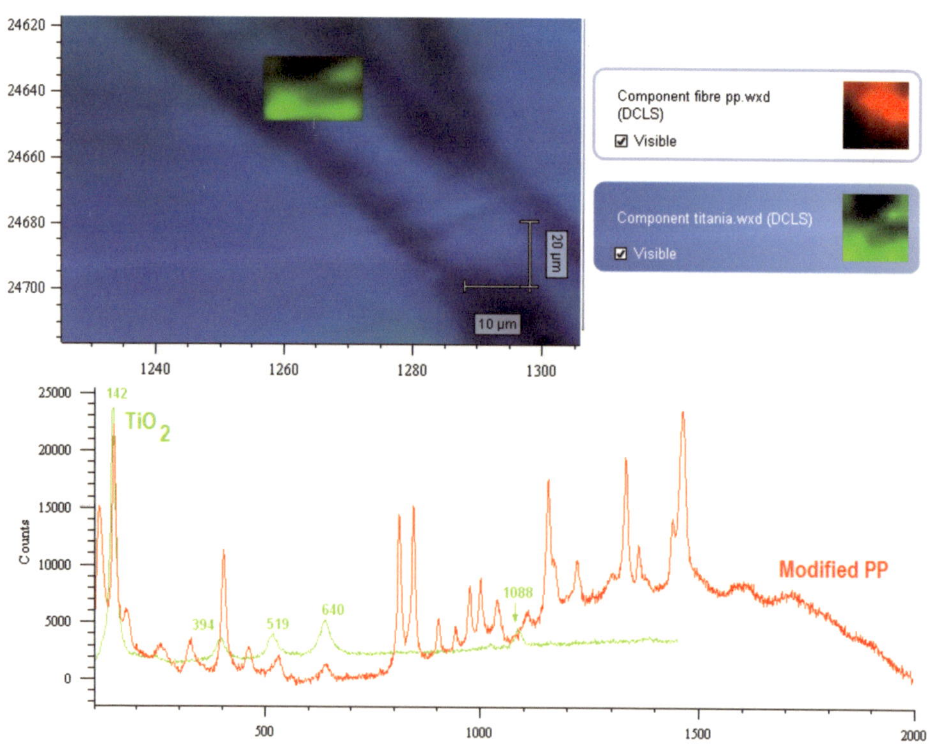

Figure 12.
Raman map of PP fiber modified by TiO_2.

smoke (ETS) sorption by the photocatalytic decomposition of ETS-derived nicotine (basic marker of tobacco smoke exposure) [50–52]. Thanks to Raman Surface mapping the TiO_2 distribution on the PP fiber can be evaluated. The analyzed area

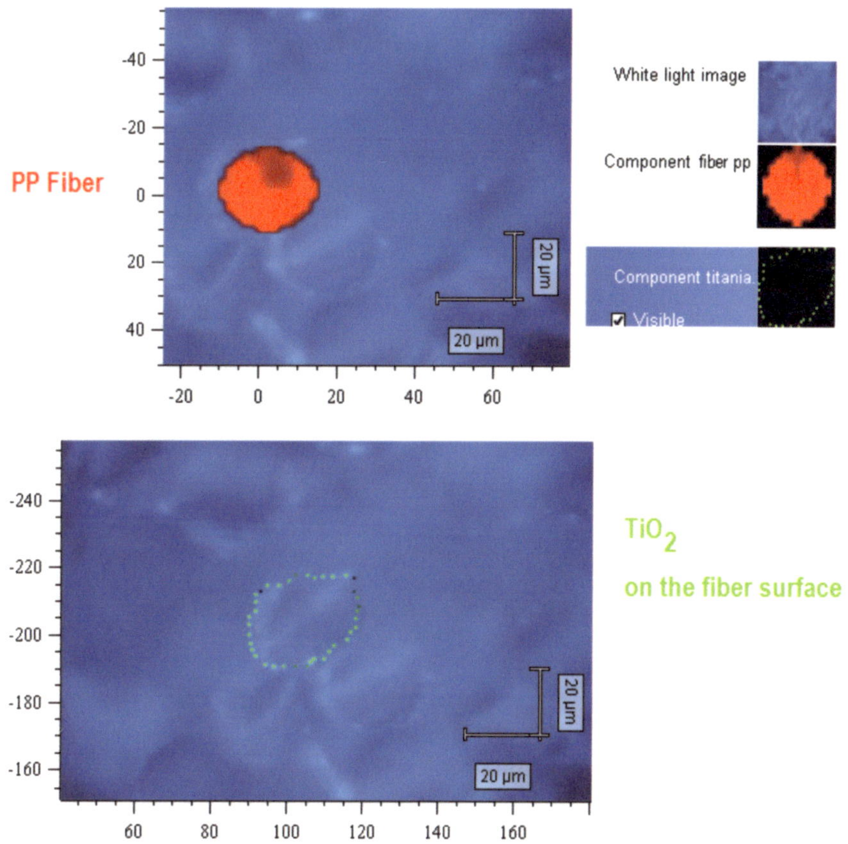

Figure 13.
Cross-sections of PP fibers modified with TiO_2.

is visualized in terms of the intensity of the 142 cm^{-1} band that is characteristic for TiO_2 in anatase form [52]. The characteristics of functional fiber can be enriched by the cross-section map as it is shown on the **Figure 13**. The cross-section map illustrates the place of modification revealing whether the modification takes place on the surface, or in the volume of the fiber.

4. SERS effect on textile fibers surface

The functionalization of textile with silver nanoparticles causes formation of metallic layer on their surfaces and in consequence possibility of Raman signal enhancement [4, 21–23]. The research on the functionalization of fibers has shown that a textile material can be a good carrier for the SERS effect [22]. SERS effect on cotton brought not only designed functionalization effects. This method turned out to be the useful tool in the identification of the reactive dyes for cotton dyed with low color intensity. In the **Figure 14** there is presented the Raman map on cotton fabric. Spectra presented below map are the reference spectra of cotton and reactive red dye. Red line is the spectrum detected on functionalized cotton surface. In addition to the band characteristic for AgNWs, the signal enhancement and in consequence increase in the intensity of the bands in the region of 1100–1600 cm is visible. This enhancement concerns main band of cotton at 1099 cm^{-1} and the bands of reactive dye. When compare this spectrum with the spectrum of cotton modified by AgNWs shown in **Figure 9**, it can be noticed that the cotton and dye

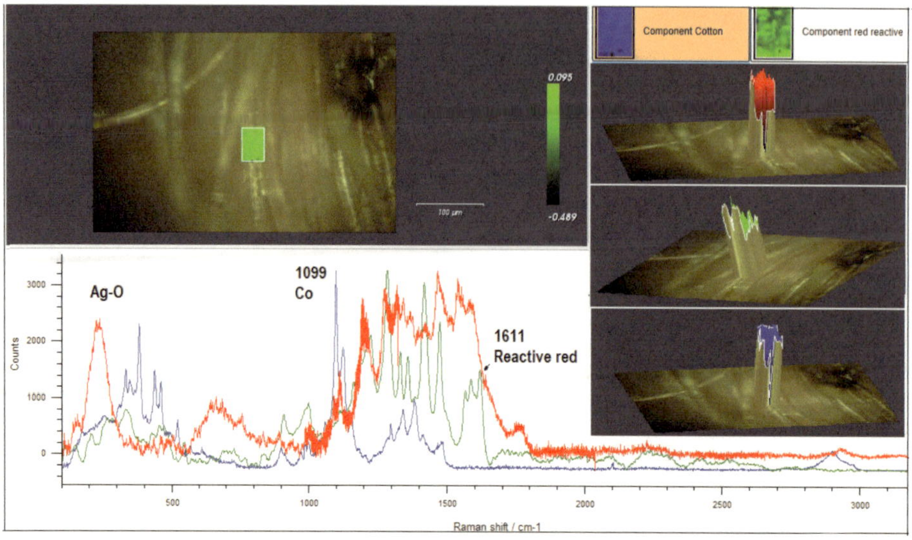

Figure 14.
Raman map of functionalized cotton fabric. Red line is the measured spectrum, blue line is the reference spectrum of cotton, green line is the reference spectrum of reactive red dye.

bands became visible thanks to this signal enhancement. SERS effect on the cotton surface was accessible only when the thin layer of AgNWs was applied [22].

The SERS effect was also recorded on the wool fibers. The Raman maps of the AgNWs modified wool surface are shown in **Figures 15** and **16**. These maps were done according to a band characteristic of Ag-O coordination and confirm the presence of the AgNWs on the wool. **Figure 16** shows a map of dyed wool and the reference spectra of wool and dye used (reactive red dye) are also presented. Raman maps collected on the wool surface, spectra made point by point, show in some places the enhancement of recorded bands.

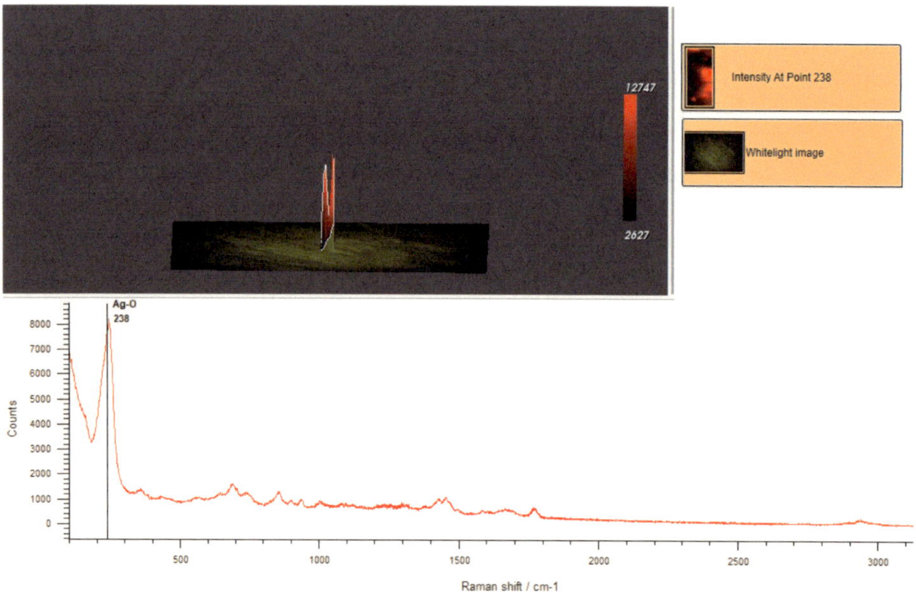

Figure 15.
Raman 3D map of functionalized wool.

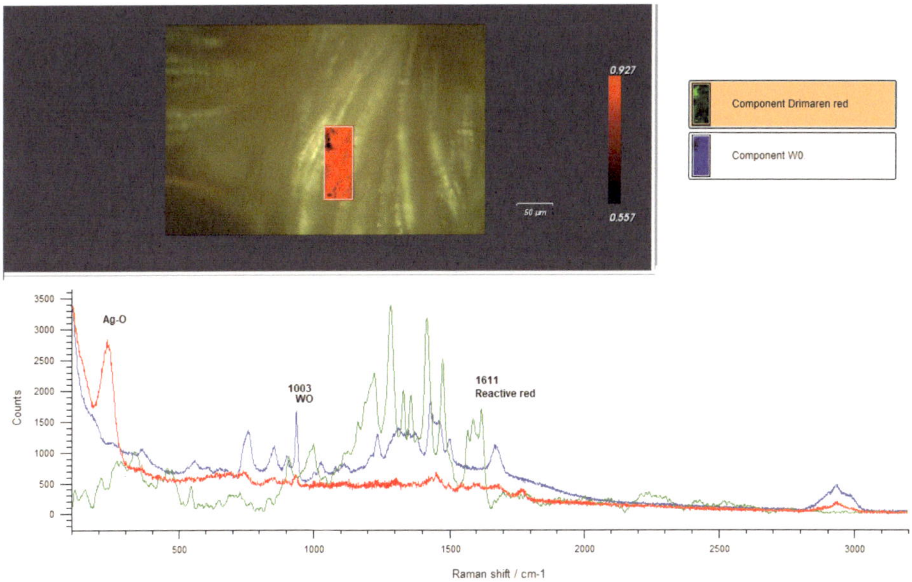

Figure 16.
Raman map of functionalized dyed wool made. Red line is the measured spectrum, blue line is the reference spectrum of cotton, green line is the reference spectrum of reactive red dye.

In the **Figure 17** there is presented one of such enhanced Raman spectrum. Stars on the **Figure 17** show strengthens bands. As the map is made on the surface of dyed wool, reinforcement of both the wool and the dye strands is observed. However, not all bands are strengthened equally, as at the same time additional SERS bands appear. Additional SERS bands that might be the effect of chemical enhancement of ring vibrations [21, 53]. The SERs effect accompanies the functionalization of

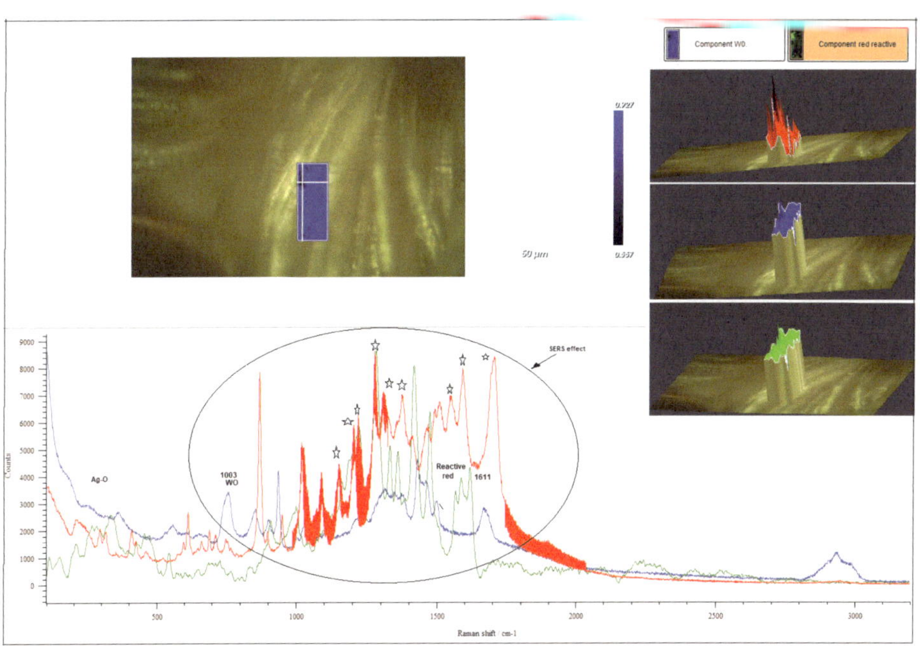

Figure 17.
Raman map with visible SERS effect of functionalized wool. Red line is the measured spectrum, blue line is the reference spectrum of wool, green line is the reference spectrum of reactive red dye.

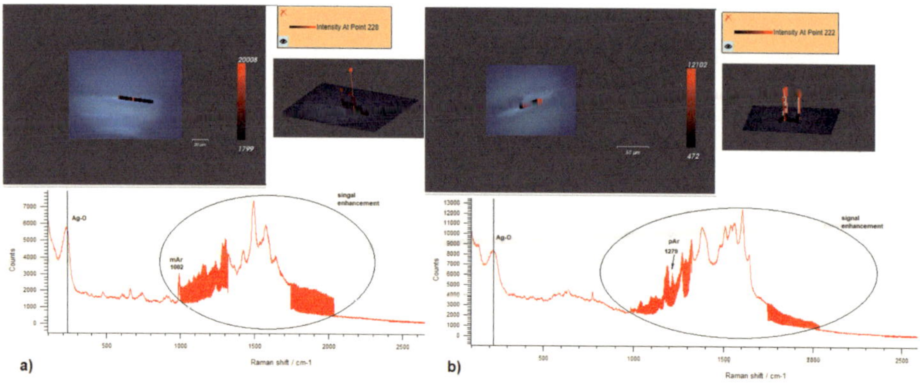

Figure 18.
Raman maps with visible SERS effect on aramid fibers: a) mAr; b) pAr.

fibers with a rough surface and should be studied more deeply as it can be useful in the analysis of textile materials i.e. in the identification of the other elements on the surface.

SERS effect was also identified on aramid fibers, as it is shown in the **Figure 18**.

5. Conclusion

Raman spectroscopy is a valuable method in the analysis of textile materials enabling fiber identification and characterization of modification effects. Identification can be carried out for natural and synthetic fiber, both by analyzing their surface and inside the structure. This technique makes also possible to distinguish between the fibers, where IR spectroscopy does not give a definite answer (wool and silk or PA 6 and PA 6.6). Raman spectrum can be also useful in assessment of the textile polymer crystallinity. The Raman spectroscopy special application has found in the study of textile materials functionalized with nanoparticles. They can be analyzed on the surface and inside the fiber. New possibilities are opened by the use of the Raman mapping system which allows the characterization of textile materials with nanoparticles, including SERS analysis. A functionalized textile structure with a noble metal on the surface can be flexible substrates for the surface enhanced Raman spectroscopy (SERS) analysis.

Acknowledgements

All presented results were carried out on Raman Renishaw InVia Spectrometer purchased the in POIG.01.03.01-00-004/08 Functional nano- and micro textile materials—NANOMITEX project, co-financed by the European Union with the financial resources of the European Regional Development Fund and the National Centre for Research and Development.

The authors would like to thank Ms. Alicja Nejman, Dr. Hubert Schmidt for functionalized fibers used in the research and Ms. Irena Kamińska for the SEM images of fibers.

Author details

Dorota Puchowicz* and Malgorzata Cieslak
Department of Chemical Textile Technologies, Textile Research Institute, LUKASIEWICZ Research Network, Lodz, Poland

*Address all correspondence to: dorota.puchowicz@iw.lukasiewicz.gov.pl

References

[1] Hashimoto K., Badarla V.R., Kawai A., Ideguchi T., Complementary vibrational spectroscopy. Nat Commun 10, 4411 (2019). DOI:10.1038/s41467-019-12442-9

[2] Giesz P., Celichowski G., Puchowicz D., Kamińska I., Grobelny J., Batory D., Cieślak M., Microwave-assisted TiO2: Anatase formation on cotton and viscose fabric surfaces, Cellulose 2016, 23 (3), 2143-2159; DOI:10.1007/s10570-016-0916-z

[3] Cieślak M., Celichowski G., Giesz P., Nejman A., Puchowicz D., Grobelny J., Formation of nanostructured TiO2-anatase films on the basalt fiber surface, Surface and Coatings Technology, 2015, 276, 686-695; DOI:10.1016/j.surfcoat.2015.05.045

[4] Giesz P., Mackiewicz E., Nejman A., Celichowski G., Cieślak M., Investigation on functionalization of cotton and viscose fabrics with AgNWs, Cellulose 24 (2017) 409-422. DOI:10.1007/s10570-016-1107-7

[5] Giesz P., Mackiewicz E., Grobelny J., Celichowski G., Cieślak M., Multifunctional hybrid functionalization of cellulose fabrics with AgNWs and TiO2, Carbohydrate Polymers, 2017, 177, 397-405, DOI:10.1016/j.carbpol.2017.08.087

[6] Boufi S., Bouattour S., Ferraria A. M, Vieira Ferreira L.F., Botelho do Rego A.M. Chehimi MM., Vilar M.R., Cotton fibres functionalized with plasmonic nanoparticles to promote the destruction of harmful molecules: An overview Nanotechnol Rev 2019; 8:671-680, DOI:10.1515/ntrev-2019-0058

[7] Syafiuddin A., Toward a comprehensive understanding of textiles functionalized with silver nanoparticles, J. Chin. Chem. Soc. 2019;66:793-814., DOI:10.1002/jccs.201800474

[8] Pakdel E., Daoud W.A., Wang X., Self-cleaning and superhydrophilic wool by TiO2/SiO2 nanocomposite, Appl. Surf. Sci., 2013, 275, 397-402, DOI:10.1016/j.apsusc.2012.10.141

[9] Radetic M., Functionalization of textile materials with TiO2 nanoparticles, J. Photochem. Photobiol. C: 2013, 16, 62-76, DOI:10.1016/j.jphotochemrev.2013.04.002

[10] Srinivas K., The role of nanotechnology in modern textiles, Journal of Chemical and Pharmaceutical Research, 8 (2016) 173-180.

[11] Yetisen A.K., Qu H., Manbachi A., Butt H., Dokmeci M.R., Hinestroza J.P., Skorobogatiy M., Khademhosseini A., Hyun Yun S., Nanotechnology in textiles, ACS Nano, 10 (2016), 3042-3068, DOI:10.1021/acsnano.5b08176

[12] Mahmud R., Nabi F., Application of nanotechnology in the field of textile, Journal of Polymer and Textile Engineering, 4 (2017), 1-6, https://www.iosrjournals.org/iosr-jpte/papers/Vol4-Issue1/A04010106.pdf

[13] Barani H., Boroumand M. N., Rafiei S., Application of silver nanoparticles as an antibacterial mordant in wool natural, Fibers and Polymers 2017, 18, 4, 658-665, DOI:10.1007/s12221-017-6473-8

[14] Sportelli M. C., Izzi M., Kukushkina E.A., Hossain S. I., Picca R. A., Ditaranto N., Cio N., Can nanotechnology and materials science help the fight against SARS-CoV-2, Nanomaterials, 2020, 10, 802; DOI:10.3390/nano10040802

[15] Arvizo R.R., Bhattacharyya S., Kudgus R., Giri K., Bhattacharya R.,

Mukherjee P., Intrinsic therapeutic applications of noble metal nanoparticles: Past, present and future, Chem. Soc. Rev. 41 (7) (2012) 2943-2970, DOI:10.1039/C2CS15355F.

[16] Mather R.R., Wardman R.H. (2011), The chemistry of textile fibres, RSC publishing, Cambridge 2011.

[17] Morton W.E., Hearle J.W.S. (1975), Practical properties of textile fibres, the textile institute Heinemann, London 1975.

[18] Pastore C.M., Kiekens P. (2001), Surface characteristics of Fibers and textiles, Marcel Deker Inc., New York 2001.

[19] Cieslak M., Schmidt H., Swiercz R., Wasowicz W., Fibers susceptibility to contamination by environmental tobacco smoke markers, Text., Res., J., 2014, 84(8), 840-853, DOI:10.1177/0040517513509850

[20] Cieslak M, Puchowicz D, Schmidt H, Evaluation of the possibility of using surface free energy study to design protective fabrics, Text Res. Journal 82 (11), 1177-1189, DOI:10.1177/0040517511426612

[21] Fateixa S., Wilhelm M., Nogueira H. I. S., Trindade T., SERS and Raman imaging as a new tool to monitor dyeing on textile fibres, journal of Raman Spectroscopy, J. Raman Spectroscopy, 2016, 47, 1239-1246, DOI:10.1002/jrs.494.

[22] Puchowicz D., Giesz P., Kozanecki M., Cieślak M., Surface-enhanced Raman spectroscopy (SERS) in cotton fabrics analysis, Talanta 2019, 195, 516-524, DOI:10.1016/j.talanta.2018.11.059

[23] Liu A., Zhang S., Guang S., Ge F., Wang J., Ag-coated nylon fabrics as flexible substrates for surface-enhanced Raman scattering swabbing applications. Journal of Materials Research, 35, 1271-1278 (2020). DOI:10.1557/jmr.2020.103

[24] Kudelski A., Analytical applications of Raman spectroscopy, Talanta 76 (2008), 1-8, DOI:10.1016/j.talanta.2008.02.042

[25] Smith E., Dent G., Modern Raman Spectroscopy – A Practical approach, applications, pp, 135-180, John Wiley And Sons, New York, Chichester 2005.

[26] https://store.textileexchange.org/wp-content/uploads/woocommerce_uploads/2019/11/Textile-Exchange_Preferred-Fiber-Material-Market-Report_2019.pdf

[27] Mather R.R., Wardman R.H. (2011), Cellulosic fibres [in:] the chemistry of textile fibres, pp. 22-60 RSC publishing, Cambridge 2011.

[28] Kavkler K., Demsar A., Examination of cellulose textile fibres in historical objects by micro-Raman spectroscopy, Spectrochim. Acta Part A 78 (2011) 740-746, DOI:10.1016/j.saa.2010.12.006.

[29] Was-Gubala J., Machnowski W., Application of Raman Spectroscopy for differentiation among cotton and viscose Fibers dyed with several dye classes, Spectroscopy Letters, 2014, 47:7, 527-535, DOI:10.1080/00387010.2013.820760

[30] Mather R.R., Wardman R.H. (2011), Protein fibres [in:] the chemistry of textile fibres, pp. 61-99, RSC publishing, Cambridge 2011.

[31] Li-Ling C., Identification of textile fiber by Raman microspectroscopy, J. Forensic Sci., 6 (2007), 55-62.

[32] Fleming G. D., Finnerty J. J., Campos-Vallette M., C'elis F., Aliaga a. E., Fredes C., Koch R., experimental and theoretical Raman and surface-enhanced Raman scattering study of

cysteine, J. Raman Spectrosc. 2009, 40, 632-638, DOI:10.1002/jrs.2175

[33] Edwards H. G. M., Farwell D. W., Raman spectroscopic studies of silk, J. Raman Spectrosc. 26, 901-909 (1995)

[34] Nejman A., Kamińska I., Jasińska I., Celichowski G., Cieślak M., Influence of low-pressure RF plasma treatment on aramid yarns properties, Molecules, 2020, 25, 3476-3499, DOI:10.3390/molecules25153476

[35] Stuart B.H., Polymer crystallinity studied using Raman spectroscopy, Vibrational Spectroscopy 10 (1996) 79-87

[36] Boerio E. J, Bahl S.K., Mc Graf G.E., Vibrational analysis of Polyethylenie terephthalate and its Deuterated derivatives, J. Polym. Sci., 1976, 14, 1029-1046.

[37] Rebollar E., Perez S., Hernandez, M., Domingo C. Martın M., a Ezquerra T.A. Garcıa-Ruiz J.P., Castillejo M., Physicochemical modifications accompanying UV laser induced surface structures on poly(ethyleneterephthalate) and their effect on adhesion of mesenchymal cells, Phys. Chem. Chem. Phys., 2014, 16, 17551, DOI:10.1039/c4cp02434f

[38] Poppe, W. A Kreklau F. H, Bergter R., Process for the identification of the pile of textile materials, in particular for the identification of PA 6 and PA 66 in carpets, DE10011254A1 Patent, 2000.

[39] Miller J. V, Bartick EG. Forensic analysis of single fibers by Raman spectroscopy. Appl Spectrosc 2001;55(12):1729-1732.

[40] Sagitova E.A., Donfack P., Nikolaeva G. Yu, Prokhorov K.A., Pashinin P.P., Nedorezova P.M., Klyamkina A.N., Materny A., New insights into the structure of polypropylene polymorphs and propylene copolymers probed by low-frequency Raman spectroscopy, Journal of Physics: Conf. Series 826 (2017) 012006 DOI:10.1088/1742-6596/826/1/012006.

[41] Andreassen E. (1999) Infrared and Raman spectroscopy of polypropylene. In: Karger-Kocsis J. (eds) Polypropylene., Pp. 320-328 Polymer Science and Technology Series, vol 2. Springer, Dordrecht. DOI:10.1007/978-94-011-4421-6_46

[42] Nielsen AS, Batchelder DA, Pyrz R. Es¬timation of crystallinity of isotactic poly¬propylene using Raman spectroscopy. Polymer 2002; 43: 2671-2676.

[43] Grieve M. C., Another look at the classification of acrylic fibers, using FTIR microscopy Science and Justice 35, 1995, 179-190.

[44] Sun Y., Gates B., Mayers B., Xia Y., Crystalline silver nanowires by soft solution processing, Nano Letters, 16 (2002) 5-168, DOI:10.1021/nl010093y

[45] Edwards H.G.M., Hakiki S., Raman spectroscopic studies of Nomex and Kevlar fibres under stress, British Polym. J. 1989, 21, 505-512.

[46] Sun Y., Xia Y., Large-scale synthesis of uniform silver nanowires through a soft, self-seeding, polyol process, Advanced Materials, 14 (2002) 833-837, DOI:10.1002/1521-4095(20020605)14:11<833::AID-ADMA833>3.0.CO;2-K

[47] Nghia N.V., Truong N.N.K., Thong N.M., Hung N.P., Synthesis of nanowire-shaped silver by polyol process of sodium chloride, International Journal of Materials and Chemistry, 2 (2012) 75-78, DOI:10.5923/j.ijmc.20120202.06

[48] Wang C.B., Deo G., Wachs I.E., Interaction of polycrystalline silver with oxygen, water, carbon dioxide, ethylene

and methanol: in situ Raman and catalytic studies, J. Phys. Chem. B 103 (1999) 5645-5656, DOI:10.1021/jp984363l.

[49] Rashid M.M., Simoncic B., Tomsic B., Recent advances in TiO2-functionalized textile surfaces, Surfaces and Interfaces 22, 2021, 100890, DOI:10.1016/j.surfin.2020.100890

[50] Cieślak M., Schmidt H., Świercz R., Wąsowicz W., „TiO2/Ag modified carpet fibres for the reduction of nicotine exposure", FIBRES and TEXTILES in Eastern Europe, 17, 2 (73), 59-65, 2009.

[51] Cieślak M., Puchowicz D., Kamińska I., SEM/EDS and Raman Micro-Spectroscopy Examination of Titanium-Modified Polypropylene Fibres., FIBRES and TEXTILES in Eastern Europe 2014; 22, 3(105): 47-53.

[52] Scepanovic M. J, Gruic-Brojcin M., Dohcevic-Mitrovic Z.D. Characterization of anatase TiO2 nanopowder by variable-temperature Raman spetroscopy. Sci. Sinter. 2009; 41: 67-73.

[53] Dong X, Gu H., Kang J., Yuan X., Wu J., Comparative study of surface-enhanced Raman scattering activities of three kinds of silver colloids when adding anions as aggregating agents, Colloids and Surfaces A: Physicochem. Eng. Aspects, 368 (2010), 142-147. DOI:10.1016/j.colsurfa.2010.07.029

Chapter 10

Application of Raman Spectroscopy in Biomedical Diagnostics

Nikiwe Mhlanga, Phumlani Tetyana, Sanele Nyembe and Lucky Sikhwivhilu

Abstract

In vivo cellular imaging and *in vitro* assays or sensors are fundamentally used to study the spatiotemporal interaction of molecules at biological interfaces. The study of these interfaces informs various applications such as diagnostics/detection of foreign materials or processes in the biological system. Raman spectroscopy, an optical, non-destructive, label-free fingerprinting tool offers a wide array of applications in both *in vitro* and *in vivo* diagnostics owing to its relatively short acquisition time, non-invasiveness and ability to provide biochemical molecular information. It has been explored in tissue imaging, *in vitro* diagnosis, DNA/RNA analysis, metabolic accretions, single cell analysis photodynamic therapy, etc. The chapter details the application of the optical Raman platform in the detection and imaging of diseases/tissues. The challenges associated with SERS applications and the future outlook as a biomedical diagnostic tool are also discussed.

Keywords: Medical diagnostics, nanoparticles, Surface Enhanced Raman Spectroscopy, Surface Plasmon Resonance, Enhancement Factor

1. Introduction

The rapid and sensitive detection, identification, and quantitative analysis of bio-species at very low concentrations for pre-symptomatic and symptomatic diagnosis represent a new frontier in biomedical research, enabled by nanomedicine [1]. The surge to achieve such diagnostics has ushered in the exploration of various tools such as Raman optical spectroscopy. The emerging Raman spectroscopy tool has the potential to provide fingerprints of bio-species, quantify and differentiate biomarkers [2].

The vibrational modes of the bio-species are measured from inelastic Raman scattering and the analysis of these modes gives a molecular picture of the bio-species [2]. However, conventional Raman spectroscopy presents a weak signal due to a small scattering cross-section. Plasmonic metal surfaces mitigate the issue of the low signal and this phenomenon is known as surface-enhanced Raman spectroscopy (SERS) [2].

Several variants of Raman besides SERS have been developed to mitigate biomedical diagnostic issues presented by the conventional Raman. Resonance Raman spectroscopy improves the low signal by the use of a laser excitation wavelength that corresponds to the electron absorption maximum of the bio-species [3].

For *in vivo* applications the penetration depth of the Raman laser to the targeted tissues is usually a bottleneck. Nonetheless, spatially offset Raman spectroscopy (SORS) measures diffuse Raman scattering from regions away from the laser excitation and thus collects vibrational modes from deeper tissues. The SORS and SERS technique are integrated into the SESORS variant, a tool that will benefit biomedical *in vivo* diagnostics [3]. The Raman spectroscopy also falls short in resolving nanostructures. To enable sub-nanometer spatial resolution and signal enhancement a Tip-enhanced Raman (TERS) that combines SERS with apertureless near-field scanning optical microscopy using a metallic tip is used [4]. This chapter presents recent developments on some of the Raman variants in biomedical diagnostics, materials developments for the application, challenges, and forecasts into the future of Raman biomedical diagnostics.

2. Materials used in Raman diagnostics

2.1 Gold and silver colloidal nanoparticles: bottom-up

It is commonplace that many condensed-matter systems exhibit collective excitation modes involving coherent oscillations of the medium. As such surface Plasmon's are known to be the collective excitation of free conduction electrons excited by electromagnetic radiation at the metal-dielectric interface [5]. Over the years the use of noble metal thin films or nanoparticles (NPs) surfaces has underpinned the success in this area. Despite all this, the study of the interaction between light and metallic nanostructures holds promise for this emerging research area of Plasmonic which is rapidly gaining traction [6–9]. Well-established targeted technologies to engineer Plasmonic nanostructures allows for better control and manipulation of visible light at the nanometer scale leading to new possible application areas with real life impact [10–12].

Although most noble metals exhibit Plasmonic properties and can potentially be used as SERS substrate gold and silver are the most widely used owing to their superior enhancement factors (EFs) and Plasmonic resonance in the visible and NIR regions [13–15]. Typically, these are used in their metallic form but also as composite with other materials. However, in order to attain the highest sensitivity and specificity these NPs materials must have the Plasmonic properties that include the resonance frequency of the surface Plasmon's and the magnitude of the electromagnetic field that is generated at the surface. Inevitably, these properties are, in the main, influenced by the type, size, shape, composition, and dielectric environment [16–19]. Despite all this, Ag has proven to have higher EF than Au making it more attractive for various applications.

Typically SERS substrates are configured in two forms and these are colloidal suspensions (NPs) and solid substrates. NPs are the most commonly used form primarily because of simplicity of preparation and relatively high EFs. Interestingly, the most significant SERS enhancement is achieved when molecular structures are bound to noble metal nanostructures with a size range of 5–15 nm (Au and Ag NPs are shown in **Figure 1**). Furthermore, it must be noted that the nature of interaction between the molecules and nanostructured surfaces, as well as the charge properties of the molecular structure play a pivotal role in terms of performance of SERS-based Raman measurements. For example, when colloidal noble metal NPs are put to use the surface charge of the NPs and that of the molecules become fundamentally important [20, 21]. This implies that maximum SERS activity is attained when the detected molecule has the opposite charge to interacting colloidal NPs. This phenomenon is ascribed to the induced aggregation resulting from reduced zeta

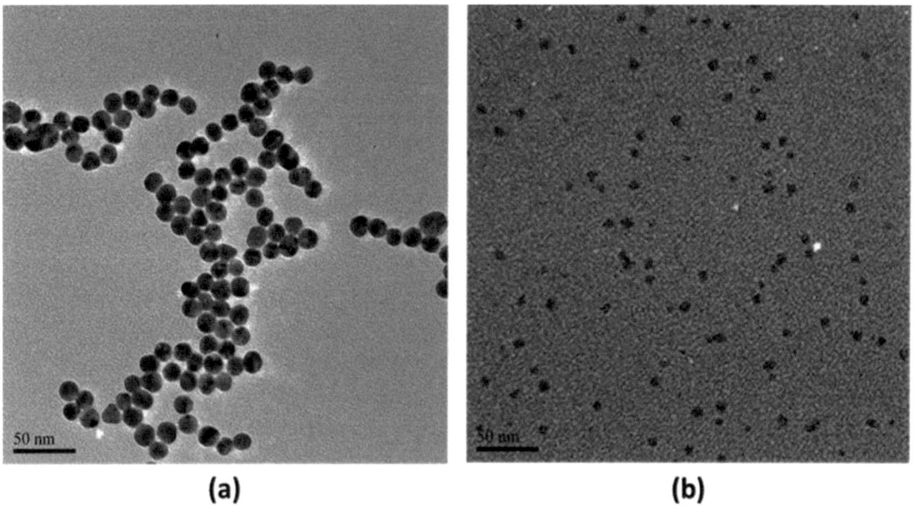

Figure 1.
TEM images of colloidal (a) Au and (b) Ag NPs.

potential of NPs [20]. Kahraman, *et al*. reported that the highest EF in the case of Ag NPs is achieved when the NPs have a pyramid-like shape [22]. On the other hand meticulously, controlling aggregation is useful for improving SERS enhancement as this could lead to a higher possibility of "hot spot" formation [23, 24]. In view of this, important consideration must be given to the fact that very large aggregates could abate the effective formation of surface Plasmons as a result of the deformations and dampening of the electron cloud within the aggregate leading to poor SERS activity. Data has shown that small-sized aggregates tend to greatly improve the EFs [25]. Furthermore, the methods of preparation are known to influence the SERS activity of Au and Ag colloidal NPs. This is because the structure of a substrate has an important role on the properties of these metal NPs and subsequently influences the applications of the substrate. Au and Ag colloidal NPs are sufficiently produced using chemical reduction method such as Citrate method.

2.2 Colloidal gold nanowires: bottom-up

Besides, the use of Au and Ag as gold standard for superior EF and possibly creating a stable platform for these metals by anchoring them on a support like graphene it is clear that the effect of changing the shape of the material on the EF is significantly large. Elongating the Au structures to very thin nanowires leads to improved enhancement owing to their large surface area and excellent molecule adsorption ability. The molecule adsorption is attributed to the fact that Au nanowires adsorb Raman active compounds via two mechanisms, namely, surface adsorption due to the surface reactivity of Au nanowires and network entrapping through their "web" network [26, 27]. The Au nanowires, as shown in **Figure 2**, reveal that ultrathin wires tend to form a spider web-like network with interstices smaller than 1 nm. These interstices have been reported to be responsible for further entrapment of adsorption molecules [27]. The spider web-like network are exclusively found in Au nanowires and not in the spherical Au NPs. Hence, Au nanowires are reported to have a relatively higher EF value compared to spherical NPs. Depending on the method of preparation for the NPs, geometrical parameters such as size distribution are not often well controlled as a result of production limitation such as random deposition or irreproducibility of size distributions leading to disordered samples.

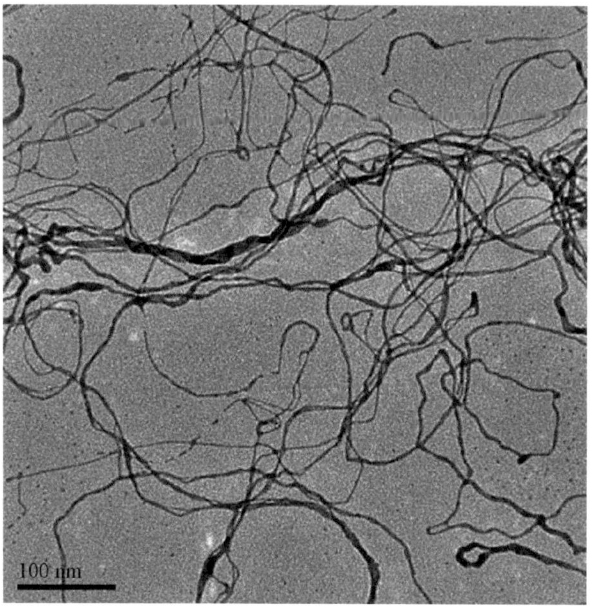

Figure 2.
The TEM image of ultrathin Au nanowires.

However, in the case of nanowires reproducibility seems to be easily controlled leading to the presence of hot spots (highly localized area of intense electric field) resulting in significant enhancement and homogeneous SERS signal (reproducibility). This phenomenon is particularly important in terms of provision of unambiguous analyte quantification as the enhancement should be highly reproducible while it is necessary to optimize SERS efficiency in relation to Plasmon excitation conditions [28]. For example, the SERS signal is directly related to the excitation wavelength and Raman scattering wavenumber [28–36] or the coupling between nanostructures [37–39]. Furthermore, aggregation of Au nanowires is less likely than in the case of colloidal Au NPs.

The size of the nanowires and its potential influence on SERS signal is considered important. It is understood that the surface of Au nanowires increases with an increase in the diameter of the wires thereby inducing an increase in the number of adsorbed molecules and consequently leading to a larger Raman signal. In order to ensure a controlled enhancement as contribution by the nanostructures orientation of the wires is important [40, 41]. Although nanowires tend to self-assemble into dense bundles resulting from the electrostatic interaction the effect of SERS activity remains insignificant. Au nanowires with a well-controlled geometry provide well controlled and reproducible SERS signal. The high aspect ratio combined with the bundling yield structures with impressive SERS activity. Ultrathin colloidal Au nanowires are produced using facile modified chemical reduction method (citrate method) making them very affordable [42, 43].

2.3 Au/Au nanowires: top-down

Whilst the use of nanowires as SERS substrate seems to be gaining momentum on all fronts Sivashanmugana *et al.* reported on the fabrication of well-oriented Au/Ag multi-layered nanorod arrays using a focused ion beam techniques [44]. The study revealed that the shape, thickness, space between nanorods and size dimensions have influence on the performance of SERS. With good experimental conditions these factors can be optimized before fabrication of the substrate and this is often done by varying the thickness of Au and Ag layers for optimum performance. Ag layer plays a pivotal role in respect to the improvement of SERS mechanism as it induces the electromagnetic effect at the Au surface [36–38, 44].

While SERS-active substrates are usually obtained either by the solvent-cast deposition of colloidal NPs [45, 46] or by lithography techniques such as electron beam lithography [46, 47], nanoimprint [48], and nano-indentation [49, 50]. These fabrication methods are based on high precision regular patterns. Consequently, most of these methods place emphasis on significant improvement on enhancement factors. However, they tend to yield low reproducibility and high cost, and they lack the essential characteristics required to make SERS a platform-enabling technology. Interestingly, the use of nanowires of a single entity or multiple has led to the construction of very reproducible SERS-active substrate, especially with gold nanorods arrays. The Localized Surface Plasmon Resonance (LSPR) from the nanorods and the lighting-rod effect of the Plasmonic field enhancement strongly depend on the shape and the spacing between the nanorods [33, 51–56].

The SERS method has great potential for the detection of Raman- active species, ranging from single molecules to biomolecules. The past five years have seen various approaches being developed for fabrication of SERS-active substrates with high sensitivity using noble metal nanostructures via bottom-up, top-down, or template-assisted routes. Nano-based substrates with relatively high SERS enhancement factors can be easily produced, with the EF largely dependent on the size and shape of the nanostructures that give rise to the effect. In order for SERS substrates to be used as a platform for applications such as bio-sensing or diagnostics various issues including sensitivity, selectivity, and intensity-concentration dependency must be addressed. The potential application of nano-based and shape-dependent (nanowires and NPs) SERS substrates in diagnostics will be discussed in the next section.

3. Application

3.1 Application of Raman spectroscopy in *In-vitro* diagnostics

In-vitro diagnostics (IVDs) are tests done on the external of the human body, on biological samples (blood or tissues) that are extracted from the human body or expressed to mimic the human body samples. IVDs serve healthcare through detection, cure, treatment and prevention of diseases. The test tube IVDs are crucial in healthcare and proffers the following merits: non-invasiveness; possible prevention of patient side-effect; a rapid diagnosis that facilitates earlier treatment even in remote areas [57]. The unique IVDs are characterized into 3 groups based on their applications: 1. Clinical laboratory IVDs – for samples that require a clinical laboratory with advanced instrumentations and specialized/trained personnel. A good example is the polymerase chain reaction (PCR) test. 2. Near-patient IVDs – advanced instrumentation is not required for these samples but a trained person is still needed to administer e.g. physician or a nurse. 3. In-home IVDs/Point of Care Testing (POCT) - test is simple, easy to use, does not require trained personnel or instrumentations. Examples of POCT devices include popular pregnancy test kits and glucose meters [57]. The last decade has propelled and translated the clinical and near-patient IVDs into the in-home (POCTs) IVDs to control and manage diseases. This translation is advantageous for early, rapid diagnosis even in remote World Health Organization (WHO) regions.

The fabrication or architecture of the IVDs devices mobilizes different scientific platforms and instruments. The optical Raman spectroscopy merited by its high chemical specificity, minimal to zero sample preparation, wide visible/near-infrared spectral range is inherently used in medical diagnostics. The capability of Raman spectroscopy to detect and quantify changes in cells, tissues and biofluids is an impetus for its application in IVDs [58]. The conventional Raman spectroscopy is however insufficient due to low inelastic Raman photons. SERS mitigates the low inelastic Raman photons. SERS uses roughened Plasmonic noble metal surfaces to amplify or enhance the Raman signal [57, 59]. Application of SERS in IVDs is driven by its high sensitivity, fingerprinting, large dynamic range, and multiplexing competencies [57, 60], molecular specificity, non-invasiveness, potential to resolve the composition of complex molecular bio-analytes [61].

SERS IVDs immunoassays started with the classical ELISA mimicking sandwich assay on a solid support or free floating and recently advanced to automated systems such as lateral flows ((LFAs) and lab on chip (microfluidics) [60]. The choice of the SERS substrate discussed in the materials section is vast and offers sufficient choices for the detection of bio-analytes [61] which in turn serves in clinical practices for the prognosis and diagnosis of diseases [5, 61].

SERS IVDs	Bioanalyte/disease	SERS substrate	Reference
Label-free SERS IVDs	Cancer (blood plasma protein	Ag NPs	Lin *et al.* [62]
	Quantification of hepatitis B DNA	Ag NPs	Batool *et al.* [63]
	HIV-1	Ag nanorods	Yada *et al.* [64]
	Nasopharyngeal cancer DNA	Negatively charged Ag NPs	Lin *et al.* [65]
	Breast cancer tissue	Ag Nps	Shen *et al.* [66]
	Sjogren's syndrome from saliva	Cl-Ag NPs	Moisoiu *et al.* [67]
	Track bi'ological injuries caused by radiation from serum and urine samples of mice	Au NPs spurted on the nano tip of a canonical anodic aluminum oxide template	Muhammad *et al.* [68]
	Huma tear uric acid	Polydimethylsiloxane film layered with polystyrene, SiO_2 and Au	Narasimhan *et al.* [69]
	Ovarian and endometrial cancer extracellular vesicles	biosilica/Ag NPs composite substrates	Rojalin *et al.* [70]
	creatinine	Nano-Au on Ag film nanostructures	Su *et al.* [71]
Labeled-SERS IVDs	DNA and microRNA	Au core/stellate shell	Wang *et al.* [72]
	Plasmodium falciparum DNA	Magnetic beads	Ngo *et al.* [73]
	mouse IgG	Au NPs	Frimpong *et al.* [74]
	SARS-coV-2 IgM/IgG	SiO_2@Ag	Liu *et al.* [75]
	Myocardial infarction biomarker	Ag@Au NPs	Zhang *et al.* [76]
	HIV-1 DNA	Au NPs	Fu *et al.* [77]
	Single prostate cancer cells	Au NPs	Willner *et al.* [78]
	Estrogen receptor alpha	Au NPs	Kapara *et al.* [79]

Table 1.
SERS IVDs.

SERS immunoassays inherit a labeled/indirect or label-free configuration. In Label-free SERS, the Raman reading comes from the fingerprint of the bio-analyte while with labeled SERS, the characteristic spectra is that of a Raman tag [61]. The label–free is simple as opposed to the labeled system which incorporates the tag on the metallic nanostructures. Both systems have been used in the detection of proteins, nucleotides, and fatty acid/lipids. Changes or alterations in these bio-samples inform the diagnosis of communicable and non-communicable diseases [61]. **Table 1** [62–79], list a few examples of bio-analytes or diseases detected using SERS IVDs.

3.1.1 Traditional sandwich SERS IVDs

The traditional sandwich SERS immunoassay is characterized by a SERS substrate and a SERS immunoprobe and inherits the ELISA principle. It is crafted with SERS surfaces (Plasmonic metals structures), which could be supported on a platform i.e., solid support (metal, non-metallic) or the new flexible supports, or free-floating in colloidal form. Several research papers have reported on the solid substrates [80–85]. We have fabricated immune SERS substrates on solid supports (glass and silicon wafer) [86] and a schematic of the preparation of the SERS substrates is shown in **Figure 3**. Noble metal NPs (Au/Ag) are self-assembled onto solid support using wet chemistry methods [86]. These SERS substrates are applied in the detection and quantification of malaria *plasmodium falciparum* antigens. A capture and detection *p. falciparum* antibodies sandwiches the *pf* antigen and it could be confirmed indirectly via a 4-mercaptobenzoic acid SERS tag [87]. **Figure 4** schematically details the SERS

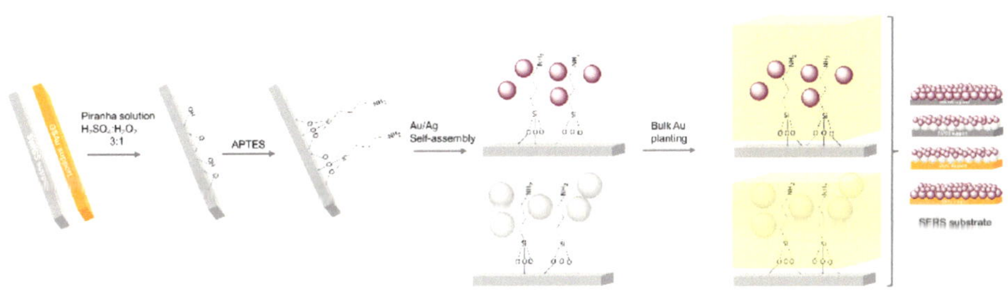

Figure 3.
Schematic representation of the preparation of SERS immune substrates on solid support. Reprinted with permission from ref. [86].

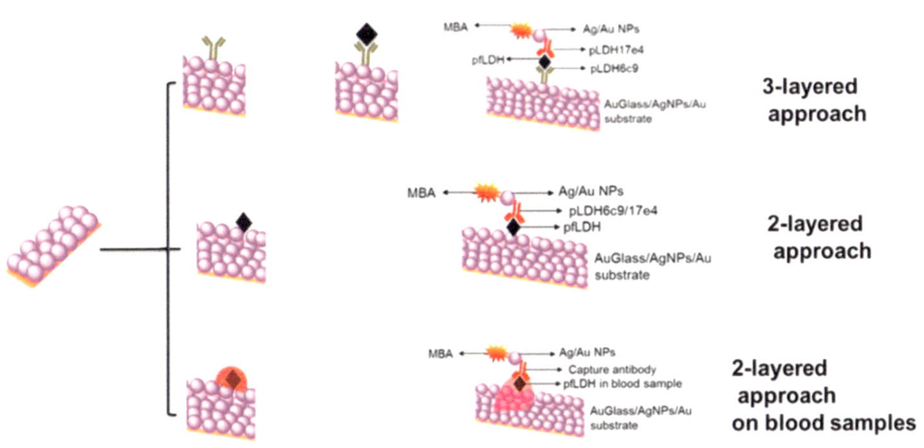

Figure 4.
SERS immunoassay schematic. Reprinted with permission from ref. [87].

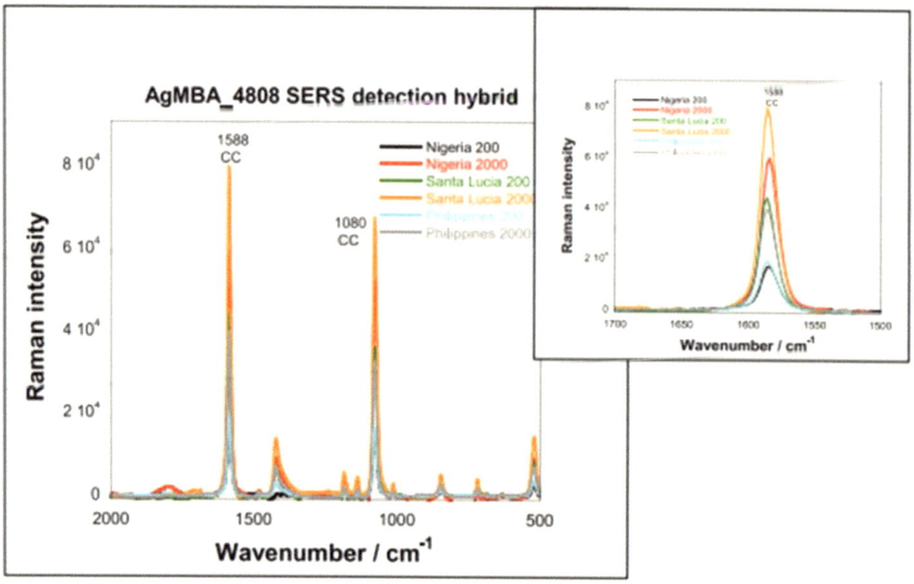

Figure 5.
Labeled-SERS spectra showing quantitative and qualitative detection of pf antigen from WHO malaria blood specimens. Reprinted with permission from ref. [87].

probe. The labeled SERS sandwich assay proves the capability of the SERS IVDs for the quantitative and qualitative detection of bio-analytes; **Figure 5** shows the SERS spectra of the SERS IVDs tested on WHO malaria specimens from different regions with known bacterial loadings.

The next section will highlight a few examples of the upcoming flexible SERS substrate. The flexible SERS substrates include polymers, graphene, graphene oxide and nanowires [88]. The lack of biodegradability and non-uniformity presented by the conventional solid SERS substrates and lithography approaches, respectively, is an impetus for the flexible substrates [89].

Fan *et al.* [90] reports on a reproducible label-free SERS hybrid of graphene oxide conjugated to popcorn shaped Au NPs. The ultrasensitive graphene oxide-Au SERS probe is used for testing of HIV DNA and MRSA bacteria and has a detection limit of 10 DFU/mL for the latter. Korkmaz *et al.* [91] demonstrated an affordable simple development of a porous biosilica Plasmonic composite flexible SERS substrates. The technique involved deposition of the materials on a regular office-grade adhesive tape. The characterization of the novel material reveals unique properties: pore size, Plasmon resonance, Raman enhancement suitable for biosensing. The usability of the platform was tested on bioanalytes; proteins and bacteria. However, due to the porous nature of the nanocomposites, smaller proteins and nanostructures disperse within the substrate and afford a reduced particle density for optical detection. Hence, low analyte concentrations cannot be detected with the exclusion of particles larger than 100 nm. The platform is useful in the detection of bacteria and other bio-analytes of nano- to micro-meter sizes.

He *et al.* [92] explored the use of graphene in SERS immunoassays. A SERS-active substrate of Au NPs grown on graphene using chemical vapor deposition is developed and used for multiplexing detection of DNA. The presence of both the Au metallic substrate and graphene leads to an enhanced signal. The Au NPs enables assemble of DNA capture probes which is normally problematic on the graphene. The SERS Au-graphene SERS platform showed extraordinarily high sensitivity and specificity for DNA detection with a detection limit of 10 pM. The multiplexing capability is tested with the simultaneous detection of two DNA targets.

3.1.2 Automated SERS IVDs

Automated SERS IVDs include LFAs and microfluidics synergized with the SERS phenomenon. LFAs are cemented for POCT due to their simplicity, user-friendliness, long-term stability, rapid detection which makes them a suitable choice for remote WHO regions. Incorporation of the SERS component into the LFAs results in SERS/LFAs IVDs. The SERS-LFAs principle is still the same as that of the classical LFAs with an addition of a Raman tag. The SERS tag is conjugated to the Plasmonic metal nanostructures, which are mostly Au NPs [93].

The SERS-LFAs improves sensitivity and enable both qualitative and quantitative detection of bio-analytes, an improvement of the traditional LFAs [93].

Ma et al. [94] capitalizing on the flexibility, portability and simplicity of polydimethylsiloxane (PDMS) used it as a matrix/support of SERS Plasmonic metallic substrates for SERS-LFAs. A novel PDMS-based SERS-LFAs for ultra-sensitive and quantitative detection of ferritin (FER), a liver cancer biomarker, is developed. The hydrophilic-hydrophobic Ag/PDMS strip is prepared and anti-FER is mobilized on the Ag strip (test line, control line 1 and control line 2). The FER bio-analyte flowed smoothly into the hydrophobic PDMS substrate coupled with SERS immonoprobes of raspberry-shaped Au NPs conjugated to 4-MBA SERS tag and anti-FER results in the detection of FER with a detection limit of 0.41 pg./ml.

Microfluidics pertains to the science and technology of handling fluids and the micro/nanofluidics devices are an ensemble of miniaturized components such as pumps, channels, valves, mixers and separators to facilitate the movement of the fluids [95].

The application of microfluidics in diagnostics proffers the following advantages: high-to-volume ratio, precise fluid control, low sample consumption and high integration with functional components [96].

For the detection of bio-analytes, microfluidics is coupled with optical, electron chemical, or electrical techniques [96]. Raman spectroscopy is one such optical technique and the synergy results in SERS-microfluidics devices [96]. The SERS/microfluidics are used for the analysis of various biospecies.

Pallaoro *et al.* [97] proved the capability of the SERS/microfluidics platform in the detection of cancer from flowing cells. A mixture of cells, cancerous (prostrate) and non-cancerous from bodily fluids are incubated with label-mediated SERS probes. The SERS probes are based on Ag NPs dimer core labeled with SERS tag and paired with an affinity-biomolecule. They are circulated continuously in the microfluidics channel and exposed to Raman laser which differentiates the cells based on their characteristic fingerprints.

3.2 Application of Raman spectroscopy in *In-vivo* diagnostics

Raman spectroscopy has gained much popularity over the years in applications that determine the biochemical composition of cells and tissues. This technique has advanced significantly and has found widespread use outside the laboratory in applications such as materials analysis, process control and environmental monitoring amongst others. Recently, Raman spectroscopy has found use in clinical applications such as in vivo diagnostics and monitoring, through coupling with other diagnostic systems. Initially, most applications of Raman spectroscopy that relate to biochemical analysis are mostly based on *ex-vivo* or

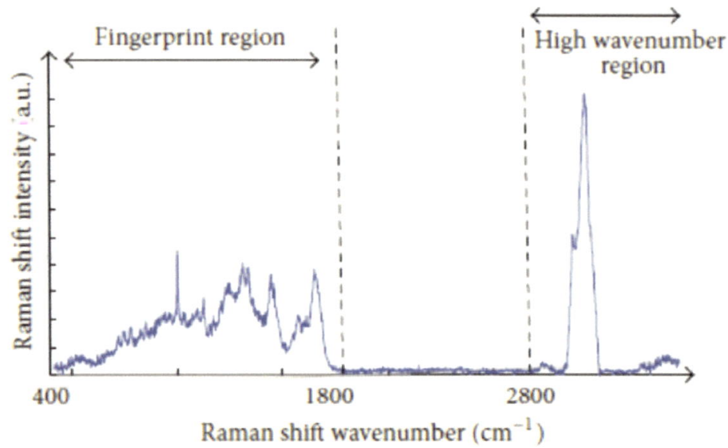

Figure 6.
Raman spectra showing the fingerprint region which depicts the area of the spectra where biological molecules and other associated molecules appear [104].

in vitro assays, but until recently, there has been a great migration from ex-vivo to in-vivo applications [98, 99].

Raman scattering occurs as a result of a change in the polarization of molecules due to light. This technique provides quantitative assessment of the biochemical composition of biological tissues. When a spectra is obtained during assessment, the peaks are highly specific to the molecular chemistry of the specimen under investigation, with intensities directly proportional to the molecular content. This allows for the investigation of concentrations and ratios of constituents within the specimen under investigation [100, 101].

Thus, Raman spectroscopy provides chemical fingerprints of biological materials such as cells, tissues or biological fluids through inelastic scattering of light by vibrating molecules [58, 102, 103]. This fingerprint represents molecular vibrations brought about by chemical bonds, thus deciphering the samples chemical or biochemical composition. The spectra is collected in the fingerprint region from 400 and 1800 cm^{-1}, as depicted in **Figure 6**. Collection of the Raman spectra does not disrupt the cellular environment. This is considered one of the advantages that have made Raman spectroscopy superior amongst other optical diagnostic tools [100, 104].

Some of the properties that have made Raman spectroscopy suitable for use in in-vivo applications include its excellent chemical specificity which result in the formation of a fingerprint like spectrum without interference from water, minimal or lack of sample preparation and its ability to employ advanced optical technologies in both the visible and near infrared spectral ranges [58, 98, 105, 106]. Water is known to be a weak Raman scatterer and shows no interference with the spectra of solutes in aqueous solution. Water absorbs in the region between 2000 and 4000 cm^{-1} in the Raman spectra, which is in the infrared (IR) and near infrared (NIR) regions, and thus falls beyond the fingerprint region in which the molecules of interest absorb [106, 107].

3.2.1 Instrumentation requirements for In vivo *applications*

Although Raman spectroscopy has been found to yield exceptional results in in-vivo applications, certain adjustments and additional components are required for optimum performance. According to Ramirez-Elias and Gonzalez [108], in vivo measurements require an integrated system capable of providing spectral

DOI: http://dx.doi.org/10.5772/intechopen.99771

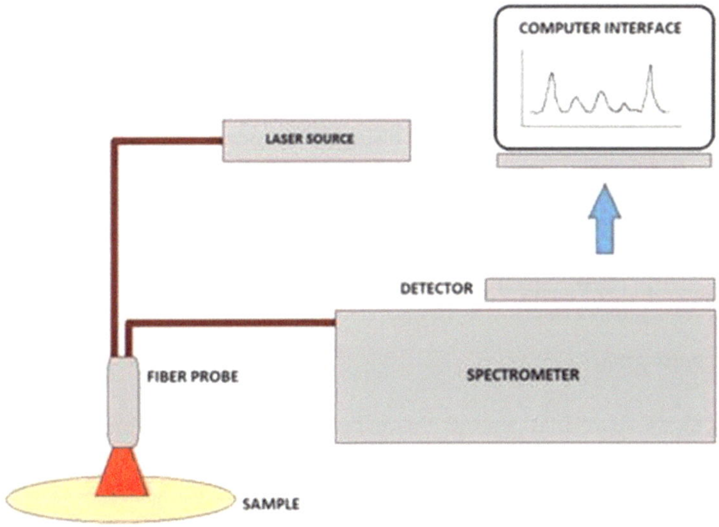

Figure 7.
A typical Raman spectroscopy setup showing the components required to perform an in vivo analysis [109].

acquisition and analysis in real time. Such a system includes a light source (lasers), light delivery system, Raman probe, signal delivery, and a signal detection system, as shown in **Figure 7** [110]. For in vivo applications, it has been reported that a spectra of sufficient signal-to-noise ratio (SNR) should be obtained within a few seconds on the signal collection time, or even faster. Thus, the instrumentation for in vivo applications should have an improved signal detection system and also be able to minimize noise contributions [111].

Raman spectroscopy has gained much popularity over the years in applications that determine the biochemical composition of cells and tissues. This technique has advanced significantly and has found widespread use outside the laboratory in applications such as materials analysis, process control and environmental monitoring amongst others. Recently, Raman spectroscopy has found use in clinical applications such as in vivo diagnostics and monitoring, through coupling with other diagnostic systems. Initially, most applications of Raman spectroscopy that relate to biochemical analysis were mostly based on ex-vivo or in vitro assays, but until recently, there has been a great migration from ex-vivo to in-vivo applications [98, 99].

3.2.2 Excitation or light source

A light source is used to deliver power or energy to the sample to generate Raman scattered photons. Raman scattering is known to be a relatively weak phenomenon and thus it is vital to ensure that sufficient power is directed towards the test sample. Laser is the most commonly used light source in the Raman spectroscopy system due to their narrow bandwidth and high power output [110]. Raman scattering is dependent on a variety of factors including signal to noise ratio, maximum permissible exposure and an increase in temperature [112]. Various lasers have been used in optical systems and only lasers offering the following characteristics are desired for use in in vivo applications; the first characteristic to consider is the line width of the emission. According to Sato et al., the line width of the emission must be narrow [99]. This requirement ensures that laser line broadening does not propagate into Raman bands through convolution, which results in irrevocable adjacent Raman peaks [113, 114].

Secondly, the wavelength of the laser light and its intensity should be stable or carefully selected. For in vivo applications, excitation with ultraviolet or visible light is not recommended. When UV light is used for excitation, there is a risk of photochemical damage to the tissue being investigated. Also, when visible is light is used for excitation, there is a strong auto-fluorescence that is generated in the biological sample under investigation. Thus, near infrared (NIR) excitation sources are commonly used. This has been chosen since only a few biological fluorophores are known to have peak emission in this region of the spectrum thereby reducing background autofluorescence and absorbance and further simplifying signal processing needed for extracting the Raman bands [107, 112].

3.2.3 Fiber optic probes

During tissue examination, excitation or illumination light needs to be delivered to the tissue and Raman signals emitted from the tissue should be transmitted back to the detectors. This is done through the use of optical fibers which enhance signal collection and lessen interfering signals. Interfering signals include signal-to-noise ratio and also includes both the Raman signal and photoluminescence generated within the optical fibers [115]. The design of the probe is mostly dependent upon the envisaged use. The following parameters are considered: Raman configuration, location of organ under investigation, microanatomy of the tissue, and the pathophysiology of the disease [112]. A probe suitable for in vivo applications should have the following characteristics; The probe should be inexpensive and sterilizable if its reusable, output of the laser delivery fiber should be filtered to prevent the Raman signal induced from the fiber reaching the sample, high signal collection efficiency, light emitted from the sample should be filtered to avoid Raman signal contributions from the fiber material [98, 111, 116].

3.2.4 Signal detection or spectrograph

A Raman detection system that is used for clinical applications consist of an imaging spectrograph linked with a charge-couple device (CCD). For in vivo Raman applications, a spectral acquisition of no more than a few seconds is required, and can be achieved through the use of a fast spectrograph and a highly sensitive detector. A suitable detection system for in vivo applications requires an appropriate imaging spectrograph that is linked to the sample interface, such as optic fibers, on one end and the CCD on the other end. Other components that form part of the detection system include rejection filters that are responsible for eliminating any laser light or elastically scattered light from the signal [112].

3.2.5 Raman spectra in In vivo *diagnosis and monitoring*

This section captures some of the biomedical applications that have employed Raman spectroscopy as an *in vivo* diagnostic tool, in conjunction with other techniques.

3.2.5.1 Breast cancer diagnosis and monitoring

Saha *et al*. demonstrated the diagnosis of early stage breast cancer by the real time detection of microcalcifications during stereotactic breast core needle biopsies. They performed their study on 159 tissue sites in 33 patients in order to detect microcalcifications in breast tissue biopsies. The authors used ordinary least squares fitting to approximate spectra that had been acquired with a breast model that

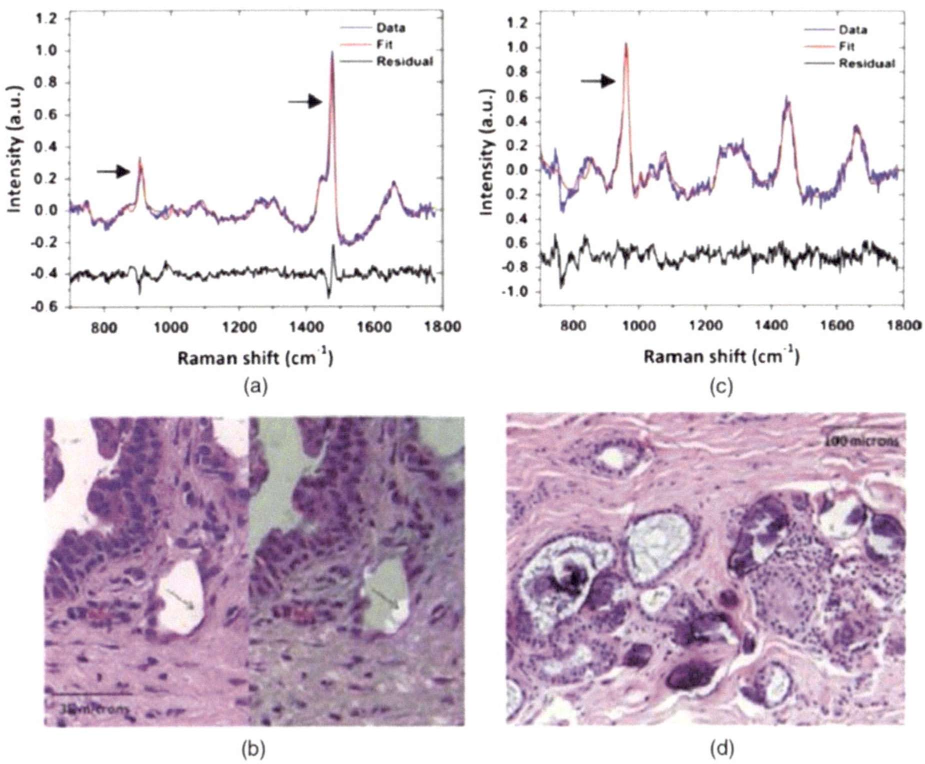

Figure 8.
Raman spectra and histopathology of breast lesion with types I and types II microcalcifications [117]. (a) and (b) depicts the Raman spectrum of type I microcalcifications characterized by calcium oxalate peaks and the spot on the biopsy where the fingerprint was collected. (c) and (d) characteristic peaks of type II microcalcifications calcium hydroxyapatite and data collection location on the biopsy, respectively.

had been developed previously. They demonstrated the possibility of distinguishing between various microcalcifications based on the appearance of vibrational bands that represent calcium oxalate in the fingerprint region at positions 912 and 1477 cm-1, as depicted in **Figure 8(a)**. The location where the spectrum was collected in the biopsy is shown in **Figure 8(b)**. **Figure 8(c)** shows the appearance of the peak or band at 960 cm-1 which represents the presence of calcium hydroxy-apatite (microcalcification type II), with **Figure 8(d)** showing the exact position on the biopsy [98, 117].

3.2.5.2 Diagnosis of skin cancer

The use of Raman spectroscopy in the in vivo diagnosis of skin cancers has been sought after due to the easy optical access to skin. Lieber and colleagues demonstrated the possibility of using Raman in the diagnosis of skin cancers through the use of a fiber optic Raman probe which recorder a sensitivity of 100% and a specificity of 91% [118]. Recently, Raman spectroscopy is coupled with autofluorescence (AF) imaging and used to diagnose basal cell carcinoma of the skin using multimodal spectral imaging, as shown in **Figure 9**. AF is used to investigate the main spatial features of skin resections for use in selecting and prioritizing sampling points in Raman spectroscopy. Skin resections are collected during Mohs surgery, and proper investigation of resection margins is obtained using 500–1500 Raman spectra and without sectioning, staining or any other form of sample or tissue preparation step [58].

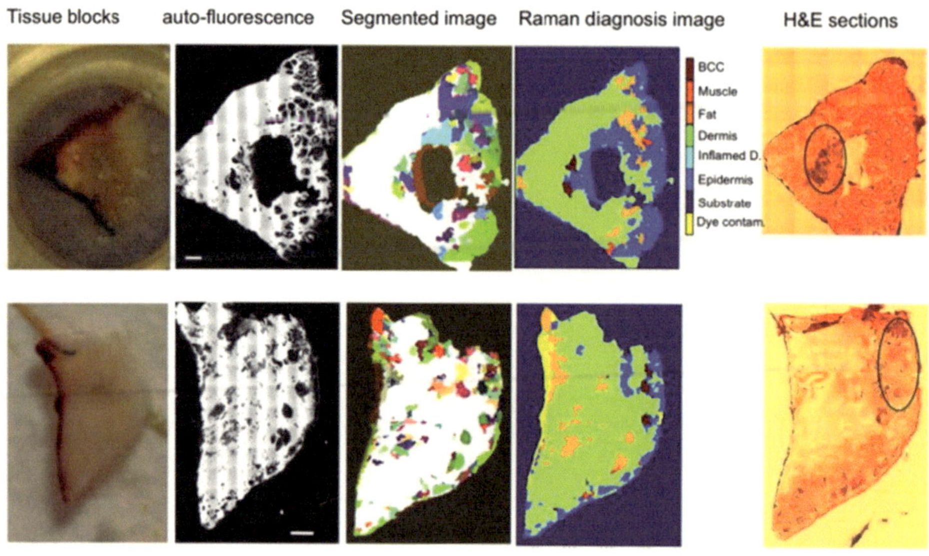

Figure 9.
Use of Raman Spectroscopy coupled with auto-fluorescence imaging in the in vivo detection of BCC in skin resections obtained during Mohs surgery [58].

4. Limitations of SERS in biomedical diagnostics

An effective SERS application for biomedical diagnostic requires a Plasmonic substrate. The metal substrates are the most superior as they produce relatively higher EFs. Plasmonic substrate play dual role in SERS technique, firstly, it interacts directly or indirectly with the targeted molecules, secondly, it amplifies the Raman signal [119]. One of the limitations often reported for SERS technique is irreproducibility of its results mostly associated with the substrate being irreproducible, quasi-ununiformed and unstable [120]. In recent decades with nanotechnology taking center stage in various areas of science, the SERS surface have shifted from bulk solid metals to colloidal nanoparticles [120]. The nanoparticles offer higher surface area and tailored optical properties over bulk solid metals leading to higher reaction "hotspots" for better enhancement of SERS signal [121].

However, the use of colloidal metal NPs does not come without its own disadvantages. Metal NPs synthesized without any stabilizing agent often have unacceptably wide particle size distribution and a mixture of undesired morphologies. Hence compromising the reproducibility of SERS application [122]. Contrary, stabilized NPs offer controlled particle size and uniformity. However, it can block the targeted molecule from adsorbing to the metal surface through steric hindrance leading to suppressed SERS signal [123].

In general the success of SERS activity is dependent on the interaction between the adsorbed molecules and the surface of Plasmonic nanostructures. This means that if the two are mismatched activity is compromised leading to poor or no enhancement factor. It is a requirement that the analyte must adsorb on the surface effectively. Thus, it should have a higher SERS cross-section than any possible interference of contaminants. It must be noted that the influence of the metal substrate on both the physical and chemical properties of the metals and the stability including the reversibility or reusability of the same material remains a major drawback in respect to applicability [122]. On the other hand many methods used to produce nanostructured materials fail to produce materials with narrow size distribution leading to poor reproducibility of the SERS substrate.

Even though stabilized metal nanoparticles offers a better Plasmonic surface to bulk metals but it compromises the interaction with the targeted analyte. Therefore, to further advance this technology commercially, the retention of Plasmonic properties in the presence of stabilizing agents of colloidal nanoparticles will be key [124].

5. Future prospects of surface enhances Raman Spectroscopy in biomedical diagnostics

SERS interest has catapulted in the research community in the past couple of decades owing to its capability of quantitative analysis of very low concentrations and fingerprints of bio-species [125–130]. Over the past 4 decades SERS technique has blossomed into various areas of research and technology due to reasons discussed in details in the previous sections. However, further efforts are still required to further advance this technique commercially [131, 132]. In this section the highlights of future prospects of SERS real world applications in selected areas of research, application and technology will be discussed. General issues hindering this technique would also be briefly highlighted in order to understand its true potential.

The new and novel areas of application of SERS in the near future include ultraviolet-SERS, tip enhanced-SERS and biological sensing [132–136].

5.1 Ultraviolet surface enhanced Raman Spectroscopy (UV-SERS)

SERS applications are largely based on the enhancement of Raman scattering molecules that are either physically or chemically adsorbed onto a Plasmonic surface. The most effective surfaces for Raman reporters has been metallic surfaces such as Ag, Au and Cu because they produce superior enhancement factors [137–139]. However, they are limited to the visible and near infrared region of the spectrum, which limits the use of SERS in the same region. The use of SERS technique in the ultraviolet region is highly desirable and remains an unexplored terrain [132, 140–143]. UV SERS would enable resonance detection of molecules in the UV wavelength range such as protein residues, DNA bases etc. Analysis of these molecules are currently analyzed with other techniques which are often not as sensitive as SERS technique. The challenge of achieving UV SERS is finding a Plasmonic material that support surface enhancement in the UV region. Research has been on going in this area, however, there has been no success in finding a material that absorbs in the UV region that outperforms or close to the enhancement factors offered by Au, Ag and Cu metals [133, 142–144]. The potential of UV SERS outweighs the current challenges and with the rapid advancement in material nanotechnology, There is hope that the ideal material will be found in the near future.

5.2 Tip enhanced Raman Spectroscopy (TERS)

Merging of SERS technique with one of the microscopic techniques such as atomic force microscopy or scanning tunneling microscopy has led to the discovery of TERS, which is an advanced and powerful imaging tool. TERS is one of a very promising spectroscopic method that can analyze a variety of samples with sub diffraction limit imaging capabilities [145–147]. In TERS technique, the electromagnetic field enhancement is located at the sharp metallic/semiconducting cantilever tip that is eradicated with a laser beam. A localized region of SERS enhancement is created when the cantilever tip is brought in close proximity to the sample of interest; this enables structural and compositional analysis of the sample [145]. Spatial resolution achieved through this hybrid technique is far superior to any other microscopic techniques commercially available. This method has been used for over a decade, initially for analysis of strong Raman scatterers such as buckyballs and dye molecules. However, the TERS technique has rapidly grown into characterization of more sophisticated and diverse samples such as single stranded RNA, individual single walled carbon nanotubes, hydrogen bonding in DNA [147], single particle dye sensitized solar cells, amongst others [146]. Few issues are still holding this technique from being routinely used commercially such as the difficulty of calculating the enhancing region for single molecules, hence the resolution and the enhancement factor. In the near future this issue would be solved since analysis of individual biomolecules is crucial for understanding and managing deadly diseases.

5.3 Biological diagnostic and sensing

SERS application has catapulted biosensing into new heights in the past decade. However, the prospects of becoming a commercial clinical diagnostic technology has stagnated. This is due to its low tissue penetration depth achievable [148, 149]. However, there are clinical areas where SERS have shown superiority and advantages over other techniques such as fluorescence imaging, MRI etc. The future of imaging single cells is likely to be via SERS based technology because of better brightness, high sensitivity and lack of photo bleaching [149]. Furthermore, SERS offers far better resolution in the micrometer range compared to commercially used MRI which offers resolution in a range of a few millimeters. The risk of using the low resolution MRI, for example, in detection of cancer cells would require 100 000 cancers cells in a tumor to be detectable via MRI technique [150]. This might render this detection too late for some aggressive tumors. Advantages offered by SERS in biosensing are likely to be leveraged in the future to make this technique a diagnostic technique with true clinical reach.

SERS is one of the most sensitive analytical methods which offers detection of molecules in low concentrations (ppb) and provides rich structural information. In this section some of the SERS applications that are likely to be commercially used outside of the laboratory while also demonstrating the versatility of this technique have been discussed. We envisage great extension of SERS technique and an advent use of UV SERS and TERS in biological diagnostic, sensing and imaging.

6. Conclusions

In this chapter recent advances relating to the use of SERS in biomedical applications were reviewed and reported. The point of departure was to concisely introduce the fundamentals of Plasmonics and SERS in respect to the enhancement of the Raman signal of molecules that are closest to the metallic nanostructures. For this reason, various metallic nanostructures that could be used to achieve the highest possible SERS enhancement factors were discussed. Discussion on the SERS efficiency of Plasmonic substrates made of gold nanowires prepared using a chemical route (bottom-up) was made. This was followed by the influence of the geometrical parameters (diameter and length of the wires or aspect ratio) on the SERS signal as a way to optimize it.

Emphasis was placed on the development of SERS as potentially dynamic technique for point-of-care diagnostics taking from its high sensitivity and multiplexing capabilities. The approach was to compare different methodologies that can be translated across various types of biomarkers. Techniques such as sandwich SERS and automated SERS for in vitro diagnostics have been elucidated. Moreover, progress made in respect to strong capture ligands such as DNA that are also specific has been unpacked and holds promise for a much needed solution. These techniques have been used in both the detection and quantification of various diseases including malaria, cancer, hepatitis B, etc. using different bioanalytes such as antigens, protein, DNA, respectively. On the other hand developments on diseases diagnosis based on label-free sensing techniques that are sensitive to low analyte concentrations found in the physiological environment were elucidated as they offer hope for therapeutic intervention. It is relatively acceptable that most disease states usually start with small changes in cellular processes that ultimately augmented as the disease progresses without medical intervention. Although SERS sensitivity can be excellent in vivo measurements require an integrated system that is capable of providing spectral acquisition and analysis in real time.

Although SERS is undoubtedly a promising technique for diagnostic purposes and for uses as POC devices the success rests on overcoming the barriers through further advancements in the fabrication of SERS substrate, assays, platforms, and making them cost-effective.

The discussion has shown that SERS allows for the continuous and highly sensitive detection and quantification of various biomarkers and end-products of disease states, making it an excellent option in the diagnosis and treatment of various diseases that are cause health concerns. However, with advances in the development of different types of novel techniques such as UV-SERS and TERS, SERS will remain an indispensable technique showing a great promise for in vitro and in vivo disease detection.

Acknowledgements

The authors wish to acknowledge the financial support of Mintek, the Department of Science and Innovation (DSI) of South Africa, and the National Intellectual Property Management Office (NIPMO) of South Africa.

Author details

Nikiwe Mhlanga[1,2]*, Phumlani Tetyana[1,2], Sanele Nyembe[1,2] and Lucky Sikhwivhilu[1,2]

1 DSI/Mintek Nanotechnology Innovation Centre, Randburg, South Africa

2 Advanced Materials Division, Mintek, Randburg, South Africa

*Address all correspondence to: nikiwem@mintek.co.za

References

[1] Henry AI, Sharma B, Cardinal MF, Kurouski D, Van Duyne RP. Surface enhanced Raman spectroscopy biosensing: in vivo diagnostics and multimodal imaging. Analytical chemistry. 2016; 88(13):6638-6647.

[2] Chalapathi D, Padmanabhan S, Manjithaya R, Narayana C. Surface-enhanced Raman spectroscopy as a tool for distinguishing extracellular vesicles under autophagic conditions: a marker for disease diagnostics. The Journal of Physical Chemistry B. 2020; 124(48):10952-10960.

[3] Ember KJ, Hoeve MA, McAughtrie SL, Bergholt MS, Dwyer BJ, Stevens MM, Faulds K, Forbes SJ, Campbell CJ. Raman spectroscopy and regenerative medicine: a review. NPJ Regenerative medicine. 2017; 2(1):1-10.

[4] He Z, Qiu W, Kizer ME, Wang J, Chen W, Sokolov AV, Wang X, Hu J, Scully MO. Resolving the Sequence of RNA Strands by Tip-Enhanced Raman Spectroscopy. ACS Photonics. 2020; 8(2):424-430.

[5] Chen C, Zhou X, Ding T, Zhang J, Wang S, Xu J, Chen J, Dai J, Chen C. Preparation and characterization of $ZnO/SiO_2/Ag$ nanoparticles as highly sensitive substrates for surface-enhanced Raman scattering. Mater. Lett. 2016;165: 55-58.

[6] Ding SY, Wu DY, Yang ZL, Ren B, Xu X, Tian ZQ. Some progresses in mechanistic studies on surface-enhanced Raman scattering, Chem. J. Chin. Univ. Chin. 2015; 29:2569-2581.

[7] Fu Q, Zhan Z, Dou J, Zheng X, Xu X, Wu M, Lei Y. Highly reproducible and sensitive SERS substrates with Ag inter-nanoparticle gaps of 5 nm fabricated by ultrathin aluminum mask technique. ACS Appl. Mat. Interfaces. 2015; 7:13322-13328.

[8] Chao Z, Song L, Zhou Y, Nie W, Chen P. Impact of PS/SiO_2 morphologies on the SERS activity of $PS/SiO_2/Ag$ nanocomposite particles. Colloid Polym. Sci. 2014; 292:2841-2848.

[9] McLellan JM, Li Z, Siekkinen AR, Xia Y. The SERS activity of a supported Ag nanocube strongly depends on its orientation relative to laser polarization. Nano Lett. 2007; 7:1013-1017.

[10] Caridad JM, Winters S, McCloskey D, Duesberg GS, Donegan JF, Krstic V. Hot-volumes as uniform and reproducible SERS-detection enhancers in weakly-coupled metallic nanohelices. Sci. Rep. UK. 2017; 7:1-7.

[11] Araujo A, Caro C, Mendes MJ, Nunes D, Fortunato DE, Franco R, Aguas H, Martins R. Highly efficient nanoplasmonic SERS on cardboard packaging substrates. Nanotechnol. 2014; 25:415202.

[12] Mikac L, Ivanda M, Gotic M, Mihelj T, Horvat L. Synthesis and characterization of silver colloidal nanoparticles with different coatings for SERS application. J. Nanopart. Res. 2014; 16:2748.

[13] Lu X, Rycenga M, Skrabalak SE, Wiley B, Xia Y. Chemical synthesis of novel plasmonic nanoparticles. Annu Rev Phys Chem 2009; 60:167-192.

[14] Sharma B, Frontiera RR, Henry A-I, Ringe E, Duyne RPV. SERS: materials, applications, and the future. Materials Today. 2012; 15:16-25.

[15] Ding T, Sigle DO, Herrmann LO, Wolverson D, Baumberg JJ. Nanoimprint lithography of Al nanovoids for deep-UV SERS. ACS Appl Mater Interfaces. 2014; 6: 17358-17363.

[16] Jain PK, Lee KS, El-Sayed IH, El-Sayed MA. Calculated absorption

and scattering properties of gold nanoparticles of different size, shape, and composition: applications in biological imaging and biomedicine. J Phys Chem B. 2006; 110:7238-7248.

[17] Kelly KL, Coronado E, Zhao LL, Schatz GC. The optical properties of metal nanoparticles: the influence of size, shape, and dielectric environment. J Phy. Chem. B. 2003; 107:668-677.

[18] Hutter E, Fendler JH. Exploitation of localized surface plasmon resonance. Adv Mater 2004; 16:1685-1706.

[19] Ly N, Foley K, Tao N. Integrated label-free protein detection and separation in real time using confined surface plasmon resonance imaging. Anal Chem 2007; 79: 2546-2551.

[20] Alvarez-Puebla RA, Arceo E, Goulet PJ, Garrido JJ, Aroca RF. Role of nanoparticle surface charge in surface-enhanced Raman scattering. J Phys Chem B 2005; 109:3787-3792.

[21] Faulds K, Littleford RE, Graham D, Dent G, Smith WE. Comparison of surface-enhanced resonance Raman scattering from unaggregated and aggregated nanoparticles. Anal Chem 2004; 76:592-598

[22] Mehmet Kahraman, Emma R. Mullen, Aysun Korkmaz, Sebastian Wachsmann-Hogiu. Fundamentals and applications of SERS-based bioanalytical sensing. Nanophotonics 2017; 6(5):831-852

[23] Kahraman M, Tokman N, Culha M. Silver nanoparticle thin films with nanocavities for surface-enhanced Raman scattering. Chem Phys Chem 2008; 9:902-910.

[24] Kahraman M, Tokman N, Türkoğlu G. Surface-enhanced Raman scattering on aggregates of silver nanoparticles with definite size. J Phys Chem C 2008; 112:10338-10343.

[25] Kahraman M, Aydin Ö, Culha M. Size effect of 3D aggregates assembled from silver nanoparticles on surface-enhanced Raman scattering. Chem Phys Chem 2009; 10: 537-542

[26] Lee SJ, Morrill AR, Moskovits M. Hot spots in silver nanowire bundles for surface-enhanced Raman spectroscopy. Journal of the American Chemical Society. 2006 Feb 22;128(7):2200-2210.

[27] Feng H, Yang Y, You Y, Li G, Guo J, Yu T, Shen Z, Wu T, Xing B. Simple and rapid synthesis of ultrathin gold nanowires, their self-assembly and application in surface-enhanced Raman scattering. Chemical Communications. 2009(15):1984-1986.

[28] Yilmaz A.; Yilmaz M. Bimetallic Core–Shell Nanoparticles of Gold and Silver via Bioinspired Polydopamine Layer as Surface-Enhanced Raman Spectroscopy (SERS) Platform. Nanomaterials. 2020; 10:688.

[29] Félidj N, Aubard, J, Lévi G, Krenn, JR., Hohenau A, Schider G, Leitner A, Aussenegg FR. Optimized surface-enhanced Raman scattering on gold nanoparticle arrays. Appl. Phys. Lett. 2003; 82:3095-3097.

[30] Haynes C.L, Van Duyne R.P. Plasmon-Sampled Surface-Enhanced Raman Excitation Spectroscopy. J. Phys. Chem. B 2003; 107:7426-7433.

[31] McFarland AD, Young MA, Dieringer JA, Van Duyne RP. Wavelength-Scanned Surface-Enhanced Raman Excitation Spectroscopy. J. Phys. Chem. B 2005; 109: 11279-11285.

[32] Laurent G, Félidj N, Aubard J, Lévi G. Krenn JR, Hohenau A, Aussenegg FR. Surface enhanced Raman scattering arising from multipolar plasmon excitation. J. Chem. Phys. 2005; 122:011102.

[33] Grand J, de la Chapelle ML, Bijeon J-L, Adam P-M, Vial A. Royer. Role of localized surface plasmons in surface-enhanced Raman scattering of shape-controlled metallic particles in regular arrays. P. Phys. Rev. B 2005; 72:033407.

[34] Billot L, Lamy de la Chapelle M, Grimault A-S, Vial A, Barchiesi D, Bijeon J-L, Adam PM, Royer P. Surface enhanced Raman scattering on gold nanowire arrays: Evidence of strong multipolar surface plasmon resonance enhancement. Chem. Phys. Lett. 2006; 422:303-307.

[35] Colas FJ, Cottat M, Gillibert R, Guillot N, Djaker N, Lidgi-Guigui N, Toury T, Barchiesi D, Toma A, Di Fabrizio E Red-Shift Effects in Surface Enhanced Raman Spectroscopy: Spectral or Intensity Dependence of the Near-Field? J. Phys. Chem. C 2016; 120:13675-13683.

[36] Guillot N, Shen H, Frémaux B, Péron O, Rinnert E, Toury T, Lamy de la Chapelle M. Surface enhanced Raman scattering optimization of gold nanocylinder arrays: Influence of the localized surface plasmon resonance and excitation wavelength. Appl. Phys. Lett. 2010; 97:023113.

[37] Gunnarsson L, Bjerneld E.J, Xu H, Petronis S, Kasemo B, Käll M. Interparticle coupling effects in nanofabricated substrates for surface enhanced Raman scattering. Appl. Phys. Lett. 2001; 78:802-804.

[38] Kessentini S, Barchiesi D, D'Andrea C, Toma A, Guillot N, Di Fabrizio E, Fzio B, Marago MO, Gucciardi GP, Lamy de la Chapelle MJ. Gold Dimer Nanoantenna with Slanted Gap for Tunable LSPR and Improved SERS. Phys. Chem. C 2014; 118: 3209-3219.

[39] D'Orlando A, Bayle M, Louarn G, Humbert B. AFM-Nano Manipulation of Plasmonic Molecules Used as "Nano-Lens" to Enhance Raman of Individual Nano-Objects. Materials 2019; 12:1372.

[40] Gillibert, R.; Sarkar, M.; Bryche, J.-F.; Yasukuni, R.; Moreau, J.; Besbes, M.; Barbillon, G.; Bartelian, B.; Canva, M.; de la Chapelle, M.L. Directional surface enhanced Raman scattering on gold nano-gratings. Nanotechnology. 2016; 27:115202.

[41] Gillibert R, Sarkar M, Moreau J, Besbes M, Canva M, de la Chapelle M.L. Near Field Enhancement Localization on Plasmonic Gratings. J. Phys. Chem. C 2016; 120: 27562-27570.

[42] Sarkar M, Bryche J-F, Moreau J, Besbes M, Barbillon G, Bartenlian B, Canva M. Generalized analytical model based on harmonic coupling for hybrid plasmonic modes: Comparison with numerical and experimental results. Opt. Express 2015; 23: 27376-27390.

[43] Sarkar M, Besbes M, Moreau J, Bryche J-F, Olivéro A, Barbillon G, Coutrot AL, Bartenlien B, Canva M. Hybrid plasmonic mode by resonant coupling of localized plasmons to propagating plasmons in a Kretschmann configuration. ACS Photonics 2015; 2:237-245.

[44] Sivashanmugan K, Liao J-D, You J-W, Wu C-L. Focused-ion-beam-fabricated Au/Ag multilayered nanorod array as SERS-active substrate for virus strain detection. Sensors and Actuators B 2013; 181:361– 367

[45] Tian ZQ, Ren B, Li JF, Yang ZL. Expanding generality of surface enhanced Raman spectroscopy with borrowing SERS activity strategy. Chemical Communications 2007; 34:3514-3534.

[46] Le F, Brandl DW, Urzhumov YA, Wang H, Kundu J, Halas NJ, Aizpurua J, Nordlander P. Metallic nanoparticle arrays: a common substrate for both

surface-enhanced Raman scattering and surface-enhanced infrared absorption. American Chemical Society Nano 2008; 2 (4):707-718.

[47] Le Ru EC, EtchegoinGrand J, Felidj N, Aubard J, Levi G, Hohenau A, Krenn JR. Surface enhanced Raman spectroscopy on nanolithography-prepared substrates. Current Applied Physics2008; 8:467-470.

[48] Das G, Mecarini F, Gentile F, De Angelis F, Kumar MHG, Candeloro P, Liberale C, Cuda G, Di Fabrizio E. Nano-patterned SERS substrate: application for protein analysis vs. temperature, Biosensor and Bioelectronics. 2009; 24:1693-1699.

[49] Chang CW, Liao JD, Lin YY, Weng C. Fabrication of nano-indented cavities on Au for the detection of chemically-adsorbed DTNB molecular probes through SERS effect. Journal of Colloid and Interface Science. 2011; 358;384-391.

[50] Bao PD, Huang TQ, Liu XM, Wu TQ. Surface-enhanced Raman spectroscopy of insect nuclear polyhedrosis virus. Journal Raman Spectroscopy. 32001; 2 (4): 227-230.

[51] Chang CW, Liao JD, Shiau AL, Yao CK. Non-labeled virus detection using inverted triangular Au nano-cavities arrayed as SERS-active substrate. Sensor and Actuators B: Chemical 2011; 156 (1):471-478.

[52] Chang CW, Liao JD, Lin YY, Weng CC. Detecting very small quantity of molecular probes in solution using nano-mechanically made Au-cavities array with SERS-active effect. Sensor and Actuators B: Chemical 2011; 153 (1):271-276.

[53] Yao CK, Liao JD, Chang CW, Lin JR. Spatially reinforced nano-cavity array as the SERS-active substrate for detecting hepatitis virus core antigen at low concentrations. Sensors and Actuators B. 2012; 174:478-484.

[54] Lin YY, Liao JD, Ju YH, Chang CW, Shiau AL. Focused ion beam-fabricated Au micro/nanostructures used as a surface enhanced Raman scattering-active substrate for trace detection of molecules and influenza virus. Nanotechnology. 2011; 22: 185308-185316.

[55] Lin YY, Liao JD, Yang ML, Wu CL. Target-size embracing dimension for sensitive detection of viruses with various sizes and influenza virus strains. Biosensors and Bioelectronics. 2012; 35:447-451.

[56] Xie Z, Tao J, Lu Y, Lin K, Yan J, Wang P, Ming H. Polymer optical fiber SERS sensor with gold nanorods. Optical Communications. 2009; 282 (3): 439-442.

[57] Zhou W, Gao X, Liu D, Chen X. Gold nanoparticles for *in vitro* diagnostics. Chemical Reviews. 2015;115(19):10575-10636.

[58] Kong K, Kendall C, Stone N, Notingher I. Raman spectroscopy for medical diagnostics—From in-vitro biofluid assays to in-vivo cancer detection. Advanced drug delivery reviews. 2015;89:121-134.

[59] Abramczyk H, Brozek-Pluska B. Raman imaging in biochemical and biomedical applications. Diagnosis and treatment of breast cancer. Chemical reviews. 2013;113(8):5766-5781.

[60] Wang Z, Zong S, Wu L, Zhu D, Cui Y. SERS-activated platforms for immunoassay: probes, encoding methods, and applications. Chemical reviews. 2017;117(12):7910-7963.

[61] Joseph MM, Narayanan N, Nair JB, Karunakaran V, Ramya AN, Sujai PT, et al. Exploring the margins of SERS in practical domain: An emerging

diagnostic modality for modern biomedical applications. Biomaterials. 2018;181:140-181.

[62] Lin J, Chen R, Feng S, Pan J, Li Y, Chen G, et al. A novel blood plasma analysis technique combining membrane electrophoresis with silver nanoparticle-based SERS spectroscopy for potential applications in noninvasive cancer detection. Nanomedicine: Nanotechnology, Biology and Medicine. 2011;7(5):655-663.

[63] Batool F, Nawaz H, Majeed MI, Rashid N, Bashir S, Akbar S, et al. SERS-based viral load quantification of hepatitis B virus from PCR products. Spectrochimica Acta Part A: Molecular and Biomolecular Spectroscopy. 2021:119722.

[64] Yadav S, Senapati S, Desai D, Gahlaut S, Kulkarni S, Singh J. Portable and sensitive Ag nanorods based SERS platform for rapid HIV-1 detection and tropism determination. Colloids and Surfaces B: Biointerfaces. 2021;198:111477.

[65] Lin D, Wu Q, Qiu S, Chen G, Feng S, Chen R, et al. Label-free liquid biopsy based on blood circulating DNA detection using SERS-based nanotechnology for nasopharyngeal cancer screening. Nanomedicine: Nanotechnology, Biology and Medicine. 2019;22:102100.

[66] Shen L, Du Y, Wei N, Li Q, Li S, Sun T, et al. SERS studies on normal epithelial and cancer cells derived from clinical breast cancer specimens. Spectrochimica Acta Part A: Molecular and Biomolecular Spectroscopy. 2020;237:118364.

[67] Moisoiu V, Badarinza M, Stefancu A, Iancu SD, Serban O, Leopold N, et al. Combining surface-enhanced Raman scattering (SERS) of saliva and two-dimensional shear wave elastography (2D-SWE) of the parotid glands in the diagnosis of Sjögren's syndrome. Spectrochimica Acta Part A: Molecular and Biomolecular Spectroscopy. 2020;235:118267.

[68] Muhammad M, Shao C, Huang Q. Label-free SERS diagnostics of radiation-induced injury via detecting the biomarker Raman signal in the serum and urine bio-samples based on Au-NPs array substrates. Spectrochimica Acta Part A: Molecular and Biomolecular Spectroscopy. 2019;223:117282.

[69] Narasimhan V, Siddique RH, Park H, Choo H. Bioinspired disordered flexible metasurfaces for human tear analysis using broadband surface-enhanced Raman scattering. ACS omega. 2020;5(22):12915-22.

[70] Rojalin T, Koster HJ, Liu J, Mizenko RR, Tran D, Wachsmann-Hogiu S, et al. Hybrid nanoplasmonic porous biomaterial scaffold for liquid biopsy diagnostics using extracellular vesicles. ACS sensors. 2020;5(9):2820-2833.

[71] Su X, Xu Y, Zhao H, Li S, Chen L. Design and preparation of centrifugal microfluidic chip integrated with SERS detection for rapid diagnostics. Talanta. 2019;194:903-909.

[72] Wang X, Xiao M, Zou Y, Lai W, Pei H, Alam MF, et al. Fractal SERS nanoprobes for multiplexed quantitative gene profiling. Biosensors and Bioelectronics. 2020;156:112130.

[73] Ngo HT, Gandra N, Fales AM, Taylor SM, Vo-Dinh T. Sensitive DNA detection and SNP discrimination using ultrabright SERS nanorattles and magnetic beads for malaria diagnostics. Biosensors and Bioelectronics. 2016;81:8-14.

[74] Frimpong R, Jang W, Kim J-H, Driskell JD. Rapid vertical flow immunoassay on AuNP plasmonic paper

for SERS-based point of need diagnostics. Talanta. 2021;223:121739.

[75] Liu H, Dai E, Xiao R, Zhou Z, Zhang M, Bai Z, et al. Development of a SERS-based lateral flow immunoassay for rapid and ultra-sensitive detection of anti-SARS-CoV-2 IgM/IgG in clinical samples. Sensors and Actuators B: Chemical. 2021;329:129196.

[76] Zhang D, Huang L, Liu B, Ni H, Sun L, Su E, et al. Quantitative and ultrasensitive detection of multiplex cardiac biomarkers in lateral flow assay with core-shell SERS nanotags. Biosensors and Bioelectronics. 2018;106:204-211.

[77] Fu X, Cheng Z, Yu J, Choo P, Chen L, Choo J. A SERS-based lateral flow assay biosensor for highly sensitive detection of HIV-1 DNA. Biosensors and Bioelectronics. 2016;78:530-537.

[78] Willner MR, McMillan KS, Graham D, Vikesland PJ, Zagnoni M. Surface-enhanced Raman scattering based microfluidics for single-cell analysis. Analytical chemistry. 2018;90(20):12004-12010.

[79] Kapara A, Findlay Paterson KA, Brunton VG, Graham D, Zagnoni M, Faulds K. Detection of Estrogen Receptor Alpha and Assessment of Fulvestrant Activity in MCF-7 Tumor Spheroids Using Microfluidics and SERS. Analytical Chemistry. 2021.

[80] Kunushpayeva Z, Rapikov A, Akhmetova A, Sultangaziyev A, Dossym D, Bukasov R. Sandwich SERS immunoassay of human immunoglobulin on silicon wafer compared to traditional SERS substrate, gold film. Sensing and Bio-Sensing Research. 2020;29:100355.

[81] Li M, Cushing SK, Zhang J, Suri S, Evans R, Petros WP, et al. Three-dimensional hierarchical plasmonic nano-architecture enhanced surface-enhanced Raman scattering immunosensor for cancer biomarker detection in blood plasma. ACS nano. 2013;7(6):4967-4976.

[82] Granger JH, Porter MD. The case for human serum as a highly preferable sample matrix for detection of anthrax toxins. ACS sensors. 2018;3(11):2303-2310.

[83] Ştiufiuc GF, Toma V, Buse M, Mărginean R, Morar-Bolba G, Culic B, et al. Solid plasmonic substrates for breast cancer detection by means of SERS analysis of blood plasma. Nanomaterials. 2020;10(6):1212.

[84] Králová ZO, Oriňak A, Oriňaková R, Petruš O, Macko J, Radoňak J, et al. Electrochemically deposited silver detection substrate for surface-enhanced Raman spectroscopy cancer diagnostics. Journal of biomedical optics. 2018;23(7):075002.

[85] Choi M, Kim S, Choi SH, Park H-H, Byun KM. Highly reliable SERS substrate based on plasmonic hybrid coupling between gold nanoislands and periodic nanopillar arrays. Optics express. 2020;28(3):3598-3606.

[86] Mhlanga N, Domfe T, Skepu A. Fabrication of surface enhanced Raman spectroscopy substrates on solid supports. Applied Surface Science. 2019;476:1108-1117.

[87] Mhlanga N, Domfe T, Skepu A, Ntho TA. Sandwich-based surface-enhanced Raman scattering probes for detection and quantification of malaria. Journal of Raman Spectroscopy. 2020;51(12):2416-2424.

[88] Xu K, Zhou R, Takei K, Hong M. Toward flexible surface-enhanced raman scattering (SERS) sensors for point-of-care diagnostics. Advanced Science. 2019;6(16):1900925.

[89] Xu K, Wang Z, Tan CF, Kang N, Chen L, Ren L, et al. Uniaxially

stretched flexible surface plasmon resonance film for versatile surface enhanced Raman scattering diagnostics. ACS applied materials & interfaces. 2017;9(31):26341-26349.

[90] Fan Z, Kanchanapally R, Ray PC. Hybrid graphene oxide based ultrasensitive SERS probe for label-free biosensing. The Journal of Physical Chemistry Letters. 2013;4(21):3813-3818.

[91] Korkmaz A, Kenton M, Aksin G, Kahraman M, Wachsmann-Hogiu S. Inexpensive and flexible SERS substrates on adhesive tape based on biosilica plasmonic nanocomposites. ACS Applied Nano Materials. 2018;1(9):5316-5326.

[92] He, S, Liu K.-K, Su S, Yan J, Mao X, Wang D, He Y, Li L.-J, Song S, Fan C. Graphene-based high-efficiency surface-enhanced Raman scattering-active platform for sensitive and multiplex DNA detection. Analytical chemistry, 2012:84(10): 4622-4627.

[93] Chen H, Das A, Bi L, Choi N, Moon J-I, Wu Y, et al. Recent advances in surface-enhanced Raman scattering-based microdevices for point-of-care diagnosis of viruses and bacteria. Nanoscale. 2020;12(42):21560-21570.

[94] Ma Y, Liu H, Chen Y, Gu C, Wei G, Jiang T. Improved lateral flow strip based on hydrophilic– hydrophobic SERS substrate for ultra– sensitive and quantitative immunoassay. Applied Surface Science. 2020;529:147121.

[95] Fallahi H, Zhang J, Phan H-P, Nguyen N-T. Flexible microfluidics: Fundamentals, recent developments, and applications. Micromachines. 2019;10(12):830.

[96] Xing Y, Zhao L, Cheng Z, Lv C, Yu F, Yu F. Microfluidics-Based Sensing of Biospecies. ACS Applied Bio Materials. 2020.

[97] Pallaoro A, Hoonejani MR, Braun GB, Meinhart CD, Moskovits M. Rapid identification by surface-enhanced Raman spectroscopy of cancer cells at low concentrations flowing in a microfluidic channel. Acs Nano. 2015;9(4):4328-4336.

[98] Cordero E, Latka I, Matthäus C, Schie IW, Popp J. In-vivo Raman spectroscopy: from basics to applications. Journal of biomedical optics. 2018;23(7):071210.

[99] Sato H, Tanaka T, Ikeda T, Wada S, Tashiro H, Ozaki Y. Biomedical applications of a new portable Raman imaging probe. Journal of Molecular Structure. 2001;598(1):93-96.

[100] Bergholt MS, Serio A, Albro MB. Raman spectroscopy: Guiding light for the extracellular matrix. Frontiers in bioengineering and biotechnology. 2019;7:303.

[101] Wachsmann-Hogiu S, Weeks T, Huser T. Chemical analysis in vivo and in vitro by Raman spectroscopy—from single cells to humans. Current opinion in biotechnology. 2009;20(1):63-73.

[102] Baraldi P, Tinti A. Raman spectroscopy in art and archaeology. Journal of Raman Spectroscopy: An International Journal for Original Work in all Aspects of Raman Spectroscopy, Including Higher Order Processes, and also Brillouin and Rayleigh Scattering. 2008;39(8):963-965.

[103] Kim JA, Wales DJ, Yang G-Z. Optical spectroscopy for in vivo medical diagnosis—a review of the state of the art and future perspectives. Progress in Biomedical Engineering. 2020;2(4):042001.

[104] Ramos IRM, Malkin A, Lyng FM. Current advances in the application of Raman spectroscopy for molecular diagnosis of cervical cancer. BioMed research international. 2015;2015.

[105] Moore TJ, Moody AS, Payne TD, Sarabia GM, Daniel AR, Sharma B. In vitro and in vivo SERS biosensing for disease diagnosis. Biosensors. 2018;8(2):46.

[106] Auner GW, Koya SK, Huang C, Broadbent B, Trexler M, Auner Z, et al. Applications of Raman spectroscopy in cancer diagnosis. Cancer and Metastasis Reviews. 2018;37(4):691-717.

[107] Fenn MB, Xanthopoulos P, Pyrgiotakis G, Grobmyer SR, Pardalos PM, Hench LL. Raman spectroscopy for clinical oncology. Advances in Optical Technologies. 2011;2011.

[108] Ramírez-Elías MG, González FJ. Raman spectroscopy for in vivo medical diagnosis. Raman Spectroscopy (April). 2018.

[109] Ramírez-Elías MG, González FJ. Raman spectroscopy for in vivo medical diagnosis. Raman Spectroscopy (April). 2018 Feb 28.

[110] Zhao J, Lui H, McLean DI, Zeng H. Real-time Raman spectroscopy for noninvasive in vivo skin analysis and diagnosis. New developments in biomedical engineering. 2010;24:455-474.

[111] Wolthuis R, Schut TB, Caspers P, Buschman H, Römer T, Bruining H, et al. Raman spectroscopic methods for in vitro and in vivo tissue characterization. Fluorescent and luminescent probes for biological activity: Elsevier; 1999. p. 433-455.

[112] Pence I, Mahadevan-Jansen A. Clinical instrumentation and applications of Raman spectroscopy. Chemical Society Reviews. 2016;45(7):1958-1979.

[113] Rzhevskii A. The recent advances in Raman microscopy and imaging techniques for biosensors. Biosensors. 2019;9(1):25.

[114] Wang W, Zhao J, Short M, Zeng H. Real-time in vivo cancer diagnosis using raman spectroscopy. Journal of biophotonics. 2015;8(7):527-545.

[115] Wei D, Chen S, Liu Q. Review of fluorescence suppression techniques in Raman spectroscopy. Applied Spectroscopy Reviews. 2015;50(5):387-406.

[116] Krafft C, Sergo V. Biomedical applications of Raman and infrared spectroscopy to diagnose tissues. Spectroscopy. 2006;20(5-6):195-218.

[117] Saha A, Barman I, Dingari N, McGee S, Volynskaya Z, Galindo L, et al. Raman spectroscopy: a real-time tool for identifying microcalcifications during stereotactic breast core needle biopsies. Biomedical optics express. 2011;2(10):2792-2803.

[118] Lieber CA, Majumder SK, Ellis DL, Billheimer DD, Mahadevan-Jansen A. In vivo nonmelanoma skin cancer diagnosis using Raman microspectroscopy. Lasers in Surgery and Medicine: The Official Journal of the American Society for Laser Medicine and Surgery. 2008; 40(7):461-467.

[119] Zhengkun, W. Ning, S. Yong, Z. Jie, Z. AgNPs decorated volcano-like Ag arrays for ultra-sensitive Raman detection. Optical Materials Express, 2020, 10 (12), 3393-3399.

[120] Morton, S. M.; Jensen, L. Understanding the Molecule-Surface Chemical Coupling in SERS. J. Am. Chem. Soc. 2009, 131, 4090– 4098.

[121] Kleinman, S. L.; Sharma, B.; Blaber, M. G.; Henry, A.-I.; Valley, N.; Freeman, R. G.; Natan, M. J.; Schatz, G. C.; Van Duyne, R. P. Structure Enhancement Factor Relationships in Single Gold Nano antennas by Surface-Enhanced Raman Excitation Spectroscopy. J. Am. Chem. Soc. 2013, 135, 301–308.

[122] C. Zapata-Urzua, M. P ´ ´erez-Ortiz, G. A. Acosta, J. Mendoza, L. Yedra, S. Estrad´e, A. Alvarez-Lueje, L. J. N ´ u´nez-Vergara, ˜ F. Albericio, R. Lavilla and M. J. Kogan, J. Colloid Interface Sci., 2015, 453, 260-269

[123] T. L. Moore, L. Rodriguez-Lorenzo, V. Hirsch, S. Balog, D. Urban, C. Jud, B. Rothen-Rutishauser, M. Lattuada and A. Petri-Fink, Chem. Soc. Rev., 2015, 44, 6287-6305.

[124] P. G. Etchegoin and E. C. Le Ru, Phys. Chem. Chem. Phys., 2008, 10, 6079-6089.

[125] Guillot, N. and Lamy de la Chapelle, M. The electromagnetic effect in surface enhanced Raman scattering: Enhancement optimization using precisely controlled nanostructures, Journal of Quantitative Spectroscopy and Radiative Transfer, 2012, 113(18), 2321-2333,

[126] Langer, J., Jimenez de Aberasturi, D., Aizpurua, J., Alvarez-Puebla, R. A., Auguié, B., Baumberg, J. J., Liz-Marzán, L. M. Present and future of surface-enhanced Raman scattering. 2020:ACS Nano, 14(1), 28-117.

[127] Dieringer, J. A., McFarland, A. D., Shah, N. C., Stuart, D. A., Whitney, A. V., Yonzon, C. R., Van Duyne, R. P. Introductory Lecture : Surface enhanced Raman spectroscopy: new materials, concepts, characterization tools, and applications. 2006: Faraday Discussions, 132, 9-26.

[128] Sivashanmugan, K., Liao, J., Liu, B.H., Yao, C., Focused-ion-beam-fabricated Au nanorods coupled with Ag nanoparticles used as surface-enhanced Raman scattering-active substrate for analyzing trace melamine constituents in solution, Analytica Chimica Acta, 2013, 800, 56-64.

[129] Stiles, P. L., Surface-enhanced Raman Spectroscopy. 2008: Annu Rev Anal Chem, 1, 601.

[130] Langer, J, Jimenez de Aberasturi, D., Aizpurua, J., Present and Future of Surface-Enhanced Raman Scattering, ACS Nano, 2020, 14 (1), 28-117

[131] Gagnon, P. Technology trends in antibody purification. 2012: Journal of Chromatography A, 1221, 57-70.

[132] Grilo, A. L., & Mantalaris, A. The increasingly human and profitable monoclonal antibody market. 2019: Trends in Biotechnology, 37(1), 9-16.

[133] Dörfer, T. Deep-UV surface-enhanced Raman scattering. 2007: J Raman Spectrosc, 38 (1),1379-1381.

[134] Abalde-Cela Sara, Aldeanueva-Potel Paula, Mateo-Mateo Cintia, Rodríguez-Lorenzo Laura, Alvarez-Puebla Ramón A. and Liz-Marzán Luis M. Surface-enhanced Raman scattering biomedical applications of plasmonic colloidal particles J. R. Soc. Interface. 2010, 7, 435-450

[135] Gao L, Zhao, H., Li, T, Huo, P., Chen D and Liu, B, Atomic Force Microscopy Based Tip-Enhanced Raman Spectroscopy in Biology, International journal of molecular sciences, 2018, 19(4), 1193-1195.

[136] Alvarez-Puebla, R. A., Liz-Marzan, L. M., & Abajo, F. J.′ Light Concentration at the Nanometer Scale. 2010:J. Phys. Chem. Lett, 1, 2428 2434.

[137] Solís, D. M., Taboada, J. M., Obelleiro, F., Liz-Marzán, L. M., & García de Abajo, F. J. Optimization of nanoparticle-based SERS substrates through large-scale realistic simulations. 2017: ACS Photonics, 4(2), 329-337.

[138] Li, K., Stockman, M. I., & Bergman, D. J. Self-similar chain of metal nanospheres as an efficient nanolens. 2003: Physical Review Letters, 91(22), 227402.

[139] Liu, Y., Tian, X., Guo, W. et al. Real-time Raman detection by the cavity

mode enhanced Raman scattering. Nano Research, 2019, 12 (7), 1643-1649.

[140] Taguchi, A. Deep-UV tip-enhanced Raman scattering 2009: J Raman Spectrosc, 40 (1), 1324-1330.

[141] Tian, Z.Q. SERS from Transition Metals and Excited by Ultraviolet Light Top. 2006: Appl Phys, 103 (1) 125-128.

[142] Sharma, B., Cardinal, M. F., Ross, M. B., Zrimsek, A. B., Bykov, S. V., Punihaole, D., Van Duyne, R.P. Aluminum Film-Over-Nanosphere Substrates for Deep-UV Surface Enhanced Resonance Raman Spectroscopy. 2016: Nano Lett, 16, 7968 7973.

[143] Ren, B., Surface-Enhanced Raman Scattering in the Ultraviolet Spectral Region: UV-SERS on Rhodium and Ruthenium Electrodes. 2003: J Am Chem Soc, 125, 9598.

[144] Stockle, R. M., Nanoscale chemical analysis by tip-enhanced Raman spectroscopy. 2000: Chem Phys Lett, 318, 131.

[145] Domke, K. F., & Pettinger, B. Studying surface chemistry beyond the diffraction limit: 10 years of TERS. 2010: Chemphyschem: A European Journal of Chemical Physics and Physical Chemistry, 11(7), 1365-1373.

[146] Bailo, E., & Deckert, V., Tip-enhanced Raman scattering. 2008: Chemical Society Reviews, 37(5), 921-930.

[147] Bailo, E., & Deckert, V., Tip-enhanced Raman spectroscopy of single RNA strands: towards a novel direct-sequencing method. 2008: Angewandte Chemie (International Ed. in English), 47(9), 1658-1661.

[148] Fabris, L., SERS tags: The next promising tool for personalized cancer detection?. 2016: ChemNanoMat: Chemistry of Nanomaterials for Energy, Biology and More, 2(4), 249-258.

[149] Indrasekara, A. S. D. S., & Fabris, L., SERS-Based Approaches toward Genetic Profiling. 2015: Bioanalysis, 7, 263 278.

[150] Ngo, H. T., Freedman, E., Odion, R. A., Strobbia, P., De Silva Indrasekara, A. S., Vohra, P., ... Vo-Dinh, T., Direct detection of unamplified pathogen RNA in blood lysate using an integrated lab-in-a-stick device and ultrabright SERS nanorattles. 2018: Scientific Reports, 8(1). doi:10.1038/s41598-018-21615-3

Section 3

Raman Spectroscopy for Low Dimensional Materials

Chapter 11

Raman Spectroscopy and Mapping Analysis of Low-Dimensional Nanostructured Materials and Systems

Karthikeyan Krishnamoorthy and Sang-Jae Kim

Abstract

This chapter describes the use of Raman spectroscopy and mapping analysis for the characterization of low dimensional nanostructures, including 2D sheets (graphene oxide, graphene sheets, MoS_2, siloxene), and one-dimensional carbyne chains. The Raman mapping analysis and their application towards understanding the molecular level interactions in these low dimensional materials, nanostructured polymer composites, and nanopaints are also discussed. The stoichiometric composition and structure of these low dimensional materials were correlated with the Raman spectral and mapping analysis. Further, Raman spectroscopy for understanding or probing the mechanism of mechanical to electrical energy harvesting properties of carbyne films via the structural transformation from cumulene to polynne networks of carbyne is demonstrated.

Keywords: Raman spectroscopy, Raman mapping, graphene oxide, graphene, MoS_2, siloxene, carbyne, nanocomposites

1. Introduction

Raman spectroscopy is a promising non-destructive testing of materials to understand their crystallinity, chemical bonding vibrations and effects of surface defects [1, 2]. The Raman spectroscopy becomes an important technique for the characterization of nanostructured materials, especially the two-dimensional (2D) materials (such as graphene, MoS_2, siloxene, metalenes), biomaterials, composites, and for understanding energy-conversion process in the recent years [3–5]. Additionally, these techniques are useful for criminological & forensic applications, biomedical applications, and as well as bio-sensors for health care sectors [6–8]. The basic principle of Raman spectroscopy relies on the "Raman effect", i.e., the inelastic scattering of light which can directly probe vibration and rotational-vibration states of any molecules and/or materials [9]. In 1923, Smekal et al. predicted the Raman scattering of light using molecules explained via classical quantum theory which was experimentally observed by Raman and Krishnan in 1928 [10]. Based on this principle, nearly 25 types of Raman spectroscopic techniques are available for characterization of different materials for various applications. Some of them are (i) spontaneous Raman, (ii) hyper-Raman scattering, (iii) Fourier-transform

Raman scattering, (iv) Raman-induced Kerr-effect spectroscopy and (v) stimulated-coherent Raman and so on [9]. There are numerous works discussed the fundamental principles and theory of Raman spectroscopy and their working mechanism were available in literature [11, 12]. In addition to these, Raman mapping is often used for identification of various information such as crystallinity, homogeneity, defect sites and molecular level bonding for materials science research [13, 14]. This book chapter focus on the use of Raman spectroscopy and mapping analysis for studying the molecular level vibrations in the 2D materials, composites, solid electrolyte entrapped in piezo-polymer matrix, low-dimensional materials such as carbyne chains, and probing method for elucidating the mechanism of energy harvesting in carbyne via mechanical stimuli.

2. Experimental section

2.1 Preparation of graphene oxide with different oxidation levels, graphene sheets and graphene nanopaint

The modified Hummers method is used for the preparation of graphene oxide sheets using the chemical reaction between graphite powders with potassium permanganate, and sulfuric acid. The oxidation degree of graphene oxide was varied via changing the concentration of potassium permanganate by keeping the remaining parameters constant as reported in our earlier work [15]. The graphene sheets were prepared via the reduction of graphene oxide using hydrazine hydrate in presence of ultrasound irradiation [16]. The graphene based nanopaint was obtained by mechanical milling process for 12 h using appropriate amount of graphene sheets (pigment) and alkyd resin binder. The graphene paint was coated on glass substrate using brush coating [17].

2.2 Preparation of 2D molybdenum disulphide nanosheets and quantum sheets

A hydrothermal method is used for the formation of MoS_2 on the surface of Mo foil (Mo source) using thiourea (sulfur source). The hydrothermal reaction process is carried out for 24 h at a temperature of 180 °C. The detailed experimental procedure can be seen from our reported work [18].

2.3 Preparation of 2D siloxene sheets

The siloxene nanosheets was obtained via topochemical de-intercalation reaction between calcium disilicide and conc. Hydrochloric acid at a temperature of 0 °C for four days [19].

2.4 Preparation of 2D antimonene

The 2D antimonene with nanodendrites structures anchored on the surface of the Ni foam was achieved via a facile electrochemical deposition process as mentioned in literature [20].

2.5 Preparation of proton conducting solid electrolyte-piezoelectric PVDF hybrids

The piezo-polymer electrolyte nanocomposite film made of phosphotungstic acid (PTA) solid electrolyte and PVDF were obtained by ultrasound irradiation followed by solvent casting method [16]. Appropriate amount of PVDF was dissolved in dimethylacetamide and acetone with the use of ultrasonication in which different weight ratios (5–25 wt%) of PTA electrolyte was added under mechanical stirring and ultrasound irradiation process. Then, the entire solution was transferred into a Petri dish and allowed to dry at 70 °C for complete evaporation of the solvents which led to the formation of PTA-PVDF piezo-electrolyte film via peel-off process.

2.6 Preparation of free standing carbyne-enriched carbon (CEC) films

The CEC film was prepared by immersing the free-standing PVDF film in a solution containing dehydrohalogenation mixture (potassium ethylate and tetrahydrofuran) in presence of ultrasound irradiation for 2 h [21, 22]. The change in color from white (PVDF) into black (CEC) confirms the occurrence of dehydrohalogenation process. Then, the CEC film was rinsed with ethanol to removal chemical impurities and dried at 60 °C. The entire reaction was performed inside an Ar-filled glove box.

2.7 Raman spectral and mapping acquisition

The Raman spectral and mapping acquisition of the samples were carried out on a LabRam HR-Evolution Raman spectrometer (Horiba Jobin-Yvon, France). The Raman system used an Ar^+ ion laser operating at a laser power of 15 mW with an excitation wavelength of 514 nm. The Raman mapping of samples were performed over the desired area to obtain spectral arrays. The results are processed and analyzed using software. The spectral arrays map was processed and analyzed using classical least squares (CLS) fitting (multivariate analysis) method on LabSpec (Ver. 6.2) software.

3. Results and discussion

3.1 Raman spectral studies of graphene oxide with various levels of oxidation

This section focused on the use of Raman spectroscopy as a promising tool for characterizing graphene-based materials and their system. It is well known that graphene sheets emerged as a material of this decade due to their wide-spread properties and applications in variety of sectors [23]. For preparation of gram-scale graphene sheets, researchers often used graphitic oxide or graphene oxide as a starting material that is originally synthesized a century ago [24]. Graphene is a one-atom thick sheet in which hexagonal carbon chains are present laterally [25]. The structure of graphene oxide is similar to that of graphene in which the carbon atoms are bonded with different functional groups (hydroxyl, carbonyl, carboxyl, and epoxide) [26]. The formation of these groups occurred because of oxidation of graphite and removal of these groups lead to the formation of chemically derived graphenes.

In general, the Raman spectrum of graphite possess G band (first order scattering of the E_{2g} mode), D band cm^{-1} (defects), and 2D band (c-axis stacking order) at 1570, 1354, and 2700 cm^{-1}, respectively [1]. **Figure 1(A)** shows the Raman spectra of graphene oxide sheets with different oxidation degrees (S1-S6 shows the low to high oxidation content). It shows that the increase in oxidation degree results in the shift of G band position (in comparison with graphite) towards higher wavenumber and reaches to 1596 cm^{-1} for the heavily oxidized graphene oxide (**Figure 1(B)**) [15]. The full-width half maximum (FWHM) of the G band raised from 45 to 124 cm^{-1} with increase in oxidation level (**Figure 1(B)**). The shift in G band position and increase in FWHM is due to the formation of different levels of sp^3 hybridized carbons with respect to the oxidation process. Likewise, the D band in graphene oxide is broadened (increased FWHM) and highly intense in comparison with graphite, i.e., due to the formation more defects and disorders as a result of oxidation process [27]. The intensity of the 2D band reduces after oxidation process because of breaking of stacking order and new band is observed around 2950 cm^{-1} that is denoted as D + G band. At much higher oxidation levels, the peak broadening of 2D band is observed. **Figure 1(C)** shows the plot of $I_{(D)}/I_{(G)}$ and $I_{(2D)}/I_{(G)}$ ratio versus the oxidation degree of graphene oxide. The $I_{(D)}/I_{(G)}$ ratio rises with initial increase in oxidation and decreased/saturated at higher degrees that is compensated by the higher FWHM. The variation of $I_{(2D)}/I_{(G)}$ ratio (**Figure 1(C)**) against the oxidation degree of graphene oxide indicating that they tend to decrease up to S-3 and increases at higher oxidation levels. These changes indicated the decrease in crystallinity and increase in amorphous nature in graphene oxide with increase in oxidation ranges. Further, the observed changes in overtone bands (2D, D + G and 2G), confirms the disruption of the stacking order in graphene oxide at higher oxidation level. The average crystallite size (L_a) of the sp^2 domains present in the prepared graphene oxides was determined in accordance with the method given by Cancodo et al. [28]. The L_a values are 18.2, 11.2, 9.6, 15.6, 14.0, and 13.7 nm for the different graphene oxides (from S-1 to S-6), and these values are lesser than the L_a of graphite (122 nm). These results indicated that there is no linear relation to quantify the amounts of sp^3 and sp^2 domains present in graphene oxide with different oxidation degrees. Altogether, the Raman analysis showed the occurrence crystalline to amorphous transition with disruption of stacking order when graphite is oxidized into graphene oxide.

3.2 Raman spectral and mapping studies on graphene-based nanopaint

This section describes the use of Raman mapping for the identification of pigment dispersion in alkyd resin-based paint coating which utilizes graphene sheets as pigment and/or conductive agent. These electrically conductive paints are

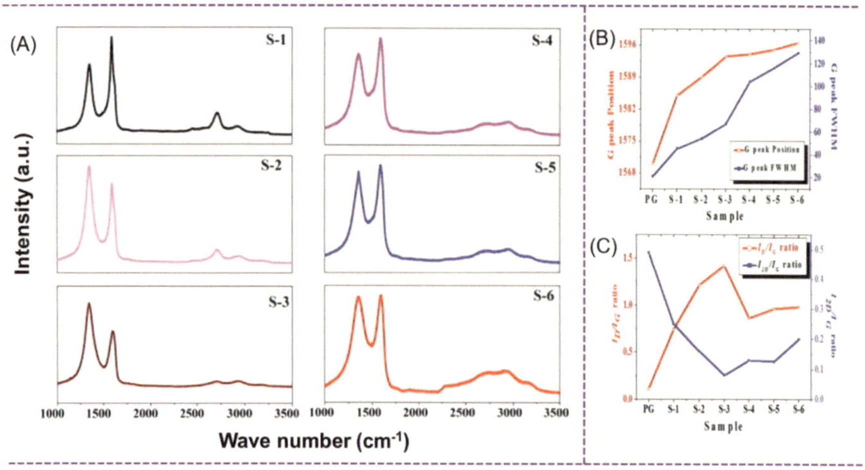

Figure 1.
(A) Raman spectra of graphene oxide with different oxidation level, (B) variation of G peak position and full width half maximum ratio with respect to the oxidation degree of graphene oxide, (C) plot of (B) $I_{(D)}/I_{(G)}$ *and* $I_{(2D)}/I_{(G)}$ *ratio versus the oxidation degree of graphene oxide.*

of high significance with applications ranging from electromagnetic interference shielding, static charge dissipation, and space [29, 30]. The comparative Raman spectrum of graphene paint and alkyd resin is given in **Figure 2(A)**. The spectrum of graphene paint coatings indicating the presence of G band (1585 cm^{-1}), and D band (1350 cm^{-1}) which confirms the presence of graphene sheets dispersed well in the alkyd resin binder [31, 32]. There were no bands related to the alkyd resin were observed in the Raman spectrum of graphene paint coating since the vibrations arise from graphene sheets overwhelms the vibrations of alkyd resin. The peak position and intensity ratios of finger imprint modes were used to study the spatial distribution of graphene sheets in the paint matrix. **Figure 2(B)–(D)** presents the peak position maps of graphene's finger imprint bands such as D band (1350–1370 cm^{-1}), G band (1584–1590 cm^{-1}) and 2D band (2710 to 2730 cm^{-1}), respectively. In comparison with the Raman spectrum of bare graphene sheets alone, the G band is red shifted as a result of molecular level bonding between the graphene sheets with the functional groups of alkyd resin [17]. Additionally, G and 2D bands of graphene sheets were seen over the entire mapped regime of the paint coatings that is responsible for the observed electrical conductivity. **Figure 2(E)** shows the intensity ratio map of I_D/I_G band in the range from 0.20 to 0.55 (blue to yellow) revealed the interconnection between of sp^2 domains of graphene with alkyd resin counterparts in the prepared paint coating. The I_{2D}/I_G ratio map (**Figure 2(F)**) is from of 0.25 to 0.65 (blue to yellow) indicating the restacking of graphene sheets (c-axis) occurred in the paint coating. This uniform distribution of restacked graphene sheets inside the alkyd resin matrix provides enough conductive channels to facilitate the electrical transport in the graphene paint [17].

3.3 Raman analysis of 2D materials directly grown on conductive substrate

In this section, we discuss about the use of Raman spectroscopy as a prominent tool understanding the crystallinity and layer numbers of 2D materials that are randomly or vertically oriented on the conductive substrates. Generally, binder-free electrodes neglecting the inclusion of insulating polymers are of great significance in electrochemical energy conversion and storage devices since they offer enhanced

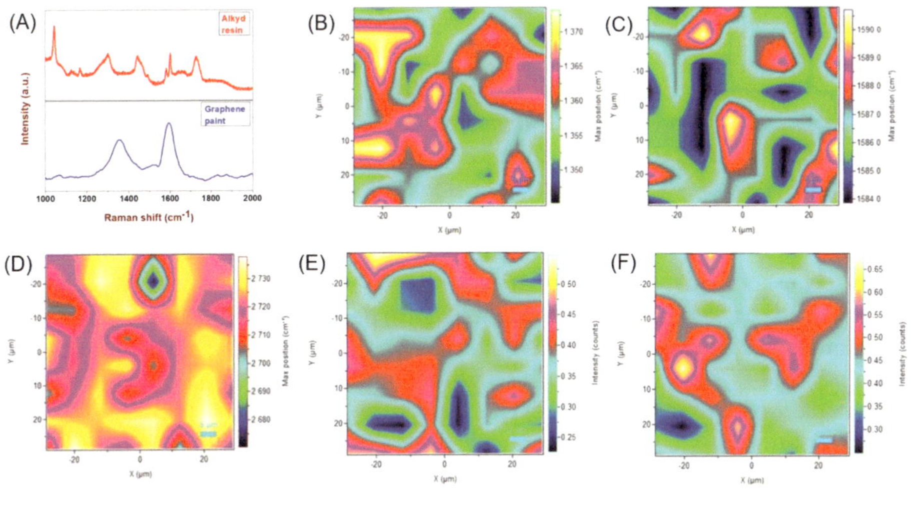

Figure 2.
Raman characterization of graphene paint coatings. (A) Raman spectrum of graphene paint in comparison with alkyd resin, peak position maps of (B) D, (C) G, and (D) 2D bands, respectively. Intensity ratio maps of (E) I_D/I_G and (F) I_{2D}/I_G bands, respectively.

electrochemical active sites [33, 34]. Usually, atomic force microscope (AFM) is used for the understanding the thickness of the 2D sheets whereas this technique is suitable only for laterally oriented sheets [35]. The AFM technique is not suitable for measurements for samples such as randomly or vertically aligned sheets on conductive substrates due to the structural issues [18]. Herein, Raman spectroscopy and mapping analysis are promising for these types of binder-free electrodes that are mainly used in electrochemical energy devices.

In recent years, MoS_2 sheets are considered as structural analogue to graphene due to their structural integrity in which the covalently bonded S-Mo-S layers were separated by Van der Waals forces [36]. These layered structure and redox properties of Mo transition metal led to superior electrochemical charge-storage properties (supercapacitors) which is extensively studied during this decade [37]. Additionally, the presence of band gap in MoS_2 that make them as an ideal candidate for optoelectronic, field effect transistor, and photoelectrochemical cells [38]. The specific capacitance of MoS_2 electrode fabricated using conventional slurry coating method is in the range of 80 to 120 F g^{-1} [39]. To boost the specific capacitance, binder-free MoS_2 electrodes were fabricated via growing MoS_2 directly on conductive substrates such as Mo foil, and Ni foam. Interestingly, the specific capacitance of MoS_2/Mo foil increases upto 192 F g^{-1} that is higher compared to many of the reported MoS_2 based planar supercapacitor electrodes and these electrodes possess better Columbic efficiency than others [18]. To understand the structural properties of MoS_2/Mo binder-free electrodes, Raman mapping analysis were performed. **Figure 3(A)** shows the Raman spectrum of MoS_2 sheets randomly aligned on Mo foil. The presence of Raman fingerprint bands of MoS_2 at 381 cm^{-1} (E^1_{2g} mode) and 405 cm^{-1} (A_{1g} mode) is observed in **Figure 3(A)**. The intensity ratio between these bands, and their band positions can be directly correlated to their thickness [40]. **Figure 3(B)** shows the integral intensity ratio map of E^1_{2g}/A_{1g} modes shows that their ratio varies from 0.1 to 0.7. Bulk MoS_2 possess an intensity ratio of 0.7 whereas tri-, bi-, and mono- layered MoS_2 have an intensity ratio of 0.6, 0.3 and 0.1 based on previous works [41]. Therefore, **Figure 3(B)** confirms the presence of minor fractions with single-layer MoS_2 (blue color), major fractions with bi-layered MoS_2 (green color), few layered ($n < 6$) MoS_2 (red color) and some bulk counterparts (yellow color) present in the MoS_2/Mo foil [18]. **Figure 3(C)** and **(D)** shows the peak position maps of A_{1g} and E^1_{2g} modes which shows their variation from 403 to 410 cm^{-1} (bulk to single-layer) and from 380 to 386 cm^{-1} (bulk to single-layer), respectively [18]. The presence of bulk and few-layered MoS_2 is due to the initial phase of the reaction due to Kirkendall effect and secondary phase resulting in the randomly oriented sheets. Altogether, the Raman mapping analysis confirmed the presence of few-layered MoS_2 sheets ($n < 5$) and some bulk counterparts were grown on the Mo foil.

Likewise, in our recent study, Raman spectroscopy coupled with mapping is effectively used to quantify the thickness of the antimonene nanodendrites grown on the surface of nickel foam via electrochemical deposition technique [20]. Antimonene is one of the important materials from the family of 2D metalenes (P, Sb, As, and Bi) due to their semi-metallic properties [42], high carrier mobility [43], oxidation resistant nature, and tuneable band gap that make them as alternative candidate for application in solar cells [44], CO_2 reduction [45], biological applications, supercapacitors, and batteries [20, 46]. Usually, antimonene sheets can be prepared via exfoliation methods (mechanical or liquid phase) like other 2D materials, whereas the yield is low. Recently, chemical/physical vapor deposition techniques were used to grow antimonene on conductive substrates [47]. In our

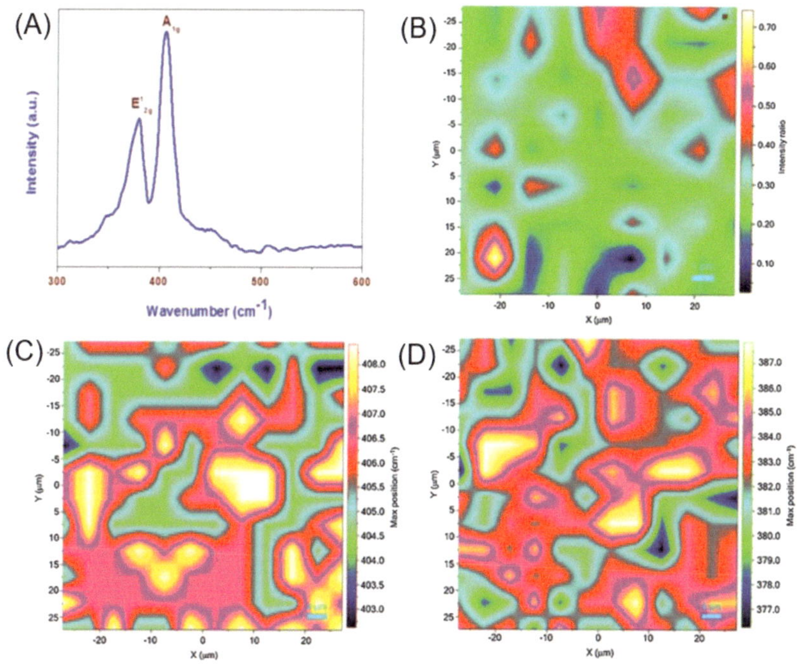

Figure 3.
Raman characterization of MoS_2 nanosheets grown on Mo foil. (A) Raman spectrum, (B) E^1_{2g}/A_{1g} intensity ratio map, (C) A_{1g} peak position map, and (D) E^1_{2g} peak position map.

recent work, we demonstrated the use of electrosynthesis route for the preparation of antimonene nanodendrites grown on nickel foam and these electrodes showed excellent charge-storage properties with a high specific capacity (1618 mA g^{-1}) than of the reported binder-free electrodes [20]. Herein, Raman mapping analysis is used for the identification of layer numbers in antimonene. Based on the E_g/A_{1g} intensity ratio map of antimonene nanodendrites that showed a ratio values vary from 0.41 to 0.49, the presence of few-layered sheets in the prepared antimonene/Ni foam binder-free electrodes were confirmed [48].

3.4 Raman analysis of siloxene sheets

Siloxene sheets are one of the emerging materials from the 2D silicon family that can be prepared via topochemical reaction (given in Section 2.3) between $CaSi_2$ (**Figure 4(A)**) with conc. HCl that results in the dissolution of calcium ions and oxidation of Si sheets (**Figure 4(B)**) [19, 49]. The structure of siloxene consists of Si_6 rings interconnected through Si–O–Si bridges with the addition of surface-terminated hydroxyl groups as seen in **Figure 4(C)** [19]. Siloxene sheets can be explained as an oxidized form of silicene sheets, and the latter is known for their excellent conductivity comparable to that of graphene [50]. Further, the 2D siloxene or silicene sheets are highly useful for micro-electronic devices since existing technology is established based on silicon [19]. Recent studies on siloxene sheets shows that they are promising candidate for applications in water splitting, Li-ion batteries, supercapacitors, electrochemical sensors, and biomedical fields [51–54]. However, the structure of siloxene is quite complicated and there are different models (such as Weiss structure, chain-like structure, and Kautsky structure) were proposed till date [54]. The Raman spectrum of siloxene sheets (**Figure 4(D)**) showed the presence of two bands viz. (i) Si-O-Si (495 cm^{-1}) and Si-Si ((520 cm^{-1})

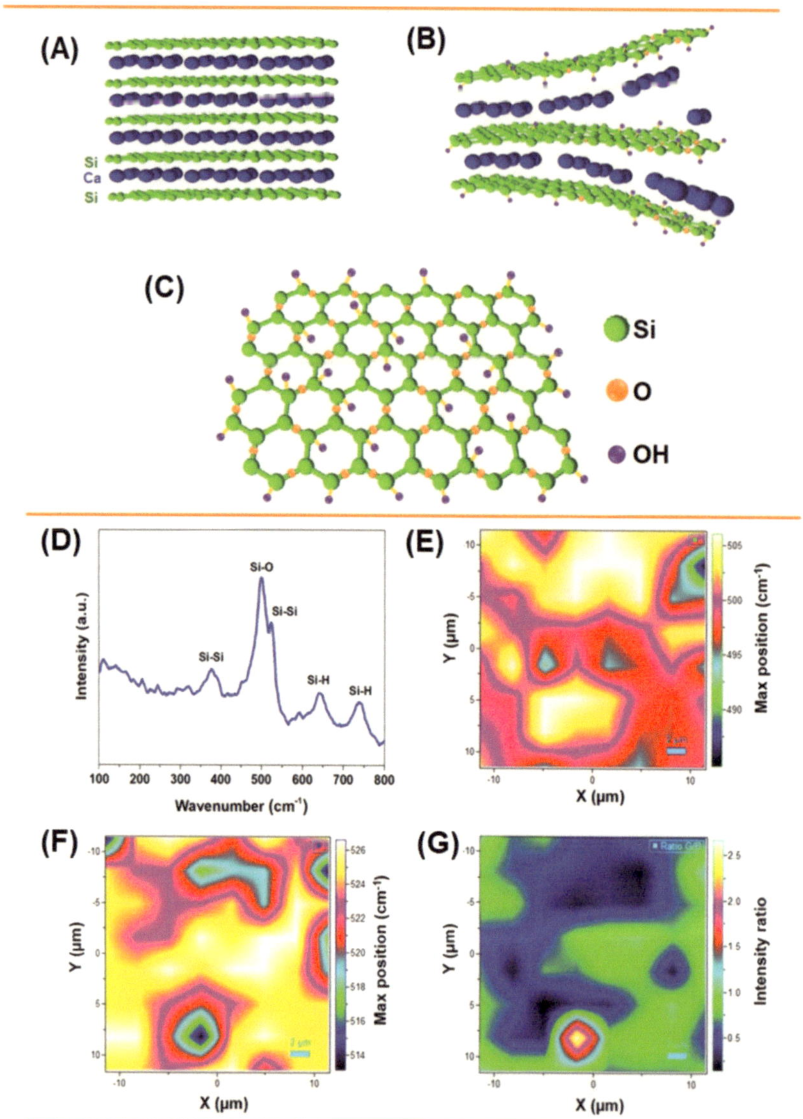

Figure 4.
(A) Structure of $CaSi_2$, (B) dissolution of Ca from $CaSi_2$ during the topochemical reaction, (C) structure of siloxene sheets, (D) Raman spectrum of siloxene sheets, (E-F) peak position maps of (E) Si-O-Si and (F) Si-Si, and (G) intensity ratio map of $I_{(Si\text{-}O\text{-}Si)}/I_{(Si\text{-}Si)}$. Reproduced from ref. 19 with permission from the Royal Society of Chemistry.

which showed that Kautsky structure of siloxene [52]. The minor bands present at 375 cm^{-1} is due to $\nu_{(Si\text{-}Si)}$ and others located at 640 and 740 cm^{-1} are originated from $\nu_{(Si\text{-}H)}$. The peak position maps of Si-O-Si and Si-Si (**Figure 4(E)** and **(F)**) vibrations shows that they vary from 495 to 505 cm^{-1} and from 518 to 526 cm^{-1}, respectively. Here, the Si-Si and Si-O-Si bonds relates to the crystalline and amorphous domains of the siloxene sheets [19], and therefore, their intensity ratio map was constructed as shown in **Figure 4(G)**. The intensity ratio map of Si-O-Si/Si-Si bands revealed that they vary from 0.3 to 2.0 over the entire mapped region. This highlighted the heterogenous distribution of Si-O-Si bridged over the Si_6 rings in the structure of siloxene and confirmed the presence of Kautsky structure [55].

3.5 Raman analysis of solid electrolyte-piezopolymer nanocomposite

Polymer electrolyte composites plays a key role in a variety of electrochemical energy conversion-, storage, and delivery devices [56]. Recently, solid electrolyte entrapped piezo polymeric separators were developed by few researchers for direct applications in self-charging power (supercapacitor or battery) cells [57]. Herein, Raman spectral and mapping analysis are used to study the distribution of electrolyte ions inside the polymer matrix and also to understand the role of electrolyte ions on the self-poling properties to yield the electroactive phase in the PVDF [58]. The piezo-polymer electrolyte nanocomposite film was prepared via the method given in Section 2.5 which contains PTA electrolyte distributed in the PVDF matrix. Initial studies on the Raman spectrum of bare and electrolyte incorporated films showed the presence of α-phase (765 & 880 cm^{-1}) and electroactive β-phase (842 cm^{-1}) in the bare PVDF whereas the presence of band due to W-O stretching (1011 cm^{-1}) were found in the composite films [16]. The Raman spectral analysis of the piezo-electrolyte film shows that there were no changes found in the electroactive phases of PVDF due to the addition of PTA. Additionally, Raman mapping analysis is performed over an area of $-1 \times 1\,\mu m^2$ with acquisition of 100 spectral arrays to study the distribution of the PTA in the PVDF polymer. **Figure 5(A)** depicts the 3D array map of PTA-PVDF revealing the presence of γ (811 cm^{-1}) and β (842 cm^{-1}) phase of PVDF and W-O vibrations (1011 cm^{-1}) of PTA electrolyte [16]. The peak position map of β-phase (**Figure 5(B)**) shows that their positions vary from 838 to 839 cm^{-1} (blue to red regime) whereas the position of the W-O vibration (**Figure 5(C)**) ranged from 1006.6 to 1007.5 (blue to red regime). **Figure 5(D)** and **(E)** presents the Raman intensity maps of β-phase PVDF (842 cm^{-1}) and (W-O 1011 cm^{-1}) respectively. The intensity ratio maps of $I_{W\text{-}O}/I_{\beta}$ (given in in **Figure 5(F)**) shows that these values are in the range from 1.198 (blue) to 1.202 (red) indicating a high order of homogeneity [16]. There is no drastic variation in the intensity ratio from which it can be concluded that PTA electrolyte is homogeneously dispersed well in the PVDF matrix [16].

3.6 Raman spectroscopy as a tool to probe the mechanical to electrical energy transduction process in CEC film

The digital photographs of the bare PVDF and CEC film (**Figure 6(A)**) shows the transformation of white to black-colored films as a result of dehydrohalogenation process [59]. The predicted structures of linear carbyne chain present in the CEC film in horizontal and vertical orientation is provided in (**Figure 6(B)** and **(C)**. The 3D Raman array map of CEC film (**Figure 6(D)**) shows the presence of β-carbyne (cumulene) and carbonoid (sp, sp^2, and sp^3 hybridized carbon) networks at 1140 and 1540 cm^{-1}, respectively [60]. The formation of sp^2 and sp^3 hybridized carbon can be explained via cross-linking reactions of adjacent sp-hybridized carbon in the carbyne chains during chemical reaction [22, 61]. The position map of the cumulene (**Figure 6(E)**) shows the width of about 40 cm^{-1} (blue to red-yellow color) as an indication of uniform distribution of $((=C=C=)_n)$ chains in the entire region of the CEC film. The carbonoid position map (**Figure 6(F)**) lies between 1525 and 1560 cm^{-1} (blue to red color) revealing the existence of sp- carbon in the mapped area. **Figure 6(G)** and **(H)** shows the intensity maps of cumulene and carbonoid bands of CEC. **Figure 6(I)** depicts the intensity-ratio map of cumulenic/carbonoid type carbon present in the CEC is in the range of 0.20 to 0.75, thus, confirmed the presence of uniformly distributed cumulene chains randomly oriented in the carbonoid species of the CEC [22].

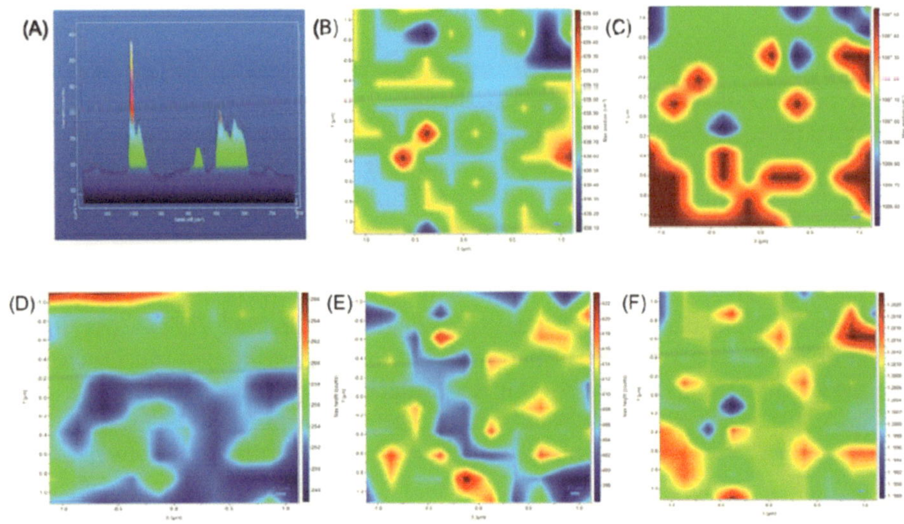

Figure 5.
Raman characterization of the PTA-PVDF film. (A) 3D Raman array map, (B-C) peak position map of (B) β-PVDF (839 cm^{-1}) and (C) W-O band (1007 cm^{-1}), (D) β-PVDF band intensity map (E) W-O band intensity map, (F) $I_{W\text{-}O}/I_{\beta}$ intensity ratio map.

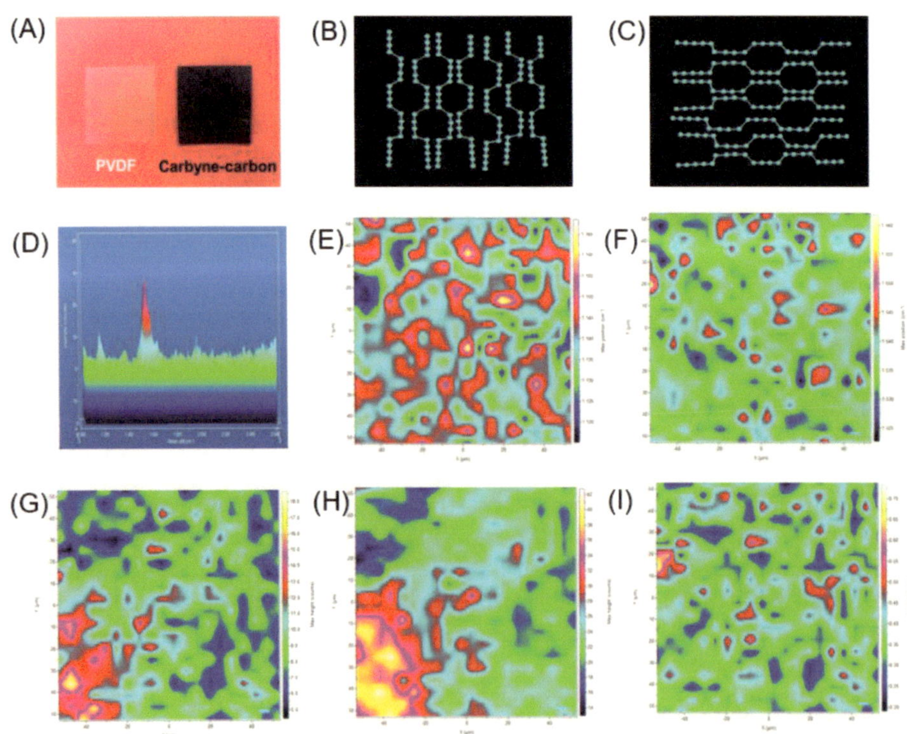

Figure 6.
(A) Digital photograph of PVDF and CEC film, (B, C) structures of carbyne orientation in the horizontal and vertical orientation, (D) 3D Raman map of CEC film, (E) Cumulene peak-position map, (F) Carbonoids peak-position map, (G and H) represents the intensity maps of (G) cumulene and (H) carbonoid-carbon bands, and (I) integral intensity ratio map (I_{1140}/I_{1540}) of CEC.

Figure 7(A) and **(B)** summarizes the mechanical to electrical energy transduction properties of the CEC film when subjected to an applied force. The CEC film generates an electrical voltage and current outputs of ~6.4 V and ~ 10 nA while

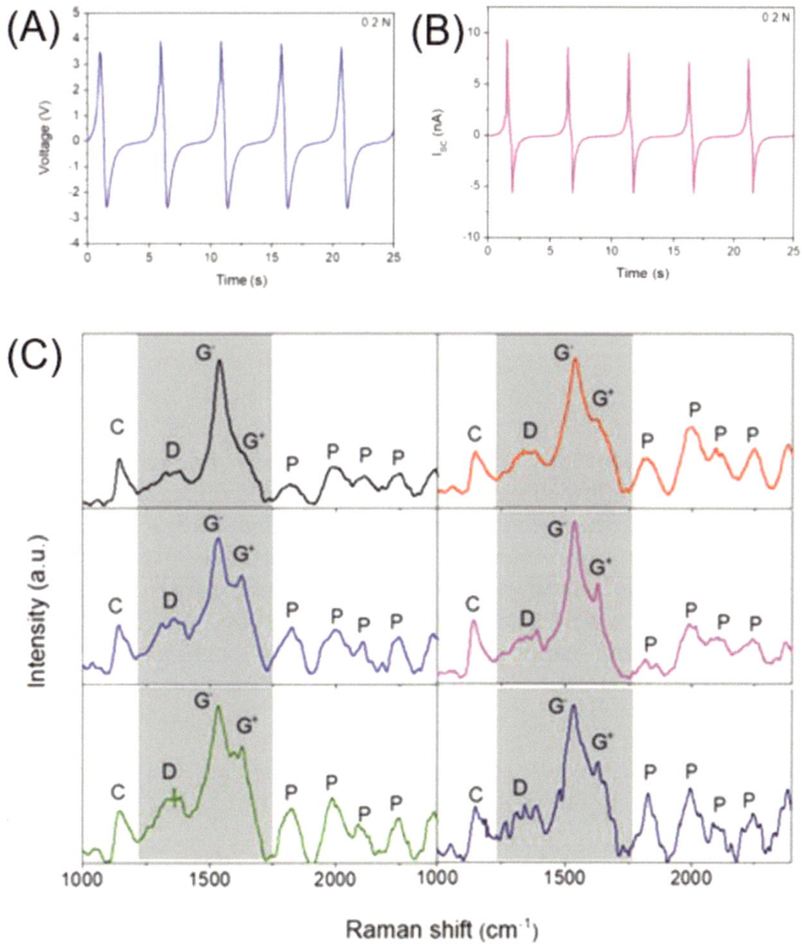

Figure 7.
(A) Voltage and (B) current outputs of CEC film when subjected to 0.1 N force, (C) ex-situ Raman spectrum of CEC film recorded at various spots subjected to mechanical stimuli.

subjected to an external applied force of 0.2 N [22]. The electrical output ranges were increased linearly with respect to an increase in forces, demonstrating their ideal electromechanical stability [62]. The observation of mechanical to electrical energy harvesting properties in carbon-based materials are not new. For instance, these types of properties were observed earlier in graphene sheets, graphene oxides, lithiated carbon fibers, carbon nanotubes (CNT), twisting CNT yarn, and graphene nitrides very recently [63–66]. However, the mechanism of energy harvesting in these carbons are different to each other. Very recently, Raman spectroscopy is used to probe the presence of alkene/alkyne transition as a cause for the electrochemical actuation process involved in graphdiyne [67]. Considering the physical properties of carbynes that can be tuned upon subjected to any form of external stimuli, and the recent theoretical studies, it is possible that chemical bonding/structural transformation can be occurred in CEC when subjected to mechanical deformation. Therefore, *ex-situ* Raman spectral analysis of CEC (in deformed state) as shown in **Figure 7(C)** is analyzed to understand their detailed mechanism of energy transduction process. In comparison to the Raman spectrum of CEC film recorded at normal states, the ex-situ Raman analysis obtained at deformed state showed interesting observations as follows: (i) The cumulenic band is retained in the CEC even at deformed condition, (ii) the origin of D band (~1370 cm^{-1}) as a result of deformation induced defects at the surface states [68, 69], (iii) occurrence of band splitting of the carbonoid band into G^- (1540 cm^{-1}) and G^+ (1590 cm^{-1}) bands due to origin of semiconducting behavior [68, 70], and (iv) the appearance of new bands over the region 1750 to 2300 cm^{-1} corresponding to the polynne type sp- carbon [71, 72]. These findings demonstrated the structural transformation of carbonoid type carbon in the CEC into semiconducting-type and polynne-type carbons via mechanical deformation which is the reason for the observed mechanical to electrical energy conversion in CEC. Moreover, these experimental results support the theoretical analysis on "electronic flexoelectricity" in low-dimensional carbon nanostructures subjected to mechanical deformation [73, 74].

4. Conclusions

This chapter describes the use of Raman spectroscopy and mapping analysis for the characterization for low-dimensional nanostructured materials, paint coatings, and solid electrolyte- piezo-polymer composites. Further, utilization of Raman spectroscopy to monitor the molecular level changes occurred in carbyne-enriched carbyne film based mechanical energy harvester when they are subjected to applied force is also described. Overall, this chapter provides a novel insight for the use of Raman spectroscopy and mapping as an important tool for characterization of low-dimensional nanostructures.

Acknowledgement

This research work was supported by the Basic Science Research Program through the National Research Foundation of Korea (NRF) grant funded by the Korean government (MSIT) (2019R1A2C3009747, 2020R1A2C2007366, and 2021R1A4A2000934).

Author details

Karthikeyan Krishnamoorthy[1*] and Sang-Jae Kim[1,2*]

1 Nanomaterials and System Laboratory, Major of Mechatronics Engineering, Faculty of Applied Energy System, Jeju National University, Jeju, Republic of Korea

2 R & D Center for New Energy Industry, Jeju National University, Jeju, Republic of Korea

*Address all correspondence to: karthi.nanotech@gmail.com and kimsangj@jejunu.ac.kr

References

[1] A.M. Herrero, Raman spectroscopy a promising technique for quality assessment of meat and fish: A review, Food Chem. 107 (2008) 1642-1651. https://doi.org/10.1016/j.foodchem.2007.10.014.

[2] M. Paillet, R. Parret, J.-L. Sauvajol, P. Colomban, Graphene and related 2D materials: An overview of the Raman studies, J. Raman Spectrosc. 49 (2018) 8-12. https://doi.org/10.1002/jrs.5295.

[3] H. Zhou, C.S. Simmons, M. Sarntinoranont, G. Subhash, Raman Spectroscopy Methods to Characterize the Mechanical Response of Soft Biomaterials, Biomacromolecules. 21 (2020) 3485-3497. https://doi.org/10.1021/acs.biomac.0c00818.

[4] B. Jalali, V. Raghunathan, D. Dimitropoulos, Ö. Boyraz, Raman-based silicon photonics, IEEE J. Sel. Top. Quantum Electron. 12 (2006) 412-421. https://doi.org/10.1109/JSTQE.2006.872708.

[5] M. Dandu, K. Watanabe, T. Taniguchi, A.K. Sood, K. Majumdar, Spectrally Tunable, Large Raman Enhancement from Nonradiative Energy Transfer in the van der Waals Heterostructure, ACS Photonics. 7 (2020) 519-527. https://doi.org/10.1021/acsphotonics.9b01648.

[6] J. Ju, C.M. Hsieh, Y. Tian, J. Kang, R. Chia, H. Chang, Y. Bai, C. Xu, X. Wang, Q. Liu, Surface Enhanced Raman Spectroscopy Based Biosensor with a Microneedle Array for Minimally Invasive in Vivo Glucose Measurements, ACS Sensors. 5 (2020) 1777-1785. https://doi.org/10.1021/acssensors.0c00444.

[7] X. Lin, Y. Wang, L. Wang, Y. Lu, J. Li, D. Lu, T. Zhou, Z. Huang, J. Huang, H. Huang, S. Qiu, R. Chen, D. Lin, S. Feng, Interference-free and high precision biosensor based on surface enhanced Raman spectroscopy integrated with surface molecularly imprinted polymer technology for tumor biomarker detection in human blood, Biosens. Bioelectron. 143 (2019) 111599. https://doi.org/10.1016/j.bios.2019.111599.

[8] S. Gulyamov, M. Shamshiddinova, W. Bae, Y.C. Park, H.J. Kim, W. Cho, Y. Lee, Identification of biomarkers on kidney failure by Raman spectroscopy, J. Raman Spectrosc. (2021) jrs.6210. https://doi.org/10.1002/jrs.6210.

[9] R.R. Jones, D.C. Hooper, L. Zhang, D. Wolverson, V.K. Valev, Raman Techniques: Fundamentals and Frontiers, Nanoscale Res. Lett. 14 (2019) 1-34. https://doi.org/10.1186/s11671-019-3039-2.

[10] K.J.I. Ember, M.A. Hoeve, S.L. McAughtrie, M.S. Bergholt, B.J. Dwyer, M.M. Stevens, K. Faulds, S.J. Forbes, C.J. Campbell, Raman spectroscopy and regenerative medicine: a review, Npj Regen. Med. 2 (2017) 1-10. https://doi.org/10.1038/s41536-017-0014-3.

[11] G.M. Do Nascimento, Introductory Chapter: The Multiple Applications of Raman Spectroscopy, in: Raman Spectrosc., InTech, 2018. https://doi.org/10.5772/intechopen.75795.

[12] G. Morari do Nascimento, Raman Spectroscopy and Imaging of Carbon Allotropes, in: Mod. Spectrosc. Tech. Appl., IntechOpen, 2020. https://doi.org/10.5772/intechopen.90867.

[13] X. Yu, Z. Zhang, F. Liu, Y. Ren, Synthesis of transfer-free graphene on cemented carbide surface, Sci. Rep. 8 (2018) 4759. https://doi.org/10.1038/s41598-018-23206-8.

[14] F. Tres, K. Treacher, J. Booth, L.P. Hughes, S.A.C. Wren, J.W. Aylott, J.C. Burley, Real time Raman imaging to

understand dissolution performance of amorphous solid dispersions, J. Control. Release. 188 (2014) 53-60. https://doi.org/10.1016/j.jconrel.2014.05.061.

[15] K. Krishnamoorthy, M. Veerapandian, K. Yun, S.-J. Kim, The chemical and structural analysis of graphene oxide with different degrees of oxidation, Carbon N. Y. 53 (2013) 38-49. https://doi.org/10.1016/j.carbon.2012.10.013.

[16] S. Manoharan, P. Pazhamalai, V.K. Mariappan, K. Murugesan, S. Subramanian, K. Krishnamoorthy, S.-J. Kim, Proton conducting solid electrolyte-piezoelectric PVDF hybrids: Novel bifunctional separator for self-charging supercapacitor power cell, Nano Energy. 83 (2021) 105753. https://doi.org/10.1016/j.nanoen.2021.105753.

[17] K. Krishnamoorthy, P. Pazhamalai, J.H. Lim, K.H. Choi, S.-J. Kim, M.P. Pazhamalai, J.H. Lim, K.H. Choi, S.-J. Kim, Mechanochemical Reinforcement of Graphene Sheets into Alkyd Resin Matrix for the Development of Electrically Conductive Paints, ChemNanoMat. 4 (2018) 568-574. https://doi.org/10.1002/cnma.201700391.

[18] K. Krishnamoorthy, G.K. Veerasubramani, P. Pazhamalai, S.J. Kim, Designing two dimensional nanoarchitectured MoS_2 sheets grown on Mo foil as a binder free electrode for supercapacitors, Electrochim. Acta. 190 (2016) 305-312. https://doi.org/10.1016/j.electacta.2015.12.148.

[19] K. Krishnamoorthy, P. Pazhamalai, S.-J.J. Kim, Two-dimensional siloxene nanosheets: novel high-performance supercapacitor electrode materials, Energy Environ. Sci. 11 (2018) 1595-1602. https://doi.org/10.1039/C8EE00160J.

[20] V.K. Mariappan, K. Krishnamoorthy, P. Pazhamalai, S. Natarajan, S. Sahoo, S.S. Nardekar, S.-J. Kim, Antimonene dendritic nanostructures: Dual-functional material for high-performance energy storage and harvesting devices, Nano Energy. 77 (2020) 105248. https://doi.org/10.1016/j.nanoen.2020.105248.

[21] V.K. Mariappan, K. Krishnamoorthy, P. Pazhamalai, S.-J. Kim, Exploring the bifunctional properties of paper-like carbyne-enriched carbon for maintenance-free self-powered systems, Mater. Adv. 1 (2020) 1644-1652. https://doi.org/10.1039/D0MA00324G.

[22] K. Krishnamoorthy, V.K. Mariappan, P. Pazhamalai, S. Sahoo, S.-J. Kim, Mechanical energy harvesting properties of free-standing carbyne enriched carbon film derived from dehydrohalogenation of polyvinylidene fluoride, Nano Energy. 59 (2019) 453-463. https://doi.org/10.1016/j.nanoen.2019.02.041.

[23] T. Yu, Z. Ni, C. Du, Y. You, Y. Wang, Z. Shen, Raman mapping investigation of graphene on transparent flexible substrate: The strain effect, J. Phys. Chem. C. 112 (2008) 12602-12605. https://doi.org/10.1021/jp806045u.

[24] D.W. Lee, J.W. Seo, Sp^2/Sp^3 carbon ratio in graphite oxide with different preparation times, J. Phys. Chem. C. 115 (2011) 2705-2708. https://doi.org/10.1021/jp107906u.

[25] U.K. Sur, Graphene: A Rising Star on the Horizon of Materials Science, Int. J. Electrochem. 2012 (2012) 1-12. https://doi.org/10.1155/2012/237689.

[26] D. Chen, H. Feng, J. Li, Graphene oxide: Preparation, functionalization, and electrochemical applications, Chem. Rev. 112 (2012) 6027-6053. https://doi.org/10.1021/cr300115g.

[27] A.C. Ferrari, J. Robertson, Interpretation of Raman spectra of

disordered and amorphous carbon, Phys. Rev. B. 61 (2000) 14095-14107. https://doi.org/10.1103/PhysRevB.61.14095.

[28] L.G. Caņado, K. Takai, T. Enoki, M. Endo, Y.A. Kim, H. Mizusaki, A. Jorio, L.N. Coelho, R. Magalhães-Paniago, M.A. Pimenta, General equation for the determination of the crystallite size la of nanographite by Raman spectroscopy, Appl. Phys. Lett. 88 (2006) 163106. https://doi.org/10.1063/1.2196057.

[29] T. Wang, G. Chen, C. Wu, D. Wu, Study on the graphite nanosheets/resin shielding coatings, Prog. Org. Coatings. 59 (2007) 101-105. https://doi.org/10.1016/j.porgcoat.2006.12.006.

[30] L.L. Vovchenko, L.Y. Matzui, V. V. Oliynyk, V.L. Launetz, The Effect of Filler Morphology and Distribution on Electrical and Shielding Properties of Graphite-Epoxy Composites, Mol. Cryst. Liq. Cryst. 535 (2011) 179-188. https://doi.org/10.1080/15421406.2011.538335.

[31] S. Kataria, A. Patsha, S. Dhara, A.K. Tyagi, H.C. Barshilia, Raman imaging on high-quality graphene grown by hot-filament chemical vapor deposition, J. Raman Spectrosc. 43 (2012) 1864-1867. https://doi.org/10.1002/jrs.4113.

[32] D. Graf, F. Molitor, K. Ensslin, C. Stampfer, A. Jungen, C. Hierold, L. Wirtz, Raman imaging of graphene, Solid State Commun. 143 (2007) 44-46. https://doi.org/10.1016/j.ssc.2007.01.050.

[33] N. Islam, M.N.F. Hoque, W. Li, S. Wang, J. Warzywoda, Z. Fan, Vertically edge-oriented graphene on plasma pyrolyzed cellulose fibers and demonstration of kilohertz high-frequency filtering electrical double layer capacitors, Carbon N. Y. 141 (2019) 523-530. https://doi.org/10.1016/j.carbon.2018.10.012.

[34] L. Wang, S. Li, F. Huang, X. Yu, M. Liu, H. Zhang, Interface modification of hierarchical Co_9S_8@NiCo layered dihydroxide nanotube arrays using polypyrrole as charge transfer layer in flexible all-solid asymmetric supercapacitors, J. Power Sources. 439 (2019) 227103. https://doi.org/10.1016/j.jpowsour.2019.227103.

[35] Y. Yao, L. Ren, S. Gao, S. Li, Histogram method for reliable thickness measurements of graphene films using atomic force microscopy (AFM), J. Mater. Sci. Technol. 33 (2017) 815-820. https://doi.org/10.1016/j.jmst.2016.07.020.

[36] X. Li, H. Zhu, Two-dimensional MoS_2: Properties, preparation, and applications, J. Mater. 1 (2015) 33-44. https://doi.org/10.1016/j.jmat.2015.03.003.

[37] M. Acerce, D. Voiry, M. Chhowalla, Metallic 1T phase MoS_2 nanosheets as supercapacitor electrode materials, Nat. Nanotechnol. 10 (2015) 313-318. https://doi.org/10.1038/nnano.2015.40.

[38] Z. Yin, H. Li, H. Li, L. Jiang, Y. Shi, Y. Sun, G. Lu, Q. Zhang, X. Chen, H. Zhang, Single-Layer MoS_2 Phototransistors, ACS Nano. 6 (2012) 74-80. https://doi.org/10.1021/nn2024557.

[39] K. Krishnamoorthy, G.K. Veerasubramani, S. Radhakrishnan, S.J. Kim, Supercapacitive properties of hydrothermally synthesized sphere like MoS_2 nanostructures, Mater. Res. Bull. 50 (2014) 499-502. https://doi.org/10.1016/j.materresbull.2013.11.019.

[40] Y. Cheng, K. Yao, Y. Yang, L. Li, Y. Yao, Q. Wang, X. Zhang, Y. Han, U. Schwingenschlögl, Van der Waals epitaxial growth of MoS_2 on SiO_2/Si by chemical vapor deposition, RSC Adv. 3 (2013) 17287. https://doi.org/10.1039/c3ra42171f.

[41] B. Chakraborty, H.S.S.R. Matte, A.K. Sood, C.N.R. Rao, Layer-dependent resonant Raman scattering of a few layer MoS_2, J. Raman Spectrosc. 44 (2013) 92-96. https://doi.org/10.1002/jrs.4147.

[42] P. Ares, J.J. Palacios, G. Abellán, J. Gómez-Herrero, F. Zamora, Recent Progress on Antimonene: A New Bidimensional Material, Adv. Mater. 30 (2018) 1703771. https://doi.org/10.1002/adma.201703771.

[43] S. Zhang, Z. Yan, Y. Li, Z. Chen, H. Zeng, Atomically Thin Arsenene and Antimonene: Semimetal-Semiconductor and Indirect-Direct Band-Gap Transitions, Angew. Chemie Int. Ed. 54 (2015) 3112-3115. https://doi.org/10.1002/anie.201411246.

[44] X. Wang, J. He, B. Zhou, Y. Zhang, J. Wu, R. Hu, L. Liu, J. Song, J. Qu, Bandgap-Tunable Preparation of Smooth and Large Two-Dimensional Antimonene, Angew. Chemie Int. Ed. 57 (2018) 8668-8673. https://doi.org/10.1002/anie.201804886.

[45] J. Medina-Ramos, R.C. Pupillo, T.P. Keane, J.L. DiMeglio, J. Rosenthal, Efficient Conversion of CO_2 to CO Using Tin and Other Inexpensive and Easily Prepared Post-Transition Metal Catalysts, J. Am. Chem. Soc. 137 (2015) 5021-5027. https://doi.org/10.1021/ja5121088.

[46] L. Liang, Y. Xu, Y. Li, H. Dong, M. Zhou, H. Zhao, U. Kaiser, Y. Lei, Facile synthesis of hierarchical fern leaf-like Sb and its application as an additive-free anode for fast reversible Na-ion storage, J. Mater. Chem. A. 5 (2017) 1749-1755. https://doi.org/10.1039/C6TA10345F.

[47] Y. Shao, Z.-L. Liu, C. Cheng, X. Wu, H. Liu, C. Liu, J.-O. Wang, S.-Y. Zhu, Y.-Q. Wang, D.-X. Shi, K. Ibrahim, J.-T. Sun, Y.-L. Wang, H.-J. Gao, Epitaxial Growth of Flat Antimonene Monolayer: A New Honeycomb Analogue of Graphene, Nano Lett. 18 (2018) 2133-2139. https://doi.org/10.1021/acs.nanolett.8b00429.

[48] M. Fortin-Deschênes, O. Waller, T.O. Menteş, A. Locatelli, S. Mukherjee, F. Genuzio, P.L. Levesque, A. Hébert, R. Martel, O. Moutanabbir, Synthesis of Antimonene on Germanium, Nano Lett. 17 (2017) 4970-4975. https://doi.org/10.1021/acs.nanolett.7b02111.

[49] H. Nakano, M. Ishii, H. Nakamura, Preparation and structure of novel siloxene nanosheets., Chem. Commun. (Camb). 2 (2005) 2945-2947. https://doi.org/10.1039/b500758e.

[50] P. Pazhamalai, K. Krishnamoorthy, S. Sahoo, V.K. Mariappan, S.-J. Kim, Understanding the Thermal Treatment Effect of Two-Dimensional Siloxene Sheets and the Origin of Superior Electrochemical Energy Storage Performances, ACS Appl. Mater. Interfaces. 11 (2019) 624-633. https://doi.org/10.1021/acsami.8b15323.

[51] K. Xu, L. Ben, H. Li, X. Huang, Silicon-based nanosheets synthesized by a topochemical reaction for use as anodes for lithium ion batteries, Nano Res. 8 (2015) 2654-2662. https://doi.org/10.1007/s12274-015-0772-4.

[52] S. Li, H. Wang, D. Li, X. Zhang, Y. Wang, J. Xie, J. Wang, Y. Tian, W. Ni, Y. Xie, Siloxene nanosheets: a metal-free semiconductor for water splitting, J. Mater. Chem. A. 4 (2016) 15841-15844. https://doi.org/10.1039/C6TA07545B.

[53] H. Imagawa, N. Takahashi, T. Nonaka, Y. Kato, K. Nishikawa, H. Itahara, Synthesis of a calcium-bridged siloxene by a solid state reaction for optical and electrochemical properties, J. Mater. Chem. A. 3 (2015) 9411-9414. https://doi.org/10.1039/C5TA00321K.

[54] R. Fu, K. Zhang, R.P. Zaccaria, H. Huang, Y. Xia, Z. Liu, Two-dimensional silicon suboxides nanostructures with Si

nanodomains confined in amorphous SiO_2 derived from siloxene as high performance anode for Li-ion batteries, Nano Energy. 39 (2017) 546-553. https://doi.org/10.1016/J.NANOEN.2017.07.040.

[55] H.D. Fuchs, M. Stutzmann, M.S. Brandt, M. Rosenbauer, J. Weber, A. Breitschwerdt, P. Deák, M. Cardona, Porous silicon and siloxene: Vibrational and structural properties, Phys. Rev. B. 48 (1993) 8172-8189. https://doi.org/10.1103/PhysRevB.48.8172.

[56] P. Yao, H. Yu, Z. Ding, Y. Liu, J. Lu, M. Lavorgna, J. Wu, X. Liu, Review on Polymer-Based Composite Electrolytes for Lithium Batteries, Front. Chem. 7 (2019) 522. https://doi.org/10.3389/fchem.2019.00522.

[57] H. He, Y. Fu, T. Zhao, X. Gao, L. Xing, Y. Zhang, X. Xue, All-solid-state flexible self-charging power cell basing on piezo-electrolyte for harvesting/storing body-motion energy and powering wearable electronics, Nano Energy. 39 (2017) 590-600. https://doi.org/10.1016/J.NANOEN.2017.07.033.

[58] C. Xu, L. Zhang, Y. Xu, Z. Yin, Q. Chen, S. Ma, H. Zhang, R. Huang, C. Zhang, L. Jin, W. Yang, J. Lu, Filling the holes in piezopolymers with a solid electrolyte: a new paradigm of poling-free dynamic electrets for energy harvesting, J. Mater. Chem. A. 5 (2017) 189-200. https://doi.org/10.1039/C6TA07297F.

[59] V. V. Korshak, Y.P. Kudryavtsev, Y. V. Korshak, S.E. Evsyukov, V. V. Khvostov, V.G. Babaev, M.B. Guseva, Formation of β-carbyne by dehydrohalogenation, Die Makromol. Chemie, Rapid Commun. 9 (1988) 135-140. https://doi.org/10.1002/marc.1988.030090304.

[60] N.N. Mel'nik, D.Y. Korobov, O.S. Plyashechnik, V. V. Savranskii, Raman spectra of carbyne-fullerene structures, Bull. Lebedev Phys. Inst. 40 (2013) 187-190. https://doi.org/10.3103/S1068335613070038.

[61] F. Cataldo, D. Capitani, Preparation and characterization of carbonaceous matter rich in diamond-like carbon and carbyne moieties, Mater. Chem. Phys. 59 (1999) 225-231. https://doi.org/10.1016/S0254-0584(99)00055-3.

[62] N. Muralidharan, M. Li, R.E. Carter, N. Galioto, C.L. Pint, Ultralow Frequency Electrochemical–Mechanical Strain Energy Harvester Using 2D Black Phosphorus Nanosheets, ACS Energy Lett. 2 (2017) 1797-1803. https://doi.org/10.1021/acsenergylett.7b00478.

[63] M.T. Ong, E.J. Reed, Engineered Piezoelectricity in Graphene, ACS Nano. 6 (2012) 1387-1394. https://doi.org/10.1021/nn204198g.

[64] L. Liu, Y. Cheng, L. Zhu, S.-T. Lee, F. Liao, M. Shao, The Surface Polarized Graphene Oxide Quantum Dot Films for Flexible Nanogenerators, Sci. Rep. 6 (2016) 32943. https://doi.org/10.1038/srep32943.

[65] E. Jacques, G. Lindbergh, D. Zenkert, S. Leijonmarck, M.H. Kjell, Piezo-Electrochemical Energy Harvesting with Lithium-Intercalating Carbon Fibers, ACS Appl. Mater. Interfaces. 7 (2015) 13898-13904. https://doi.org/10.1021/acsami.5b02585.

[66] M. V Ilina, Y.F. Blinov, O.I. Ilin, N.N. Rudyk, O.A. Ageev, Piezoelectric effect in non-uniform strained carbon nanotubes, IOP Conf. Ser. Mater. Sci. Eng. 256 (2017) 012024. https://doi.org/10.1088/1757-899X/256/1/012024.

[67] C. Lu, Y. Yang, J. Wang, R. Fu, X. Zhao, L. Zhao, Y. Ming, Y. Hu, H. Lin, X. Tao, Y. Li, W. Chen, High-performance graphdiyne-based electrochemical actuators, Nat. Commun. 9 (2018) 752. https://doi.org/10.1038/s41467-018-03095-1.

[68] S. Takabayashi, K. Okamoto, T. Nakatani, Influence of post-annealing on a diamondlike carbon film analyzed by Raman spectroscopy, Surf. Interface Anal. 50 (2018) 441-447. https://doi.org/10.1002/sia.6387.

[69] S. Thangavel, K. Krishnamoorthy, V. Krishnaswamy, N. Raju, S.J. Kim, G. Venugopal, Graphdiyne–ZnO Nanohybrids as an Advanced Photocatalytic Material, J. Phys. Chem. C. 119 (2015) 22057-22065. https://doi.org/10.1021/acs.jpcc.5b06138.

[70] H. Telg, M. Fouquet, J. Maultzsch, Y. Wu, B. Chandra, J. Hone, T.F. Heinz, C. Thomsen, G- and G+ in the Raman spectrum of isolated nanotube: a study on resonance conditions and lineshape, Phys. Status Solidi. 245 (2008) 2189-2192. https://doi.org/10.1002/pssb.200879658.

[71] J. Kastner, H. Kuzmany, L. Kavan, F.P. Dousek, J. Kuerti, Reductive Preparation of Carbyne with High Yield. An in Situ Raman Scattering Study, Macromolecules. 28 (1995) 344-353. https://doi.org/10.1021/ma00105a048.

[72] K. Akagi, M. Nishiguchi, H. Shirakawa, Y. Furukawa, I. Harada, One-dimensional conjugated carbyne - synthesis and properties, Synth. Met. 17 (1987) 557-562. https://doi.org/10.1016/0379-6779(87)90798-3.

[73] S. V. Kalinin, V. Meunier, Electronic flexoelectricity in low-dimensional systems, Phys. Rev. B. 77 (2008) 033403. https://doi.org/10.1103/PhysRevB.77.033403.

[74] A.G. Kvashnin, P.B. Sorokin, B.I. Yakobson, Flexoelectricity in Carbon Nanostructures: Nanotubes, Fullerenes, and Nanocones, J. Phys. Chem. Lett. 6 (2015) 2740-2744. https://doi.org/10.1021/acs.jpclett.5b01041.

Chapter 12

Tip-Enhanced Raman Spectroscopy of 2D Semiconductors

Mahfujur Rahaman and Dietrich R.T. Zahn

Abstract

Two-dimensional (2D) semiconductors are one of the most extensively studied modern materials showing potentials in large spectrum of applications from electronics/optoelectronics to photocatalysis and CO_2 reduction. These materials possess astonishing optical, electronic, and mechanical properties, which are different from their bulk counterparts. Due to strong dielectric screening, local heterogeneities such as edges, grain boundaries, defects, strain, doping, chemical bonding, and molecular orientation dictate their physical properties to a great extent. Therefore, there is a growing demand of probing such heterogeneities and their effects on the physical properties of 2D semiconductors on site in a label-free and non-destructive way. Tip-enhanced Raman spectroscopy (TERS), which combines the merits of both scanning probe microscopy and Raman spectroscopy, has experienced tremendous progress since its introduction in the early 2000s and is capable of local spectroscopic investigation with (sub-) nanometer spatial resolution. Introducing this technique to 2D semiconductors not only enables us to understand the effects of local heterogeneities, it can also provide new insights opening the door for novel quantum mechanical applications. This book chapter sheds light on the recent progress of local spectroscopic investigation and chemical imaging of 2D semiconductors using TERS. It also provides a basic discussion of Raman selection rules of 2D semiconductors important to understand TERS results. Finally, a brief outlook regarding the potential of TERS in the field of 2D semiconductors is provided.

Keywords: TERS, Raman spectroscopy, 2D semiconductors, TMDC, MMC, plasmonics, nanoscale chemical imaging, local heterogeneities

1. Introduction

The ability of isolating stable, atomically thin monolayers of layered materials stimulated a new field of atomic-scale interface physics, with tremendous potential for novel quantum optoelectronic applications [1]. Graphene, the first isolated 2D material consisting of an atomically thin carbon sheet provides much stronger mechanical strength and electrical or heat conductivity compared to its bulk counterpart graphite [2–4]. Thus, graphene was predicted to have great impact on devices with ballistic charge transport or quantum anomalous Hall effect [5]. However, the lack of a bandgap coupled with challenges and consequences associated to the attempts of bandgap opening hinders its application as an active material in semiconducting devices [6]. In this context, 2D semiconductors have attracted significant attention owing to their suitable bandgap required for optoelectronic applications. Some of these semiconductors already possess carrier mobilities that can outperform existing Si CMOS technology at the scaling limit, although they are affected by some other limiting factors [7]. Therefore, extensive research is going on to gain both fundamental understanding of these materials and to explore new 2D semiconductors for potential applications.

One of the novel aspects of 2D semiconductors is that their optical properties can be tuned as a function of layer number [8]. The most famous 2D semiconductors in this respect are the transition metal dichalcogenides (TMDCs). For example, in bulk TMDCs are indirect bandgap semiconductors, which can be tuned upward ~1 eV with decreasing layer number down to monolayer. More importantly, the nature of the bandgap also changes from indirect to direct at the monolayer thickness [9]. Density functional theory (DFT) predicts that the direct excitonic transition energy at the Brillouin zone (BZ) K point in TMDCs hardly shows any dependence with respect to the layer thickness [9]. However, the indirect transition along the $\Gamma - Q$ (valence band – conduction band) direction (the smallest energy gap in the bulk, *i.e.* the bandgap) increases monotonically as the layer number decreases. Therefore, at monolayer thickness the direct transition becomes smaller than the indirect gap and hence the TMDC becomes a direct bandgap semiconductor. Black phosphorous (BP) also exhibits a similar behavior (from 0.3 eV in bulk to 1.5 eV in monolayer) whereas the nature of the bandgap remains direct due to strong interlayer coupling [10]. Other 2D semiconductors such as, metal monochalcogenides (MMCs) are direct bandgap semiconductors in bulk and turn into indirect bandgap semiconductor at some thicknesses as the layer number decreases [11, 12].

Even though these semiconductors are few atoms thick, some of them can absorb up to 15% of light in the visible range due to strong light-matter interaction [13]. The dramatically reduced dielectric screening in the out-of-plane direction creates strongly bound excitons with binding energies in the range of few hundreds of meV [14, 15]. Therefore, their optical response is dominated by excitons or multi-particle complexes at room temperature (RT) [16, 17]. Atomically flat dangling bond free surfaces of these materials are free from career scattering caused by surface roughness, which can lead to high performance optoelectronic devices with large on–off ratio ($\sim 10^8$), photoresponsivity ($\sim 10^7$ mAW^{-1}), and career mobility (10^3 cm^2V^{-1} s^{-1}) at RT [7, 18, 19]. Both few layer BP and InSe based field effect transistors show comparable career mobility (10^3 cm^2V^{-1} s^{-1}) at RT suggesting they can outperform Si based CMOS at the scaling limits [20]. However, BP suffers from poor stability in air [21] and the polar nature of InSe leads to layer dependent Fröhlich interaction [22] and thus optical phonon limited career mobility.

For all the extraordinary optoelectronic properties of 2D semiconductors, the influence of local heterogeneities such as edges, grain boundaries, defects, strain, doping, chemical bonding *etc.* are very crucial due to the strong Coulomb screening in these systems [23, 24]. Therefore, a powerful analytical technique, which is capable of studying chemical structures, phase, stress, impurities, molecular interaction *etc.* at the nanoscale, is essential to understand the underlying physics important for device applications. Raman spectroscopy meets all criteria when it comes to measure these heterogeneities at the micro-scale. It has been frequently applied as a non-destructive and straightforward to measure technique to study 2D semiconductors over the last decade [25]. However, the major disadvantages of this technique are its low sensitivity and optical diffraction limited spatial resolution. While typically one in every 10^5–10^7 scattered photons is counted for Raman scattering, the typical spatial resolution of conventional micro-Raman scattering is in the range of 0.5 to 1 μm. Hence, both of these characteristics limit Raman spectroscopy from acquiring the local heterogeneous information stated above. Therefore, tip-enhanced Raman spectroscopy (TERS) has become a popular analytical technique for studying 2D semiconductors in recent years [26]. It combines the benefits of both scanning probe microscopy (SPM) and Raman spectroscopy, thus permits a very good spatial resolution and high sensitivity for local spectroscopic investigation and chemical imaging. In this chapter, we start with discussing the principle of TERS by explaining the technical aspects of it. Then, recent developments and applications of TERS of 2D semiconductors are discussed. Finally, conclusions and future directions of TERS of 2D semiconductors are addressed.

2. Principle of TERS

Raman scattering is at the core of TERS, which is light inelastically scattered by elementary excitations such as vibrational modes (phonons) in the material probed. Consequently photons are emitted that are shifted in energy with respect to the energy of the exciting light. The shifts are directly correlated to the phonon frequencies of the material. This effect was first observed by C. V. Raman in 1928 and thus named after him [27]. The fundamentals and the various aspects of Raman scattering are comprehensively covered *e.g.* in the book series "Light Scattering in Solids I-IX" [28, 29]. Here, we would like to point out a few facts, which are relevant for TERS. First, the Raman process in matter involves the instantaneous absorption of a photon, excitation of the electronic system, scattering of electron or hole by a vibrational mode or phonon, and relaxation of the excited state accompanied by the emission of a photon. This higher order process has an extremely weak Raman scattering cross section (typically 10^{-26}–10^{-31} cm^2) [30, 31]. The intensity of the inelastically scattered Raman light is thus orders of magnitude weaker than the elastically scattered Rayleigh light, which imposes the need of strong reduction of the elastically scattered light by the Raman spectrometers. The weakness of Raman scattering is especially critical if the amount of matter and/or the available scattering volume are very small as is the case for low-dimensional structures such as quantum dots or 2D semiconductors. For such samples, enhancement mechanisms are employed. An important one is resonance enhancement, for which the exciting light is tuned to a real electronic transition of the sample leading to strong optical absorption and consequently stronger Raman scattering (typically two orders of magnitude). However, this may not be sufficient for low-dimensional systems and thus further enhancement *e.g.* by using surface-enhanced Raman scattering (SERS) is required. SERS has seen tremendous increase in applications in recent years [32]. Compared to conventional Raman cross sections SERS cross section can be up to ten orders of magnitude larger allowing *e.g.* single molecule detection [33–35]. SERS uses metallic nanostructures to create a very strong electric field via localized surface plasmon resonances (LSPR) or lightening rod effects also known as plasmonic effects [36]. Besides a local electromagnetic (EM) field enhancement, photo-driven charge transfer (CT) mechanisms can also lead to an additional enhancement factor of up to 10^3 [34]. Despite the remarkable improvement compared to conventional Raman scattering, SERS still suffers from the optically diffraction limit and is restricted to resolve spatial chemical information beyond 200 nm [26]. TERS can also be esteemed as the ultimate case of SERS with the SPM tip acting as the single metallic nanostructure enhancing the Raman cross section. The photon emitted in the Raman process is shifted in energy with respect to the photon energy of the exciting light towards higher (anti-Stokes) or lower energy (Stokes) by the energy of the elementary excitation (phonon) depending on whether a phonon is absorbed or emitted. The Stokes-shifted part of the Raman spectrum is predominantly probed in Raman experiments. As the Raman shift directly reflects the energy of the elementary excitation (phonon), it can directly be used to identify materials. Moreover, the intensity (under non-resonant conditions) is related to the number of scatterers (scattering volume), while the energy position and the lineshape can be influenced by parameters such as temperature, strain, doping, and crystallinity. Finally, the Raman scattering process obeys symmetry selection rules, which are represented by the so-called Raman tensors (derivative of the polarization tensors with respect to the phonon elongation). By choosing the polarization of the incident and scattered light it is possible to probe distinct components of the Raman tensor. The Raman selection rules can be altered by *e.g.* electric fields, strain, and by reduced dimensionality as well as by the presence of the SPM tip inducing strong electric field gradients.

TERS works using the same principal as SERS, *i.e.* it utilizes the plasmonic enhancement around a sharp metallic tip. However, unlike SERS it employs a single SPM tip to collect local phonon information by taking advantage of the lateral resolution of SPM [37, 38]. When excited by a suitable photon energy, the collective oscillation of the conduction band electrons in the sharp metallic tip amplifies and confines the optical field in the vicinity of the tip apex. The resulting enhancement of the EM field then produces Raman scattering from a nanoscopic volume of the sample under the apex. The general schematic of the TERS mechanism is shown in **Figure 1a**. For simplification the tip apex can be assumed as shown in the schematic to be a metal sphere acting as the metal nanostructure prerequisite for the plasmonic enhancement. When an EM wave with ***E*** field parallel to the tip long axis excites the tip, a strong plasmonic field (also known as local electric field or simply local field) is created around the apex since the metal nanostructure can facilitate both LSPR and lightening rod effect. A finite element method (FEM) simulation of the plasmonic field enhancement around the Au tip apex with a radius of 10 nm at 638 nm excitation is presented in **Figure 1b**. The local field amplitude is enhanced by a factor of 5 at the tip apex. Since, the Raman scattering is approximately proportional to the fourth power of the excitation field, TERS can thus amplify the local Raman information by a factor of 625 under these simulated conditions.

The lightning rod effect is introduced by the anisotropy of the tip geometry (the larger the anisotropy, the stronger the field enhancement) and thus independent of the excitation wavelength. However, the LSPR is created due to the collective oscillation of conduction band electrons at the metal surface. Therefore, several factors influence the LSPR energy and oscillator strength, such as material, size, shape, and dielectric interface [36, 39, 40]. Gold and silver are the two most widely used noble metals for TERS tips. Both metals reveal small dielectric loss, stability in air, and tunability of the LSPR in the visible spectrum. Importantly, they both have some advantages and disadvantages, which become critical deciding factors for

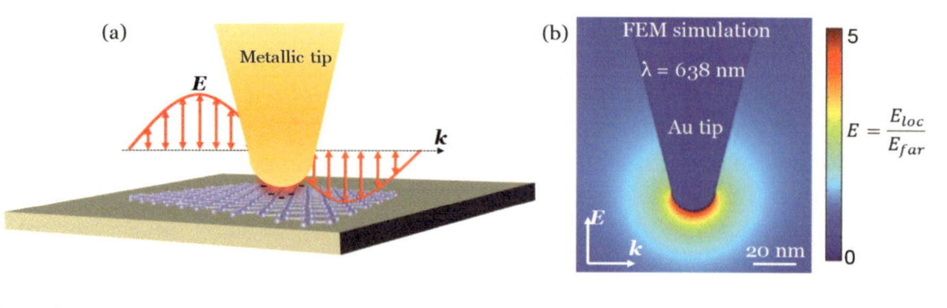

Figure 1.
Schematic representation of TERS (a) and FEM simulation of the plasmonic field enhancement at the tip apex for an Au tip (b).

TERS applications. For example, silver shows a more pronounced plasmonic effect but gold provides a better environmental stability. Thus, the latter makes gold the more popular material for TERS tips. Additionally, for the same size and shape of the apex, the LSPR of a tip made of Ag lies at a higher energy than that of an Au tip [36]. Therefore, Ag tips are more commonly in use for green excitation and Au tips are more suitable for red and near-IR TERS applications.

The TERS schematic shown in **Figure 1a** is also known as conventional TERS, in which the local electric field is created by the tip dipole oscillation. The TERS sensitivity (*i.e.* enhancement factor, EF, and spatial resolution, SR) is moderate in a conventional TERS geometry. Hence, a usual practice of increasing TERS sensitivity multifold is to introduce a metal substrate, which then creates an image dipole inside the substrate [41]. Thus, the combined tip-substrate system acts as a dimer and enhances the sensitivity to a great extent such that the EF can reach more than three order of magnitude higher values than that in the conventional TERS [42]. This way of enhancing the TERS sensitivity is also known as gap-mode TERS configuration.

Figure 2a displays the schematic of the gap-mode TERS configuration. The strength of the dimer coupling and consequently the gap-mode TERS sensitivity depend on the tip-substrate distance d. A comparison of generated local electric field distribution around an Au tip apex in conventional TERS and gap-mode TERS is presented in **Figure 2b** and **c**. As can be seen, the enhancement of the local electric field amplitude in the case of conventional TERS is almost 20 times smaller than that in the gap-mode TERS configuration. The spatial resolution of gap-mode TERS is also superior to that of the conventional TERS. The full width at half maximum (FWHM) of the local electric field E_{loc} distribution around the tip apex in conventional TERS is proportional to ($\sim 1.345(R + d)$), where R is the tip radius and d is the distance between the sample surface and the tip apex [41]. However, in the case of gap-mode TERS the FWHM of the E_{loc} distribution is given by $2\sqrt{Rd}$ as shown in **Figure 2d** [37]. Since the TERS intensity is proportional to the fourth power of E_{loc}, the TERS SR also turns out to be much smaller than that in the conventional geometry. The analysis of SR can be performed using the schematic shown in **Figure 2a**. The potential drop between the tip sphere and the image sphere can be written as $\Delta V = |E_{loc}|d$; whereas the potential difference between two

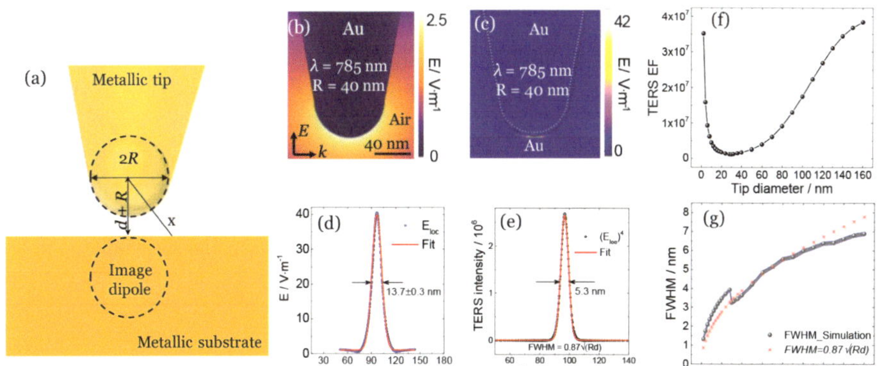

Figure 2.
Schematic of the gap-mode TERS configuration (a). Comparison of local electric field distribution between conventional (b) and gap-mode TERS (c). Local electric field amplitude (d) and TERS enhancement (e) in a gap-mode configuration shown in (c) calculated for a tip-substrate distance of 1 nm. TERS EF (f) and SR (g) in a gap-mode TERS configuration as a function of tip diameter. The tip-substrate distance was kept at 1 nm. Figure (a) to (e) reproduced from ref. [37] with permission from the Royal Society of Chemistry and (f) and (g) from ref. [38].

sites in the absence of the these two spheres can be expressed as $\Delta V = |E_0|(2R + d)$, E_0 is the incident field. Since, the spheres are equipotential, we can thus write

$$\left.\begin{aligned} \Delta V &= |E_0|(2R + d) = |E_{loc}|d \\ |E_{loc}| &= \frac{2R + d}{d} E_0 \end{aligned}\right\} \quad (1)$$

For a certain tip-substrate geometry both R and d are fixed at the smallest distance from the tip apex to the sample surface being d, which is along the center of the electric field lines. Therefore, the lateral offset of the electric field from the center can be written as [38].

$$|E_{loc}(x)| = \frac{\Delta V}{2R + d - 2\sqrt{R^2 - x^2}} \quad (2)$$

For the TERS intensity Eq. (2) becomes

$$|E_{loc}(x)|^4 = \left(\frac{\Delta V}{2R + d - 2\sqrt{R^2 - x^2}}\right)^4 \quad (3)$$

At $x = 0$, $|E_{loc}|^4$ has a maximum of $(\Delta V/d)^4$. Hence, at half maximum we can write

$$2d^4 = \left(2R + d - 2\sqrt{R^2 - x^2}\right)^4 \quad (4)$$

For a very small d and $R> > d$, we can neglect the terms containing d^2. Thus, the SR of gap-mode TERS becomes

$$SR = 2\sqrt{\left(\sqrt[4]{2} - 1\right)Rd} \approx 0.87\sqrt{Rd} \quad (5)$$

A more detailed theoretical study of the TERS enhancement and spatial resolution in gap-mode TERS as a function of tip diameter is shown in **Figure 2f–g**. The tip-substrate distance d was kept fixed at 1 nm in the simulation. Both the calculated TERS EF and SR under 638 nm excitation show two distinctive regimes, one below and the other above a tip size of 28 nm. The sharp increase of the TERS EF and SR below 28 nm of the tip diameter is due to the lightning rod effect. On contrary, the slow increment of both EF and SR above 28 nm is the result of the LSPR of the tip apex. The red asterisks in the SR graph (**Figure 2g**) represent the results of Eq. (5) in very good agreement to the SR calculated from the LSPR contribution of the tip apex. Eq. (5) clearly indicates that for a given tip radius both EF and SR can be amplified by reducing the distance d. However, at very small d, the quantum mechanical effects such as non-local screening effects and electron tunneling weaken the TERS sensitivity and impose an unavoidable limiting factor avoiding a singularity at $d = 0$ [43, 44].

3. Surface selection rules of TERS in 2D semiconductors

According to the FEM simulations, the overall electric field intensity beneath the tip apex is zero for an incident beam polarized along the XY plane as shown in **Figure 3a**. Therefore, for an incident beam polarized along the z-direction (along the tip long axis) the intensity enhancement matrix can be written as [26].

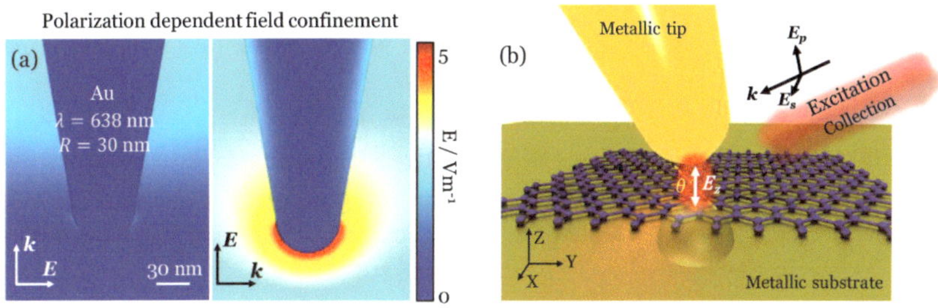

Figure 3.
Incident light polarization dependent E_{loc} *distribution around the tip apex (a). Schematic of the gap-mode TERS in the side-illumination geometry (b). The dashed line projects the normal to the basal plane of 2D semiconductor at any given geometrical situation, which is inclined at an angle* θ *to the z-axis.*

$$G = \begin{pmatrix} G_{xx} & 0 & 0 \\ 0 & G_{yy} & 0 \\ 0 & 0 & G_{zz} \end{pmatrix} \approx \begin{pmatrix} 0 & 0 & 0 \\ 0 & 0 & 0 \\ 0 & 0 & G_{zz} \end{pmatrix} \tag{6}$$

The above enhancement condition is also applicable for the gap-mode TERS configuration. Hence, for a side-illumination geometry at an incident angle, which satisfies the tip polarization condition, a tip in gap-mode TERS configuration confines and enhances the E_{loc} along z-direction as shown in **Figure 3b**. Thus, the TERS intensity can be expressed as

$$I_{TERS} \propto \left| G_{\omega_{in}} \cdot \alpha'_{zz} \cdot G_{\omega_{sc}} \right|^2 \tag{7}$$

$G_{\omega_{in}}$ and $G_{\omega_{sc}}$ are the intensity enhancement matrices for incident and scattered light. α'_{zz} is the Raman tensor involved in a vibrational mode along the z-direction, which can be derived from the 3 x 3 matrix of the anisotropic polarizability tensor associated to the phonon vibrations as follows

$$\alpha = \begin{pmatrix} \alpha_{xx} & \alpha_{xy} & \alpha_{xz} \\ \alpha_{yx} & \alpha_{yy} & \alpha_{yz} \\ \alpha_{zx} & \alpha_{zy} & \alpha_{zz} \end{pmatrix} \tag{8}$$

The idea of surface selection rules was first conceptualized by Moskovits on flat metallic surfaces [45]. Due to the adsorption of molecules on a flat metal surface, they will experience a different electric field, since incident light is reflected off the metal substrate. Assuming both s- and p-polarized incident light causing Raman scattering, the scattered radiation at any given angle of incidence, ϕ consists of superposition of two waves as given by [45]

$$\left. \begin{aligned} E'_s &= (1 + r'_s) p_x \\ E'_p &= p_y \left(1 - r'_p\right) \cos\phi' + p_z \left(1 + r'_p\right) \sin\phi' \end{aligned} \right\} \tag{9}$$

The primes are associated to the scattered radiation, r_s and r_p are the respective Fresnel reflection coefficient, and p_i is the polarization inside the molecule expressed by $p = \alpha E$, where α can be derived from Eq. (8). The four components associated to the Raman experiment are ss, sp, ps, and pp, the first and second letters refer to the incident and scattered light polarization, respectively. In the limit of a very good reflector, all the components except pp reduce to zero, since $r_s \approx -1$ and $r_p \approx 1$. Therefore, the effective Raman scattering intensity becomes

$$I_{sc}^{eff} \approx \left|\alpha'_{zz}\left(1+r_p\right)\left(1+r'_p\right)\right|^2 \sin^2\phi \, \sin^2\phi' \tag{10}$$

According to Eq. (10) the effective Raman scattering depends on the z-component of the Raman tensor, normally not measured in ordinary Raman spectroscopy. However, in the gap-mode TERS configuration the coupling between the phonon with tensor elements having z-component and the E_{loc} changes the TERS intensity dramatically leading to so-called TERS selection rules.

In a real situation, the normal to the 2D material basal plane is parallel to the direction of E_{loc} (*i.e.* $\theta = 0$ in **Figure 3b**) in a Raman backscattering geometry. Therefore, one can expect that the TERS signal is only associated to the α'_{zz} component. However, if the normal to the 2D basal plane is tilted by an angle θ due to local structural deformation, other components of the Raman tensor also contribute to the overall TERS response.

It is also worth noting that above mentioned assumption is a simplified model, which does not consider interface chemistry. For example, in the case of chemisorption the molecular geometry or orientation may alter significantly leading to the breaking of the Raman selection rules. In such cases, some Raman inactive modes can become Raman active [46]. Moreover, light-plasmon coupling in a nanocavity between the tip and the metal substrate can lead to the alteration of classical Raman selection due to photon tunneling through perturbation of the evanescent field [47].

In polarization dependent Raman measurements we observe the scattered light intensity as a function of polarization directions of both the incident and scattered light. The Raman intensity can be written as

$$I \propto \left|\boldsymbol{e}_s \mathcal{R} \boldsymbol{e}_i\right|^2 \tag{11}$$

Here, $\boldsymbol{e}_s$ and $\boldsymbol{e}_i$ are the unit vectors of the electric field of the scattered and incident light and $\mathcal{R}$ is a second rank tensor also called Raman tensor derived from Eq. (8). There is plenty of literature including text books available discussing polarization and Raman tensors in conventional Raman measurements [25, 50–52]. In this chapter, we revisit the Raman tensors of first order Raman modes in 2D semiconductors, which are important to understand the TERS selection rules. Group theory analysis tells us that the Raman active phonon modes belong to the irreducible representations, the basis function of which contain quadratic terms of x, y, z. Thus, to have non-zero Raman intensity at a given $\boldsymbol{e}_s$ and $\boldsymbol{e}_i$ geometry, $\mathcal{R}$ should have a non-zero matrix element. The Raman tensors of three first order Raman modes in 2D semiconductors (D_{3h} point group) are presented below [25, 50].

$$\left.\begin{aligned} R(E', xy) &= \begin{pmatrix} 0 & d & 0 \\ d & 0 & 0 \\ 0 & 0 & 0 \end{pmatrix} \\ R\left(E', x^2 - y^2\right) &= \begin{pmatrix} -c & 0 & 0 \\ 0 & -c & 0 \\ 0 & 0 & 0 \end{pmatrix} \\ R\left(A'_1, x^2 + y^2 \, \text{and} \, z^2\right) &= \begin{pmatrix} a & 0 & 0 \\ 0 & a & 0 \\ 0 & 0 & b \end{pmatrix} \end{aligned}\right\} \tag{12}$$

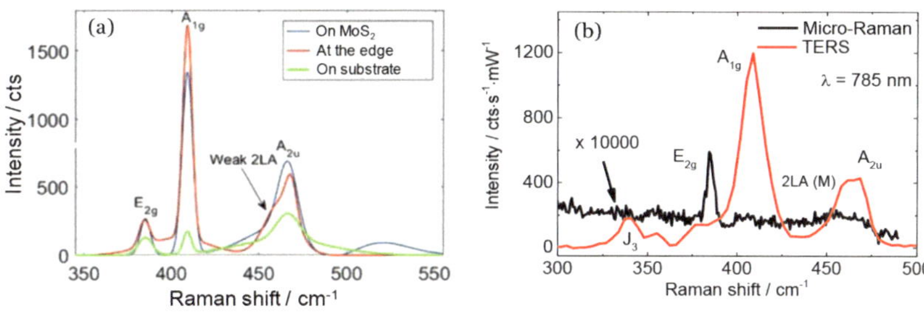

Figure 4.
TERS spectra of few layer MoS_2 deposited on an Au coated glass slide acquired using 638 nm excitation (a). Micro-Raman and TERS spectra of monolayer MoS_2 deposited on an Au nanodisc array on a Si substrate acquired using 785 nm excitation (b). The micro-Raman spectrum was obtained over an area of ~1 μm² (spot size) and the TERS spectrum was taken locally on top of Au nanodisc with a spatial resolution of 2.3 nm. In both cases, together with the A_{1g} Raman mode, the infra-red active A_{2u} is enhanced significantly. Figure (a) is adapted from ref. [48] (Copyright © 2017, IEEE) and (b) is adapted from ref. [49] with permission from the Royal Society of Chemistry.

There is one more doubly degenerate Raman mode with E'' (E_{1g}) symmetry available in 2D semiconductors, which is forbidden in backscattering geometry. Therefore, the discussion of this mode is omitted here. Interested readers can learn more about this phonon feature in this review [50]. From Eq. (12) it is clear that the out-of-plane, A_1' (A_{1g}) phonon mode has a Raman tensor component along z-direction (α'_{zz}). Therefore, the A_1' Raman mode is strongly sensitive to the TERS geometry. Since other modes are active in cross polarization geometry, they can still be observed in side-illumination TERS configuration, even though the TERS intensity is not as enhanced as for the out-of-plane Raman mode. An experimental demonstration of TERS selection rules on MoS_2 is shown in **Figure 4** [48, 49]. In the case of few layer MoS_2 on an Au surface and resonantly excited by a 638 nm laser the far-field contribution on the substrate close to the flake shows comparable intensity of the in-plane E_{2g} and out-of-plane A_{1g} mode (**Figure 4a**). However, as the tip approaches the flake, the out-of-plane A_{1g} mode enhances significantly [48]. Similarly, a marked TERS enhancement of the A_{1g} mode is observed from the monolayer MoS_2 deposited on Au nanodisc arrays on Si substrate and excited by 785 nm excitation (**Figure 4b**). Due to the below bandgap excitation (bandgap of monolayer MoS_2 is ~660 nm) only the in-plane E_{2g} mode is observed in micro-Raman spectra. In contrast, TERS spectra of monolayer MoS_2 on Au nanodisc show a significant modification with a giant enhancement of the A_{1g} mode. Together with other factors (as discussed later in this chapter), the TERS selection rules amplify multifold the out-of-plane phonon modes [49].

Another important aspect of TERS selection rules is demonstrated by the enhancement of the infra-red (IR) active out-of-plane A_{2u} mode in both works. As shown in **Figures 2** and **3**, the tip-substrate dimer system creates a strong dipole moment in the nanocavity (or even sub-nanocavity) resulting in a strong out-of-plane E_{loc}. Therefore, IR active phonon modes, which induce a transition dipole moment parallel to the E_{loc}, can also be strongly enhanced. Since the IR active A_{2u} phonon vibration creates a transition dipole moment perpendicular to the MoS_2 basal plane, it couples to the E_{loc} at the tip-substrate nanocavity polarized along z-direction. This is remarkable since the A_{2u} phonon mode is Raman forbidden and only observed in resonant Raman conditions. Therefore, Voronine *et al.* [48] could still see this mode on the substrate as a far field contribution. However, the strong plasmonic field created at the tip-nanodisc nanocavity is able to not only excite this Raman forbidden IR mode, but also amplify it significantly under non-resonant 785 nm excitation as shown in **Figure 4b** [49].

4. Local phonon properties in 2D semiconductors

4.1 Strain induced local bandgap modulation

Strain plays an important role in manipulating optoelectronic properties of 2D semiconductors. The remarkable strength of 2D semiconductors such as TMDCs compared to other conventional semiconductors makes them well suited for straintronic applications. For example, a MoS_2 monolayer can sustain a large biaxial strain >10%; whereas bulk silicon usually breaks at a strain of ~1.5% [53]. This extraordinary property of 2D semiconductors has led to burgeoning research of strain engineered optoelectronic properties in recent times [53–55]. Among them, most of the studies provide macro/microscale information of strain induced optoelectronic manipulation using conventional optical spectroscopy. In contrast, the local modification of band structures due to spatially inhomogeneous strain distribution on the nanoscale is much less explored. Like the band structure, phonons in 2D semiconductors are directly affected by a wide range of parameters including strain [56, 57]. Therefore, Raman spectroscopy, one of the most frequently used non-destructive spectroscopies, becomes a powerful technique to investigate the different components of strain and their effect on optoelectronic properties of these semiconductors. Due to energy and momentum conservation, phonons participating in first order Raman scattering processes need to satisfy the wavevector condition $\boldsymbol{q} \approx 0$ (*i.e.* near the Γ point in the crystal). Though, phonons with non-zero $\boldsymbol{q}$ can also contribute to the Raman signals. However, they require two or more scattering processes assisted by a real electronic transition, a process called resonant-Raman scattering [58]. Hence, monitoring both the first order and the higher order Raman modes can provide quantitative knowledge of strain and its impact on the electronic structure of 2D semiconductors. Additionally, both tip enhanced photoluminescence (TEPL) and TERS can be performed simultaneously and can be used as complementary techniques to investigate local strain effects on the excitonic response. Though, special care needs to be taken to interpret the results since PL intensity, lineshape, and spectral position also depend on other local heterogeneities [24].

The two well-studied first order Raman modes in TMDCs are in-plane E_{2g} and out-of-plane A_{1g}. Using the combination of both experiments and *ab initio* calculations shows that the in-plane E_{2g} mode is more sensitive to uni- or bi-axial strain than the out-of-plane A_{1g} mode [56, 60]. TERS reveals similar behavior for multilayer and monolayer TMDCs when probing highly localized strain at the nanoscale. Rahaman *et al*. studied local strain heterogeneities in 3 L MoS_2 caused by a nanostructured gold substrate using TERS with a spatial resolution of less than 25 nm [59]. TERS measurements were performed in a side-illumination geometry under 638 nm excitation. When deposited on hexagonal periodic arrays of gold nanotriangles, 3 L MoS_2 underwent local deformations on top and in between the nanotriangles as shown in **Figure 5a**. Hence, this is an ideal system to study the local flexibility at the nanoscale. Looking at the AFM topography and the 2nd order derivative of it, which represents the local curvature image (**Figure 5b**), it is clear that a variety of deformations caused strain inhomogeneities in MoS_2. The center of the valley is unstrained and the twisted areas between two gold nanotriangles is most strained. The corresponding TERS map together with selective TERS spectra of the MoS_2/Au heterostructure are presented in **Figure 5c** and **d**. As explained in the previous section the out-of-plane A_{1g} mode is found to be more enhanced than the in-plane E_{2g} mode due to TERS selection rules. After deconvoluting all the spectra using voigt functions, the frequency shift of the in-plane E_{2g} mode agrees

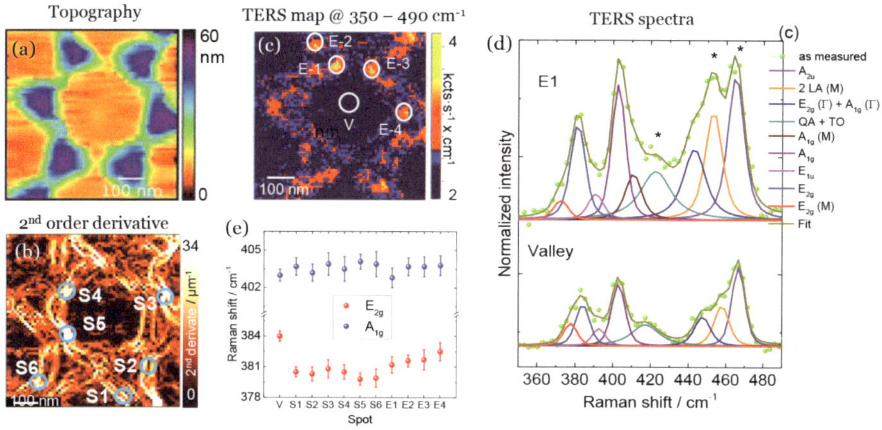

Figure 5.
AFM topography (a) and 2nd order derivative of the topography (b). Both topography and 2nd derivative images show heterogeneous biaxial strain distribution in 3 L MoS_2. The corresponding TERS map created for the spectral range of 350–490 cm^{-1} (c) and two representative TERS spectra (d) taken from two respective areas shown in (c). Site dependent peak position of two first order Raman modes for the quantification of strain (e). Adapted from ref. [59] (Copyright © 2017 American Chemical Society).

well with the local curvature image and a maximum strain of 1.4% is determined in the twisted regions from the frequency shift of 4.2 cm^{-1}. However, at the corner of the nanotriangles the strain is calculated to be ~0.9% from a frequency shift of 2.6 cm^{-1} of the E_{2g} mode. The same authors then studied monolayer MoS_2 on top of gold nanodiscs and observed a similar amount of strain at the rim of the discs [49].

It is important to note that the above described TERS experiments were performed in resonant Raman condition with 638 nm excitation (excitation close to the A optical transition in MoS_2) [61]. Therefore, there are as many as seven Raman features observed in the TERS spectra together with the two first order Raman modes (E_{2g} and A_{1g}). The Raman band around 378 cm^{-1} is the in-plane $E_{2g}(M)$ [62]. However, some reports argued that Davydov splitting of the in-plane $E_{2g}(\Gamma)$ should appear around this frequency position as well [13]. The feature around 390 cm^{-1} is a Raman inactive E_{1u} excited due to the resonant Raman condition in the TERS configuration [63]. The Raman mode around 410 cm^{-1} is an out-of-plane A_{1g} phonon in the vicinity of the M point of the BZ and analogous to the mode $E_{2g}(M)$. The frequency around 420 cm^{-1} is attributed to a two phonon process involving a successive emission of a dispersive longitudinal quasi-acoustic (QA) phonon and a dispersionless transverse optical (TO) phonon along the c-axis [64]. The broad feature around 450 cm^{-1} is a convolution of minimum three modes. The first one around 440 cm^{-1} is $E_{2g}(\Gamma) + A_{1g}(\Gamma)$ [62]. The frequency centered around 450 cm^{-1} involves two phonons of the longitudinal acoustic branch with opposite momenta at the M point of BZ ($2LA$ (M)) [65], and the third feature around 460 cm^{-1} is the infrared active A_{2u} mode as discussed in the previous section.

The rich variety of second order Raman features in MoS_2 in resonant Raman condition also yields important information about its physical properties and the electronic band structure. Zhang *et al*. performed similar TERS experiments on monolayer MoS_2 deposited on gold nanopyramids and by combining TERS and TEPL they probed the nanoscale variation of the electronic band structure induced by strain [60]. Under resonance excitation, a few of the second order Raman features in MoS_2 were assigned to longitudinal acoustic (LA) and/or transverse acoustic (TA) phonons. These Raman bands scatter light through intervalley resonant scattering of the excited electrons by phonons when excited near the B exciton energy. The process is known as double resonance Raman scattering (DRRS) and is similar to the process that leads to the creation of the *D* and 2*D* band in graphene/graphite [66, 67]. The DRRS process is explained by the schematic shown in **Figure 6a**. After the excitation of an e-h pair near the *K* valley of BZ, the excited electron is inelastically scattered from *K* to the *K′* valley by the emission of a phonon. In the relaxation process the excited electron is then inelastically scattered back to the *K* valley by the emission of a second phonon, where the e-h pair recombines and emits a photon. In MoS_2, the Raman mode around 420 cm^{-1} and the 2 *LA* mode around 450 cm^{-1} are the results of DRRS processes. Zhang *et al.* observed that together with the shift in peak position of these second order features, the peak intensities are also inhomogeneously modified due to local strain [60]. Their DFT calculations of the phonon dispersion for acoustic branches show that DRRS processes are sensitive to the changes in the momentum and energy conservation constraints that govern which transitions and states in the BZ participate in the resonant interactions. Thus any distortion in the band structure caused by strain is reflected in the DRRS processes. With positive hydrostatic strain, the calculated phonon dispersion indicates a negative shift in peak position for both *LA* and *TA* modes at the *K* point in the BZ due to increasing lattice constants. On the other hand, the increase in peak areas associated to the DRRS processes depends on the relative position of the conduction band minima at the *K* and *Q* point in the BZ. In an unstrained monolayer MoS_2, the *K* point conduction band minima is lower in energy than the *Q* point conduction band minima. Therefore, the $K - K'$ transition mediated by the $q_{\sim K}$ acoustic phonons are much more prominent than the $K - Q$ transition mediated by $q_{\sim M}$ phonons. However, when tensile strain is applied, the difference in energy between the *K* and *Q* point conduction band minima decreases. Hence, more transitions mediated by the $q_{\sim M}$ phonons satisfy the constraints on a DRRS process, which in turn increases the respective peak area.

Like externally induced strain via transferring TMDCs monolayers on nanostructured substrates as discussed above, direct evaporation of a metal, for example gold on monolayer MoS_2, can cause large strain ($\sim$ 5%) at the interface due to lattice mismatch. Combining various SPM techniques and TERS Jo *et al.* demonstrated that

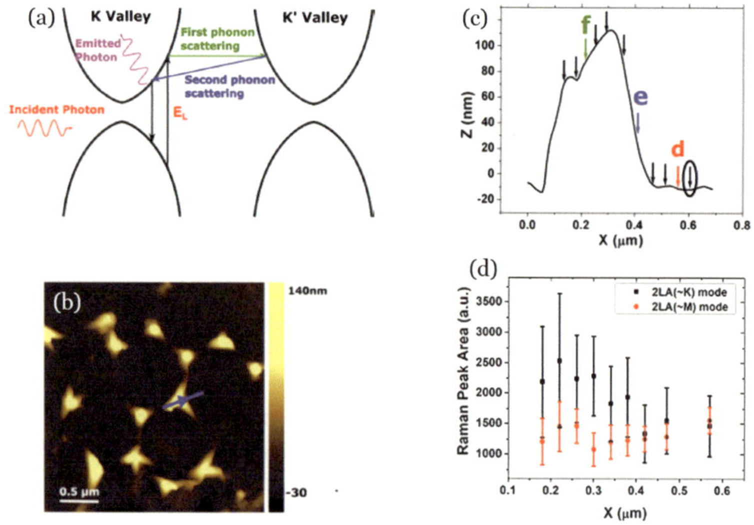

Figure 6.
Illustration of the DRRS process in 2D semiconductors (a). AFM topography of monolayer MoS_2 deposited on gold NTs (b). A line profile is created from the topography along the dashed line and presented in (c). Integrated Raman peak area of the DRRS triggered 2LA mode in MoS_2 (d) taken at the arrow positions along the line profile shown in (c). Adapted from ref. [60] (© 2018 American Physical Society).

large tensile strain induced at the interface in this process reduces the bandgap of MoS_2 and thus influences the contact resistance as a consequence [68].

TERS has become a powerful technique to probe local strain variation in 2D semiconductors, which in turn helps us to understand the inhomogeneities in optoelectronic response of these materials.

4.2 Effect of variance in atomic registry on electronic properties

The reduced dimensionality of 2D semiconductors makes them susceptible to structural variations near the internal and external boundaries. Parameters, such as impurities, defects, and grain boundaries interfere strongly with intrinsic properties resulting in highly inhomogeneous optoelectronic response [23, 69, 70]. While conventional optical spectroscopy leaves a confusing picture regarding the effect of these heterogeneities [71], near-field optical studies can help us to understand the correlation of local structural heterogeneities with nano-optical response. In recent years, the combination of TEPL and TERS has been successfully introduced to investigate these heterogeneities. In TMDCs, the out-of-plane A_{1g} Raman mode is particularly sensitive to changes in carrier concentration (or doping) via strong electron–phonon coupling [72]. Therefore, by comparing the peak positions of both in-plane E_{2g} and out-of-plane A_{1g} modes it is possible to separate the local doping effect from other heterogeneities [49]. Additionally, the Raman intensity of the out-of-plane A_{1g} mode is also affected by the physical and chemical irregularities such as doping, atomic vacancies, bond strain, grain boundaries (GB) at the atomic level [73]. Understanding the variations of atomic registries are especially important for 2D semiconductors as Park *et al*. showed that the excitonic behavior differs at the edges, nucleation sites (NS), and twin boundaries (TB) [24]. Using the combination of TERS, TEPL, and atomic force local strain microscopy they studied the PL response of PVD grown WSe_2 and the results suggest that the defect concentration and stoichiometry of W and Se atoms in different sites can result in different excitonic mechanisms.

Due to their extraordinary physical properties, many groups are now synthesizing 2D semiconductors by CVD and other deposition techniques with the vision of industry scale production. However, in terms of crystal quality these deposition processes still require further optimization to be able to use them in electronic applications. CVD grown 2D semiconductors often reveal several structural heterogeneities as discussed in this section. Therefore, proper understanding is the key for the optimization of the deposition process and hence improving the crystal quality. While confocal micro-Raman spectroscopy is unable to obtain locally heterogeneous information, TERS can uncover them with nanometer precision. Smithe *et al.* studied the growth mechanism, GB, and defects in CVD grown monolayer $MoSe_2$ using TERS [74]. Their investigation revealed that CVD $MoSe_2$ monolayers have nano-domains of MoO_3, which are invisible in confocal Raman spectroscopy. They also observed that the TERS intensity was strongly suppressed at GBs, which was attributed to the fast degradation of the GBs in ambient conditions due to the presence of higher defect concentration.

For optoelectronic devices the physics of semiconductor–metal interfaces is critical, since they are at the core of charge separation and collection. Jariwala *et al.* directly probed interfaces between WSe_2 and noble metals such as gold and silver to investigate the influence of intrinsic doping on electronic properties using the cross-correlated spatial mapping of contact potential difference (CPD), photocurrent, TERS, and TEPL [75]. Their TERS map acquired with 638 nm excitation showed two distinct regions with domains of 10–100 nm lateral size, which correlated with

CPD images as well. The regions of lower CPD exhibited non-resonant TERS spectra with a single Raman mode around 250 cm^{-1}. This single Raman band corresponds to the combination of E_{2g} and A_{1g} modes of WSe_2. In contrast, TERS spectra representing higher CPD regions exhibited resonant Raman scattering with a number of additional features. It is important to note that the laser line 638 nm overlaps with the broad shoulder on the high-energy side of the *A* exciton in WSe_2. Therefore, Raman spectra acquired with this laser line induces resonant Raman response in this material with a number of additional peaks including DRR features, which are not observed with other excitation wavelengths. Later, by performing photocurrent measurement they observed that both higher and lower CPD areas generated higher photocurrents, but with opposite signs. Therefore, they concluded that the areas showing resonant Raman behavior are intrinsically n-type, while non-resonant areas are p-type.

4.3 Semiconducting to metallic phase transition

The stable crystal structure of TMDCs is the 2*H* semiconducting phase. However, these materials are known to form also metallic 1*T* or 1*T*′ phases (though not stable) due to high density of charge donation or electron doping [76, 77]. Due to strong light matter interaction TMDC monolayers, when integrated with plasmonic antennas, show great potential via modulation of their optoelectronic properties [78]. Generated hot electrons from plasmonic nanostructures can be injected into the conduction band of these 2D semiconductors due to favorable Schottky barrier heights at the interface and a large oscillator strength of these hot electrons [22, 79]. The concept of the doping induced structural phase change is illustrated in **Figure 7a**. According to the crystal field theory, the semiconducting 2*H* phase consists of M (M = Mo/W) outermost *d* orbitals of a D_{3h}- MX_6 (X = S, Se) unit into three groups creating the stable 2*H* lattice structure [80]. The doping via hot electrons first destabilizes the 2*H* phase and induces a splitting of the metal

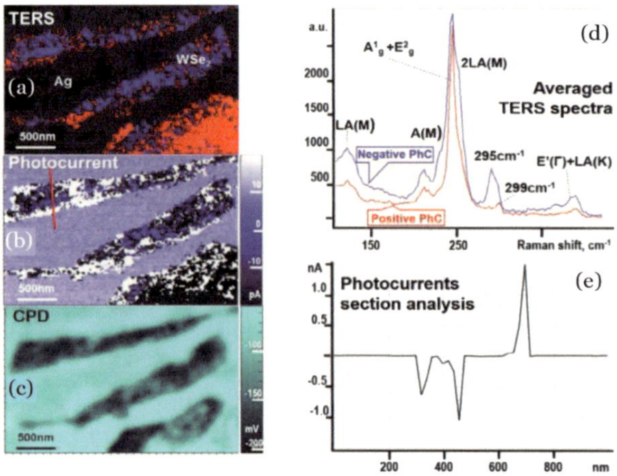

Figure 7.
Single layer flakes of WSe_2 exfoliated onto template stripped silver. Combined TERS map (intensity of 295 cm^{-1} peak in blue and of the 250 cm^{-1} ($A_{1g} + E_{2g}$) peak in red) (a). Photocurrent map at zero bias voltage collected simultaneously with TERS map (b), CPD image of the same area (c) and averaged TERS spectra showing resonant and non-resonant behavior (d). Section analysis of the photocurrent map again showing that the value of the photocurrent generated over differently doped domains was similar in absolute value, but of the opposite direction (e). The image is reproduced from ref. [75] (© 2018 IOP Publishing Ltd).

outermost *d* orbitals to form metallic O_h-MX_6 units into two groups as shown in **Figure 7a** [81–83]. The unoccupied orbitals then accommodate the extra electrons crossing the Schottky barrier (**Figure 8**).

Milekhin *et al.* studied the extremely localized site dependent transient reversible 2*H* to 1*T* phase change in monolayer MoS_2 using TERS. Monolayer MoS_2 was deposited on periodic arrays of gold nanodiscs, thus sandwiched between the nanodisc and the tip to conduct this experiment [49]. Hence, the tip-nanodisc system creates a strongly coupled dimer, which then pumps a high concentration of electrons into the conduction band of MoS_2. As demonstrated by the high resolution TERS images and later by FEM simulations by Rahaman *et al.* [37] the coupling is the strongest between the tip and the rim of the nanodiscs. Therefore, the rim of the nanodiscs became the strongest source of hot electron doping in MoS_2, which consequently showed the transition from the 2*H* to 1*T* phase via the appearance of an additional Raman mode in TERS spectra. Metallic MoS_2 has three additional Raman modes around 156, 226, and 333 cm^{-1} known as J_1, J_2, and J_3, respectively [84]. Due to strong increasing Rayleigh background, Milekhin *et al.* [49] were able to observe only the J_3 Raman mode around 333 cm^{-1}. As can be seen in **Figure 7b**, the transition from 2*H* to 1*T* is extremely site dependent with an area limited by the spatial resolution, which was $\sim$ 2 nm.

4.4 Probing edge related properties

TMDCs have two distinct edges in the form of armchair and zigzag, which possess different electronic properties. For example, the zigzag edge in MoS_2 is metallic, whereas the armchair is semiconducting [85]. Therefore, quantitative measurements or characterization of these edges are important for effective edge engineering, especially for edge contacts for electronic device applications. Huang *et al.* studied edge related properties of atomically thin MoS_2 using TERS [69]. Their local Raman measurements aided by strong plasmonic enhancement of either gold or silver tips probed different Raman features of the edge defects in mono- and bilayer MoS_2 corresponding to the unique lattice vibration and electronic properties of the respective edge.

A broad Raman feature around 220 cm^{-1} was observed along the edges (full of dangling bonds) and absent on the wrinkles and main body (**Figure 9a**). The peak was assigned to the defect (dangling bonds) induced acoustic *LA* with a DRRS feature as explained in Section 4.1. More interestingly, the TERS spectra taken along

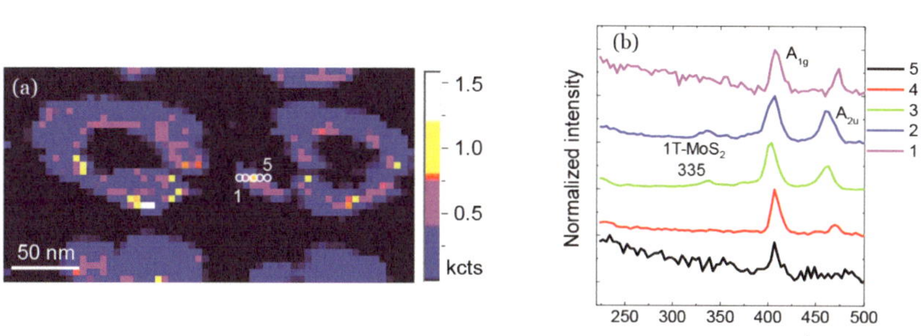

Figure 8.
High resolution TERS map of monolayer MoS_2 on gold nanodiscs acquired with a gold tip and 785 nm excitation (a). Corresponding TERS spectra along the circles shown in the map (b). Thanks to the very high spatial resolution the structural phase change from 2H to 1T due to highly localized plasmonic hot electron doping can be observed. The image is reproduced from ref. [49] with permission from the Royal Society of Chemistry.

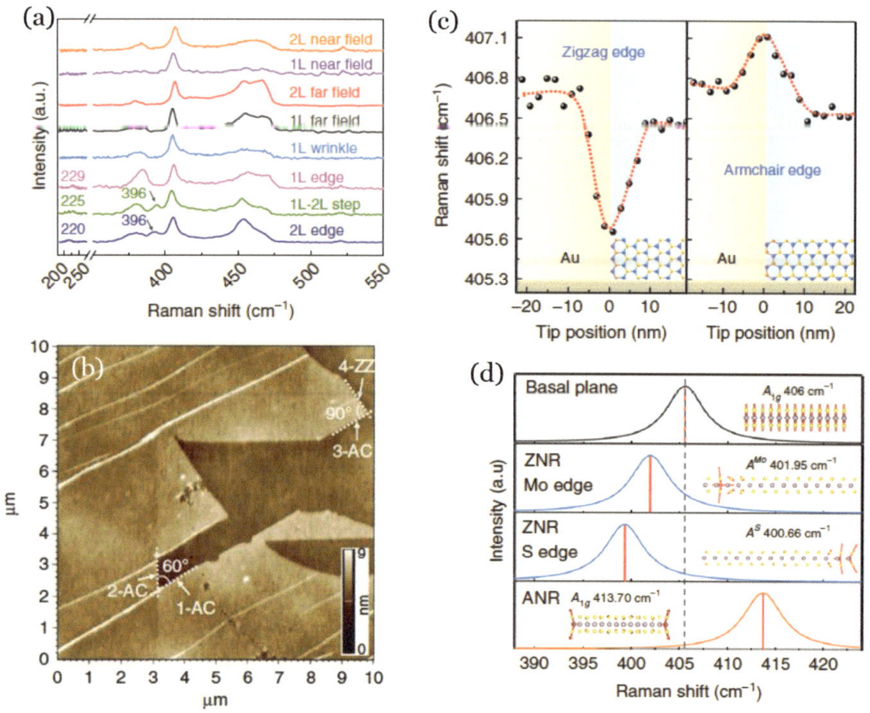

Figure 9.
TERS of mono- and bilayer MoS_2 acquired along the edges and the step between mono- and bilayer (a). AFM topography of a monolayer MoS_2 showing different terminated edges (b). Fitted peak position of A_{1g} mode acquired along a line crossing both zigzag and armchair edges (c). DFT calculated Raman spectra of at different sites in MoS_2 (d). Reproduced from ref. [69].

the edges of the bilayer and the step between mono- and bilayer exhibits an additional feature around 396 cm^{-1}. This particular feature is explicitly observed in bulk at cryogenic temperature and assigned to the $LA(M) + TA(M)$ acoustic mode [86]. This Raman mode involves a DRRS process between the K and Q valleys with wavevector $\boldsymbol{q}_M$ and thus the Raman cross section of this mode depends on the relative energy position of the conduction band minima at the K and Q points in the BZ. Since in bulk the conduction band minima at the Q point is lower than the K point, the Raman scattering efficiency of this mode becomes larger. Therefore, this mode is absent in the basal plane of mono- and bilayer MoS_2. However, along the edges, especially in the case of bilayer MoS_2 the modification of the electronic band structure leads to the considerable lowering of the conduction band energy at the Q point resulting in the activation of the DRRS process of the $LA(M) + TA\ (M)$ mode.

Another important observation of this study is the edge dependent local Raman sensitivity. The frequency position of the out-of-plane A_{1g} mode showed either upward or downward shift compared to the basal plane when probed along the armchair or zigzag edges, respectively. The AFM topography of monolayer MoS_2 presented in **Figure 9b** shows edges terminated with different angles. For an angle of 60° both edges have the same structure (either armchair or zigzag) and for an angle of 90° the edges are different. The TERS spectra acquired along a line crossing both zigzag and armchair edges show a clear pattern, namely that the A_{1g} mode downshifts at one of the edges and upshifts at other one (**Figure 9c**). According to the DFT calculations the local strain introduced by the two edge structures cause this opposite direction of the frequency shift as shown in **Figure 9d**.

5. Conclusion and outlook

TERS has developed into a versatile characterization technique for a wide range of materials. In recent years we have witnessed chemical mapping of single molecules with a spatial resolution of the bond length using TERS. For 2D semiconductors, an excellent SR of ~2 nm in TERS helped us to understand local heterogeneous behavior of monolayer MoS_2 and demonstrate its tremendous capabilities. In this chapter, we explained the basic principal of TERS and described the surface selection rules in regards to the 2D semiconductors important for understanding the near-field Raman spectra. We also reviewed the recent progress of this technique in the 2D semiconductor field. The potential of TERS certainly guarantees new breakthroughs in the 2D semiconductor field in the near future. One of the recent direction of 2D semiconductors is to create heterostacks like Lego, which promises exotic physics due to the creation of moiré superlattices. The size of a moiré unit cell varies from nm to few 10s nm depending on the lattice mismatch and rotation angle. The moiré superlattice forces the constituent monolayers into phonon renormalization, also known as moiré phonons. TERS can be used to probe these moiré phonons, thus resolving the moiré supercell critical for understanding the moiré physics. Besides, there is a lot of information yet to be resolved on how local heterogeneities control the optical response such as strain induced exciton funneling, defects induced excitonic response, or single photon emitters *etc*. in 2D semiconductors, for which TERS can be a powerful tool to achieve a thorough understanding.

Acknowledgements

This work was supported by the German research council funded project DFG-ZA 146/44-1.

Conflict of interest

The authors declare no conflict of interest.

Author details

Mahfujur Rahaman* and Dietrich R.T. Zahn
Semiconductor Physics, Technische Universität Chemnitz, Chemnitz, Germany

*Address all correspondence to: mahfujur.rahaman@physik.tu-chemnitz.de

References

[1] Wang Q H, Kalantar-Zadeh K, Kis A, Coleman J N and Strano M S 2012 Electronics and optoelectronics of two-dimensional transition metal dichalcogenides *Nature Nanotechnology* 7 699-712

[2] Geim A K and Novoselov K S 2007 The rise of graphene *Nature Materials* 6 183-191

[3] Lee C, Wei X, Kysar J W and Hone J 2008 Measurement of the Elastic Properties and Intrinsic Strength of Monolayer Graphene *Science* 321 385

[4] Seol J H, Jo I, Moore A L, Lindsay L, Aitken Z H, Pettes M T, Li X, Yao Z, Huang R, Broido D, Mingo N, Ruoff R S and Shi L 2010 Two-Dimensional Phonon Transport in Supported Graphene *Science* 328 213

[5] Högl P, Frank T, Zollner K, Kochan D, Gmitra M and Fabian J 2020 Quantum Anomalous Hall Effects in Graphene from Proximity-Induced Uniform and Staggered Spin-Orbit and Exchange Coupling *Physical Review Letters* 124 136403

[6] Jiao L, Zhang L, Wang X, Diankov G and Dai H 2009 Narrow graphene nanoribbons from carbon nanotubes *Nature* 458 877-880

[7] Bandurin D A, Tyurnina A V, Yu G L, Mishchenko A, Zólyomi V, Morozov S V, Kumar R K, Gorbachev R V, Kudrynskyi Z R, Pezzini S, Kovalyuk Z D, Zeitler U, Novoselov K S, Patanè A, Eaves L, Grigorieva I V, Fal'ko V I, Geim A K and Cao Y 2017 High electron mobility, quantum Hall effect and anomalous optical response in atomically thin InSe Nature Nanotechnology 12 223-227

[8] Mak K F, Lee C, Hone J, Shan J and Heinz T F 2010 Atomically Thin MoS_2: A New Direct-Gap Semiconductor Physical Review Letters 105 136805

[9] Splendiani A, Sun L, Zhang Y, Li T, Kim J, Chim C-Y, Galli G and Wang F 2010 Emerging Photoluminescence in Monolayer MoS_2 *Nano Letters* 10 1271-5

[10] Xu Y, Shi Z, Shi X, Zhang K and Zhang H 2019 Recent progress in black phosphorus and black-phosphorus-analogue materials: properties, synthesis and applications *Nanoscale* 11 14491-527

[11] Rahaman M, Bejani M, Salvan G, Lopez-Rivera S A, Pulci O, Bechstedt F and Zahn D R T 2018 Vibrational properties of GaSe: a layer dependent study from experiments to theory *Semiconductor Science and Technology* 33 125008

[12] Sun Y, Luo S, Zhao X-G, Biswas K, Li S-L and Zhang L 2018 InSe: a two-dimensional material with strong interlayer coupling *Nanoscale* 10 7991-7998

[13] Tonndorf P, Schmidt R, Böttger P, Zhang X, Börner J, Liebig A, Albrecht M, Kloc C, Gordan O, Zahn D R T, Michaelis de Vasconcellos S and Bratschitsch R 2013 Photoluminescence emission and Raman response of monolayer MoS_2, $MoSe_2$, and WSe_2 *Opt.* Express 21 4908-4916

[14] Chernikov A, Berkelbach T C, Hill H M, Rigosi A, Li Y, Aslan O B, Reichman D R, Hybertsen M S and Heinz T F 2014 Exciton Binding Energy and Nonhydrogenic Rydberg Series in Monolayer WS_2 *Physical Review Letters* 113 076802

[15] Zhu B, Chen X and Cui X 2015 Exciton Binding Energy of Monolayer WS_2 *Scientific Reports* 5 9218

[16] Mak K F, He K, Lee C, Lee G H, Hone J, Heinz T F and Shan J 2013

Tightly bound trions in monolayer MoS_2 *Nature Materials* 12 207-11

[17] Drüppel M, Deilmann T, Krüger P and Rohlfing M 2017 Diversity of trion states and substrate effects in the optical properties of an MoS_2 monolayer *Nature* Communications 8 2117

[18] Illarionov Y Y, Smithe K K H, Waltl M, Knobloch T, Pop E and Grasser T 2017 Improved Hysteresis and Reliability of MoS_2 Transistors With High-Quality CVD Growth and Al_2O_3 Encapsulation *IEEE Electron Device Letters* 38 1763-6

[19] Buscema M, Groenendijk D J, Blanter S I, Steele G A, van der Zant H S J and Castellanos-Gomez A 2014 Fast and Broadband Photoresponse of Few-Layer Black Phosphorus Field-Effect Transistors *Nano Letters* 14 3347-3352

[20] Feng W, Zheng W, Gao F and Hu P 2017 Atomically thin InSe: A high mobility two-dimensional material *Science China Technological Sciences* 60 1121-1122

[21] Island J O, Steele G A, Zant H S J v d and Castellanos-Gomez A 2015 Environmental instability of few-layer black phosphorus *2D Materials* 2 011002

[22] Rahaman M, Aslam M A, He L, Madeira T I and Zahn D R T 2021 Plasmonic hot electron induced layer dependent anomalous Fröhlich interaction in InSe *Communications Physics* 4 172

[23] Chow P K, Jacobs-Gedrim R B, Gao J, Lu T-M, Yu B, Terrones H and Koratkar N 2015 Defect-Induced Photoluminescence in Monolayer Semiconducting Transition Metal Dichalcogenides *ACS Nano* 9 1520-1527

[24] Park K-D, Khatib O, Kravtsov V, Clark G, Xu X and Raschke M B 2016 Hybrid Tip-Enhanced Nanospectroscopy and Nanoimaging of Monolayer WSe_2 with Local Strain Control *Nano Letters* 16 2621-7

[25] Tan P-H 2019 *Raman Spectroscopy of Two Dimensional Materials*: *Springer Series in Materials Science*, vol. 276, Springer

[26] Shao F and Zenobi R 2019 Tip-enhanced Raman spectroscopy: principles, practice, and applications to nanospectroscopic imaging of 2D materials *Analytical and Bioanalytical Chemistry* 411 37-61

[27] Raman C V 1928 A Change of Wave-length in Light Scattering *Nature* 121 619-

[28] Cardona M 1995 *Light Scattering in Solids: Topics in Applied Physics*, vol 8: Springer, Berlin, Heidelberg

[29] Manuel Cardona R M 2007 *Light scatteriing in Solid IX*: *Topics in Applied Physics*, Springer, Berlin, Heidelberg

[30] Penney C M, Goldman L M and Lapp M 1972 Raman Scattering Cross Sections *Nature Physical Science* 235 110-112

[31] Aggarwal R L, Farrar L W, Saikin S K, Aspuru-Guzik A, Stopa M and Polla D L 2011 Measurement of the absolute Raman cross section of the optical phonon in silicon *Solid State Communications* 151 553-556

[32] Samir Kumar P K, Anamika Das and Chandra Shakher Pathak 2020 *Surface-Enhanced Raman Scattering: Introduction and Applications*: IntechOpen

[33] Blackie E J, Le Ru E C and Etchegoin P G 2009 Single-Molecule Surface-Enhanced Raman Spectroscopy of Nonresonant Molecules *Journal of the American Chemical Society* 131 14466-14472

[34] Langer J, Jimenez de Aberasturi D, Aizpurua J, Alvarez-Puebla R A,

Auguié B, Baumberg J J, Bazan G C, Bell S E J, Boisen A, Brolo A G, Choo J, Cialla-May D, Deckert V, Fabris L, Faulds K, García de Abajo F J, Goodacre R, Graham D, Haes A J, Haynes C L, Huck C, Itoh T, Käll M, Kneipp J, Kotov N A, Kuang H, Le Ru E C, Lee H K, Li J-F, Ling X Y, Maier S A, Mayerhöfer T, Moskovits M, Murakoshi K, Nam J-M, Nie S, Ozaki Y, Pastoriza-Santos I, Perez-Juste J, Popp J, Pucci A, Reich S, Ren B, Schatz G C, Shegai T, Schlücker S, Tay L-L, Thomas K G, Tian Z-Q, Van Duyne R P, Vo-Dinh T, Wang Y, Willets K A, Xu C, Xu H, Xu Y, Yamamoto Y S, Zhao B and Liz-Marzán L M 2020 Present and Future of Surface-Enhanced Raman Scattering *ACS Nano* 14 28-117

[35] Le Ru E C and Etchegoin P G 2009 *Principles of Surface-Enhanced Raman Spectroscopy,* ed E C Le Ru and P G Etchegoin (Amsterdam: Elsevier) pp 185-264

[36] Rahaman M, Moras S, He L, Madeira T I and Zahn D R T 2020 Fine-tuning of localized surface plasmon resonance of metal nanostructures from near-Infrared to blue prepared by nanosphere lithography *Journal of Applied Physics* 128 233104

[37] Rahaman M, Milekhin A G, Mukherjee A, Rodyakina E E, Latyshev Alexander V, Dzhagan V M and Zahn D R T 2019 The role of a plasmonic substrate on the enhancement and spatial resolution of tip-enhanced Raman scattering *Faraday Discussions* 214 309-323

[38] He L, Rahaman M, Madeira T I and Zahn D R T 2021 Understanding the Role of Different Substrate Geometries for Achieving Optimum Tip-Enhanced Raman Scattering Sensitivity *Nanomaterials* 11

[39] Hulteen J C, Treichel D A, Smith M T, Duval M L, Jensen T R and Van Duyne R P 1999 Nanosphere Lithography: Size-Tunable Silver Nanoparticle and Surface Cluster Arrays *The Journal of Physical Chemistry B* 103 3854-3863

[40] Murray W A, Suckling J R and Barnes W L 2006 Overlayers on Silver Nanotriangles: Field Confinement and Spectral Position of Localized Surface Plasmon Resonances *Nano Letters* 6 1772-1777

[41] Richard-Lacroix M, Zhang Y, Dong Z and Deckert V 2017 Mastering high resolution tip-enhanced Raman spectroscopy: towards a shift of perception *Chemical Society Reviews* 46 3922-3944

[42] Kumar N, Su W, Veselý M, Weckhuysen B M, Pollard A J and Wain A J 2018 Nanoscale chemical imaging of solid–liquid interfaces using tip-enhanced Raman spectroscopy *Nanoscale* 10 1815-1824

[43] Savage K J, Hawkeye M M, Esteban R, Borisov A G, Aizpurua J and Baumberg J J 2012 Revealing the quantum regime in tunnelling plasmonics *Nature* 491 574-577

[44] Esteban R, Borisov A G, Nordlander P and Aizpurua J 2012 Bridging quantum and classical plasmonics with a quantum-corrected model *Nature Communications* 3 825

[45] Moskovits M 1982 Surface selection rules The Journal of Chemical Physics 77 4408-4416

[46] Chen X, Liu P, Hu Z and Jensen L 2019 High-resolution tip-enhanced Raman scattering probes sub-molecular density changes *Nature Communications* 10 2567

[47] Poliani E, Wagner M R, Vierck A, Herziger F, Nenstiel C, Gannott F, Schweiger M, Fritze S, Dadgar A, Zaumseil J, Krost A, Hoffmann A and Maultzsch J 2017 Breakdown of

Far-Field Raman Selection Rules by Light–Plasmon Coupling Demonstrated by Tip-Enhanced Raman Scattering *The Journal of Physical Chemistry Letters* 8 5462-5471

[48] Voronine D V, Lu G, Zhu D and Krayev A 2017 Tip-Enhanced Raman Scattering of MoS2 *IEEE Journal of Selected Topics in Quantum Electronics* 23 138-43

[49] Milekhin A G, Rahaman M, Rodyakina E E, Latyshev A V, Dzhagan V M and Zahn D R T 2018 Giant gap-plasmon tip-enhanced Raman scattering of MoS2 monolayers on Au nanocluster arrays *Nanoscale* 10 2755-63

[50] Saito R, Tatsumi Y, Huang S, Ling X and Dresselhaus M S 2016 Raman spectroscopy of transition metal dichalcogenides *Journal of Physics: Condensed Matter* 28 353002

[51] Loudon R 1964 The Raman effect in crystals Advances in Physics 13 423-482

[52] Yu P Y and Cardona M 2010 *Fundamentals of Semiconductors: Physics and Materials Properties*, ed P Y Yu and M Cardona, Springer Berlin Heidelberg, pp 345-426

[53] Peng Z, Chen X, Fan Y, Srolovitz D J and Lei D 2020 Strain engineering of 2D semiconductors and graphene: from strain fields to band-structure tuning and photonic applications *Light: Science & Applications* 9 190

[54] Deng S, Sumant A V and Berry V 2018 Strain engineering in two-dimensional nanomaterials beyond graphene *Nano Today* 22 14-35

[55] Dadgar A M, Scullion D, Kang K, Esposito D, Yang E H, Herman I P, Pimenta M A, Santos E J G and Pasupathy A N 2018 Strain Engineering and Raman Spectroscopy of Monolayer Transition Metal Dichalcogenides *Chemistry of Materials* 30 5148-5155

[56] Lloyd D, Liu X, Christopher J W, Cantley L, Wadehra A, Kim B L, Goldberg B B, Swan A K and Bunch J S 2016 Band Gap Engineering with Ultralarge Biaxial Strains in Suspended Monolayer MoS_2 *Nano Letters* 16 5836-41

[57] Frisenda R, Drüppel M, Schmidt R, Michaelis de Vasconcellos S, Perez de Lara D, Bratschitsch R, Rohlfing M and Castellanos-Gomez A 2017 Biaxial strain tuning of the optical properties of single-layer transition metal dichalcogenides *npj 2D Materials and Applications* 1 10

[58] Loudon R 1965 Theory of the resonance Raman effect in crystals J. Phys. France 26 677-683

[59] Rahaman M, Rodriguez R D, Plechinger G, Moras S, Schüller C, Korn T and Zahn D R T 2017 Highly Localized Strain in a MoS2/Au Heterostructure Revealed by Tip-Enhanced Raman Spectroscopy *Nano Letters* 17 6027-33

[60] Zhang Z, De Palma A C, Brennan C J, Cossio G, Ghosh R, Banerjee S K and Yu E T 2018 Probing nanoscale variations in strain and band structure of MoS_2 on Au nanopyramids using tip-enhanced Raman spectroscopy *Physical Review B* 97 085305

[61] Chakraborty B, Matte H S S R, Sood A K and Rao C N R 2013 Layer-dependent resonant Raman scattering of a few layer MoS2 *Journal of Raman Spectroscopy* 44 92-6

[62] Peña-Álvarez M, del Corro E, Morales-García Á, Kavan L, Kalbac M and Frank O 2015 Single Layer Molybdenum Disulfide under Direct Out-of-Plane Compression: Low-Stress Band-Gap Engineering *Nano Letters* 15 3139-3146

[63] Molina-Sánchez A and Wirtz L 2011 Phonons in single-layer and few-layer

MoS_2 and WS_2 *Physical Review B* 84 155413

[64] Sekine T, Uchinokura K, Nakashizu T, Matsuura E and Yoshizaki R 1984 Dispersive Raman Mode of Layered Compound 2H-MoS_2 under the Resonant Condition *Journal of the Physical Society of Japan* 53 811-8

[65] Pimenta M A, del Corro E, Carvalho B R, Fantini C and Malard L M 2015 Comparative Study of Raman Spectroscopy in Graphene and MoS_2-type Transition Metal Dichalcogenides Accounts of Chemical Research 48 41-47

[66] Ni Z, Liu L, Wang Y, Zheng Z, Li L-J, Yu T and Shen Z 2009 G-band Raman double resonance in twisted bilayer graphene: Evidence of band splitting and folding *Physical Review B* 80 125404

[67] Carvalho B R, Wang Y, Mignuzzi S, Roy D, Terrones M, Fantini C, Crespi V H, Malard L M and Pimenta M A 2017 Intervalley scattering by acoustic phonons in two-dimensional MoS_2 revealed by double-resonance Raman spectroscopy *Nature Communications* 8 14670

[68] Jo K, Kumar P, Orr J, Anantharaman S B, Miao J, Motala M J, Bandyopadhyay A, Kisslinger K, Muratore C, Shenoy V B, Stach E A, Glavin N R and Jariwala D 2021 Direct Optoelectronic Imaging of 2D Semiconductor–3D Metal Buried Interfaces *ACS Nano* 15 5618-30

[69] Huang T-X, Cong X, Wu S-S, Lin K-Q, Yao X, He Y-H, Wu J-B, Bao Y-F, Huang S-C, Wang X, Tan P-H and Ren B 2019 Probing the edge-related properties of atomically thin MoS2 at nanoscale *Nature Communications* 10 5544

[70] Fan X, Siris R, Hartwig O, Duesberg G S and Niklaus F 2020 Rapid and Large-Area Visualization of Grain Boundaries in MoS_2 on SiO_2 Using Vapor Hydrofluoric Acid *ACS Applied Materials & Interfaces* 12 34049-57

[71] Ogletree D F, Schuck P J, Weber-Bargioni A F, Borys N J, Aloni S, Bao W, Barja S, Lee J, Melli M, Munechika K, Whitelam S and Wickenburg S 2015 Revealing Optical Properties of Reduced-Dimensionality Materials at Relevant Length Scales *Advanced Materials* 27 5693-5719

[72] Chakraborty B, Bera A, Muthu D V S, Bhowmick S, Waghmare U V and Sood A K 2012 Symmetry-dependent phonon renormalization in monolayer MoS_2 transistor *Physical Review B* 85 161403

[73] Kato R, Umakoshi T, Sam R T and Verma P 2019 Probing nanoscale defects and wrinkles in MoS_2 by tip-enhanced Raman spectroscopic imaging *Applied Physics Letters* 114 073105

[74] Smithe K K H, Krayev A V, Bailey C S, Lee H R, Yalon E, Aslan Ö B, Muñoz Rojo M, Krylyuk S, Taheri P, Davydov A V, Heinz T F and Pop E 2018 Nanoscale Heterogeneities in Monolayer MoSe2 Revealed by Correlated Scanning Probe Microscopy and Tip-Enhanced Raman Spectroscopy *ACS Applied Nano Materials* 1 572-9

[75] Jariwala D, Krayev A, Wong J, Robinson A E, Sherrott M C, Wang S, Liu G-Y, Terrones M and Atwater H A 2018 Nanoscale doping heterogeneity in few-layer WSe_2 exfoliated onto noble metals revealed by correlated SPM and TERS imaging *2D Materials* 5 035003

[76] Ambrosi A, Sofer Z and Pumera M 2015 2H $\rightarrow$ 1T phase transition and hydrogen evolution activity of MoS2, MoSe2, WS2 and WSe2 strongly depends on the MX_2 composition Chemical Communications 51 8450-8453

[77] Ma Y, Liu B, Zhang A, Chen L, Fathi M, Shen C, Abbas A N, Ge M,

Mecklenburg M and Zhou C 2015 Reversible Semiconducting-to-Metallic Phase Transition in Chemical Vapor Deposition Grown Monolayer WSe_2 and Applications for Devices ACS Nano 9 7383-7391

[78] Schneider C, Glazov M M, Korn T, Höfling S and Urbaszek B 2018 Two-dimensional semiconductors in the regime of strong light-matter coupling *Nature Communications* 9 2695

[79] Lee M-K, Kim T G, Kim W and Sung Y-M 2008 Surface Plasmon Resonance (SPR) Electron and Energy Transfer in Noble Metal–Zinc Oxide Composite Nanocrystals The Journal of Physical Chemistry C 112 10079-10082

[80] Enyashin A N, Yadgarov L, Houben L, Popov I, Weidenbach M, Tenne R, Bar-Sadan M and Seifert G 2011 New Route for Stabilization of 1T-WS_2 and MoS_2 Phases The Journal of Physical Chemistry C 115 24586-24591

[81] Py M A and Haering R R 1983 Structural destabilization induced by lithium intercalation in MoS_2 and related compounds *Canadian Journal of Physics* 61 76-84

[82] Mattheiss L F 1973 Band Structures of Transition-Metal-Dichalcogenide Layer Compounds *Physical Review B* 8 3719-40

[83] Kang Y, Najmaei S, Liu Z, Bao Y, Wang Y, Zhu X, Halas N J, Nordlander P, Ajayan P M, Lou J and Fang Z 2014 Plasmonic Hot Electron Induced Structural Phase Transition in a MoS_2 Monolayer *Advanced Materials* 26 6467-71

[84] Jiménez Sandoval S, Yang D, Frindt R F and Irwin J C 1991 Raman study and lattice dynamics of single molecular layers of MoS_2 *Physical Review B* 44 3955-62

[85] Li Y, Zhou Z, Zhang S and Chen Z 2008 MoS_2 Nanoribbons: High Stability and Unusual Electronic and Magnetic Properties Journal of the American Chemical Society 130 16739-16744

[86] Livneh T and Sterer E 2010 Resonant Raman scattering at exciton states tuned by pressure and temperature in 2H-MoS_2 *Physical Review B* 81 195209

Chapter 13

Raman Features of Linear-Carbon-Chain and Multiwall Carbon Nanotube Composites

Yahachi Saito and Koji Asaka

Abstract

Structural and electronic properties of multiwall carbon nanotubes (MWCNTs) containing linear carbon chains (LCCs), which were produced by arc-discharge between carbon electrodes in an atmospheric pressure, have been studied by Raman spectroscopy as well as electron microscopy. Spectral features of Raman scattering from the LCC/MWCNT composites were reviewed with emphasis on the spectra obtained with a low energy photon (1.58 eV, 785 nm) excitation, which have not been described in detail so far. Characteristic frequencies of LCC stretching modes with the 785 nm laser excitation are observed at around 1740, 1759, and 1789 cm^{-1}. In a low frequency region, radial breathing modes (RBMs) of the innermost tube within MWCNTs are observed at specific frequencies of 293, 341, 402, and 510 cm^{-1}; the highest RBM frequency is tentatively assigned to a tube with the chiral index (4,3), whose diameter is expected to 0.50 nm. LCC bands observed with various excitation wavelengths from 488 to 785 nm show that the band consists of several peaks, and the relative intensities of constituent peaks change with the excitation wavelengths due to the resonance effect; the higher the excitation photon energy is, the higher the intensity of high-frequency LCC modes.

Keywords: carbyne, linear carbon chain, carbon nanotube, Raman, resonance

1. Introduction

Carbon exists in a variety of allotropes: e.g., diamond, graphite, fullerene, and carbon nanotube (CNT). Besides these well-known allotropes, carbyne, an infinite carbon chain, has attracted much interest with significant controversy since the late 1960s [1–5]. Linear carbon chains (LCCs) in its own forms are unstable in an ambient environment. In solution, LCCs up to $C_{12}H_2$ are formed [6]. When both ends of carbon chains are anchored by bulky molecules, polyyne consisting of carbon atoms up to 44 have so far been chemically synthesized [7, 8]. Recently, LCCs confined inside CNTs have been discovered in multiwall CNT (MWCNT) produced by arc discharge [9], in high-temperature annealed double-wall CNTs (DWCNT) [10], and in single-wall CNT (SWCNT) suffered from electric discharge [11]. The structural and electronic properties of the encapsulated LCCs have been mainly measured by transmission electron microscopy (TEM) and

Raman scattering spectroscopy, and it has been revealed that the length of LCC inside CNTs exceeds 30 nm (i.e., more than 230 carbon atoms).

LCC corresponds to a long polyyne with effectively infinite length, in which the sp-hybridized orbitals form alternated single and triple bonds between carbon atoms in the one-dimensional (1D) chain, and thus an electronic band gap is open at the Fermi level. The electronic energy gap of hydrogen-capped polyynes $C_{2m}H_2$ is expected to be in a range from 4.1 eV for $2m = 8$ to 2.2 eV for an infinitely long polyyne [12]. Reflecting the energy gap in 1D materials, strong resonance Raman scatterings are expected. Actually, Raman scattering characteristic to LCC (called "LCC bands") has been observed from 1750 to 1855 cm^{-1}, depending on incident photon energies (1.6–2.6 eV). The LCC bands originate from one-dimensional stretching vibration of triple-bonded carbon atoms [6, 13–15].

Raman spectroscopy is a powerful and effectual technique to characterize nano-carbon materials including LCC and CNT. In measuring LLCs encapsulated in CNTs by Raman, the wavelengths of incident lasers employed so far were mainly in a visible region [9, 10, 16–18], because strong scatterings from LCCs are observed owing to resonance Raman. In this report, focusing to Raman spectra obtained with a long wavelength (near infrared) incident laser of 785 nm, which have not been reported in detail so far, characteristic features of LCC bands and radial breathing modes of MWCNTs at various incident wavelengths (785, 633, 532 and 488 nm) are presented. The materials studied here is MWCNTs encapsulating LCCs, which were produced by carbon arc-discharge being blown with inactive gases in the atmosphere [19].

2. MWCNT encapsulating LCC prepared by arc discharge

MWCNTs encapsulating LCC were synthesized by arc-discharge between carbon electrodes in the atmospheric pressure. The LCC/MWCNT composite materials investigated in the present study was supplied from JFE Engineering Corporation. The details of the production procedure have been described in the public patent documents [19]. Briefly, the anode is a hollow carbon rod (outer diameter 10 mm, inner diameter 4 mm), and the cathode is a fat carbon rod (diameter 35 mm) which executes a motion of rotation around its axis (1.5 rotation/min), as shown in **Figure 1(a)**. The anode translates (35 mm/min) parallel to the axis of the cathode, keeping a gap between the anode end and the cathode surface constant (ca. 1 mm). Arc discharging (DC 100A, 20 V) was fired in the gap between the anode and the

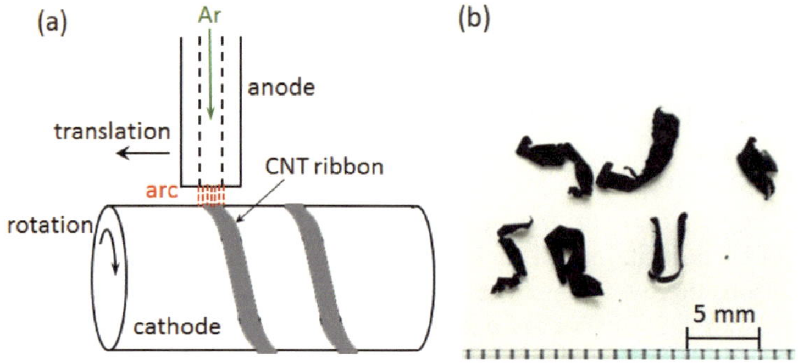

Figure 1.
(a) Schematic diagram of an arc-discharge apparatus for MWCNT production, and (b) photograph of produced ribbons consisting of MWCNT encapsulating LCC.

cathode. During the discharge, Ar gas was flowed through the inner hole of the anode toward the arcing gap, and also a cooling gas (e.g., N_2) was blown to the gap in order to cool and remove carbonaceous deposits formed on the cathode surface. The carbon deposit produced after the discharge had a form of thin tape (2–5 mm wide, and 10–20 μm thick) as shown in **Figure 1(b)**.

3. Surface morphology observed by SEM

As shown later by micro-Raman spectroscopy, in a tape of LCC/MWCNT composite, there are two types of regions; one contains abundant LCCs while the other scarce LCCs. Hereafter, the former region is called the "A" region, and the latter the "B" region. The surface morphology of the MWCNT tape at various places are observed by a scanning electron microscope (SEM) using a ThermoFisher Prisma-E operated at 15–20 kV. **Figure 2(a)** and **(b)** show SEM images of surfaces of the "A" and "B" regions, respectively. In both regions, abundant MWCNTs are observed, and no outstanding difference between the SEM images is apparently found. By close-inspection, however, the surface of the "A" region (**Figure 2(a)**) is found to be rugged and show a blanket-like surface, while the "B" region (**Figure 2(b)**) looks smooth and contains small particles (possibly, carbon nanoparticles).

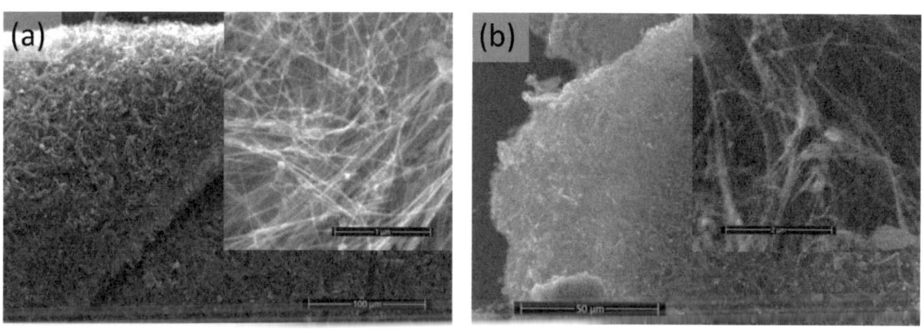

Figure 2.
SEM images of (a) the area "A" with the blanket-like surface morphology and (b) the area "B" with smooth appearance. Insets show high magnification images of each area.

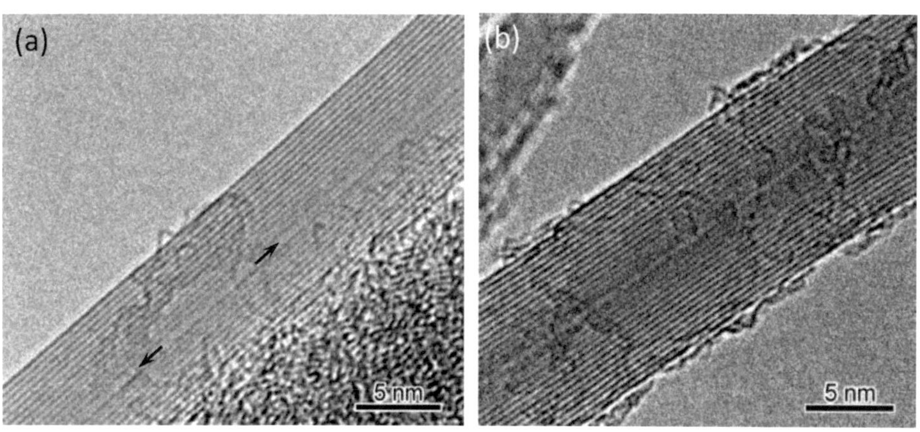

Figure 3.
TEM images of (a) MWCNT encapsulating LCCs inside the innermost tube, and (b) MWCNT with the empty innermost tube with diameter of ca. 0.47 nm. Arrows in (a) point LCCs encapsulated in the core of the MWCNT.

4. TEM observation of LCC encapsulated in MWCNT

Transmission electron microscopy (TEM) was also performed to investigate the detailed inner structure of MWCNTs encapsulating LCCs by using a JEOL JEM-2010 operated at 120 kV. The TEM samples were prepared by dropping MWCNTs dispersed in isopropyl alcohol onto a microgrid for TEM. The diameter of individual MWCNTs is in a range from 7 to 20 nm, as revealed by high resolution transmission electron microscopy, and the length exceeds 1 μm. **Figure 3(a)** shows a MWCNT encapsulating a LCC inside the innermost tube. Arrows point LCCs encapsulated in the core of the MWCNT. The length of the LCC is over 20 nm, and the diameter of the innermost tube is about 0.7 nm. **Figure 3(b)** shows a MWCNT with an empty innermost tube with diameter of ~ 0.47 nm. The innermost diameters in the present sample is smaller than those (4–10 nm) of ordinary MWCNTs [20] prepared by other methods, e.g., ordinary arc-discharge in He gas, catalyst-assisted thermal decomposition of hydrocarbon gases and so on.

5. Raman spectra of LCC/MWCNT composite at different excitation lasers

Raman spectroscopy was carried out using a Renishaw inVia confocal Raman microscope with 488, 532, 633, and 785 nm excitation lasers. A piece of the LCC/MWCNT composite tape adhered to an Al plate was placed on the microscope stage, and Raman spectra were measured at room temperature. Objective lenses employed were mainly 50× and 20×, and the spot size of laser on a sample was approximately 1 μm. Laser power on sample surfaces was in a range from 0.2 to 3.0 mW for 488, 532, and 633 nm, and in a range from 0.25 to 5.2 mW for 785 nm.

Typical Raman scattering spectra from the region “A” of the LCC/MWCNT composite obtained with four different excitation lasers, 488, 532, 633, and 785 nm are shown in **Figure 4**. In addition to D and G bands, strong peaks at higher frequencies (1750–1850 cm^{-1}) labeled LCC, attributed to stretching vibrations of triple bonds in one-dimensional carbon chains, are observed; the peak height is much stronger than the G-band for 488, 532, and 633 nm excitations. The enhanced peaks of LCC are attributed to resonance of the band gaps of LCCs with the incident photon energies.

A band of LCC observed from the tape-shaped sample (coagulated dense sample) is relatively wide in frequency (~50 cm^{-1}) and exhibits multiple peaks and/or shoulders, as found conspicuously in **Figure 4(b)** and **(c)**. Micro-Raman investigation of a finely dispersed thin sample showed that only one peak with a narrow FWHM (full width at half-maxima) of about 10 cm^{-1} was observed [21] when one or a few MWCNTs containing a LCC were shined by a focused incident laser.

In a low frequency region (< 600 cm^{-1}), radial breathing modes (RBMs) are clearly observed even for MWCNTs. RBM, whose frequency is inversely proportional to the diameter of CNT [22], is a characteristic signature for SWCNT and thin DWCNT, but even MWCNT shows this mode when the inner-most tube in it is extremely small, e.g., less than 1 nm [23]. Observable RBM frequencies depend on the excitation photon energies; CNTs which are in resonance with the incident photon energy (i.e., when optically-allowed transition gaps between van Hove singularities match with the photon energy) expose RBM peaks in Raman spectra. **Figure 5** shows a Raman spectrum extending up to 6000 cm^{-1} with the 532 nm

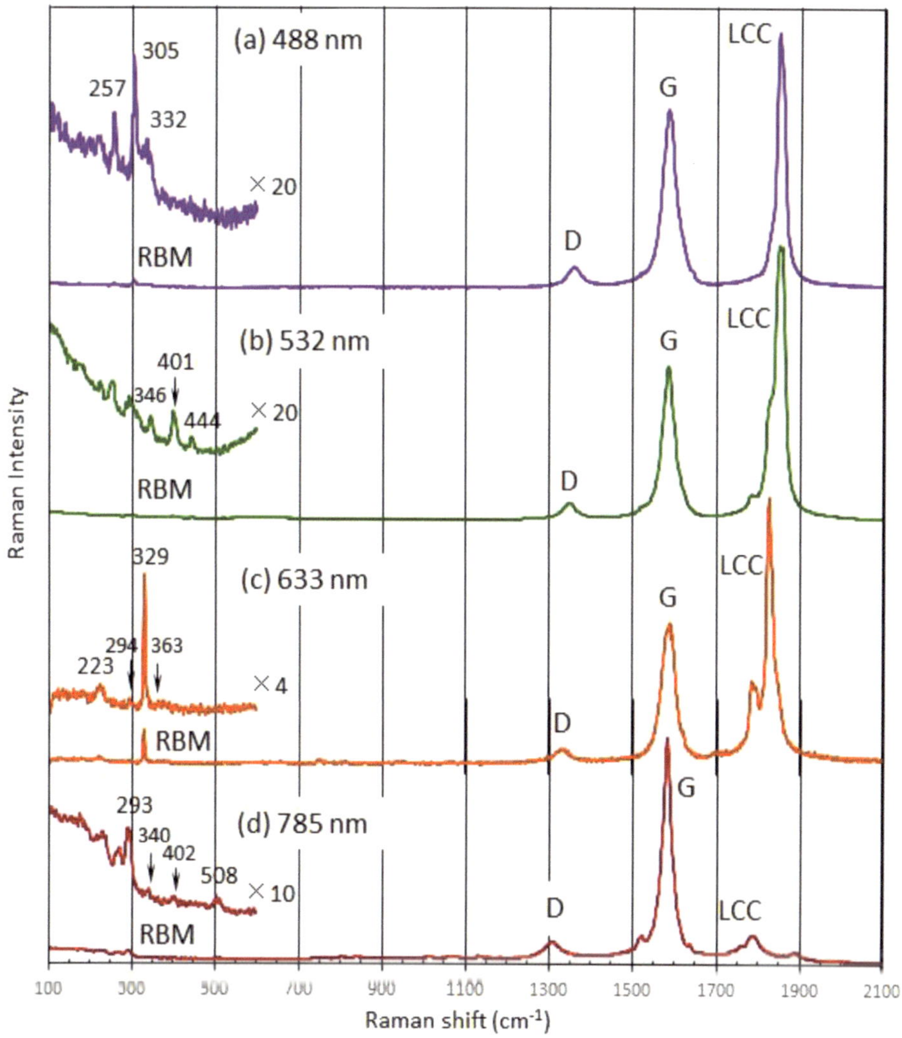

Figure 4.
Raman scattering spectra of LCC@MWCNT composite with 488, 532, 633, and 785 nm laser excitations.

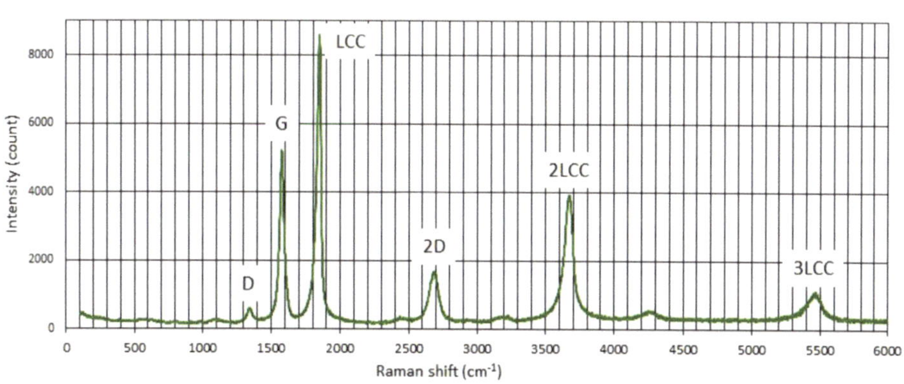

Figure 5.
Raman spectrum showing overtones of LCC with 532 nm laser excitation.

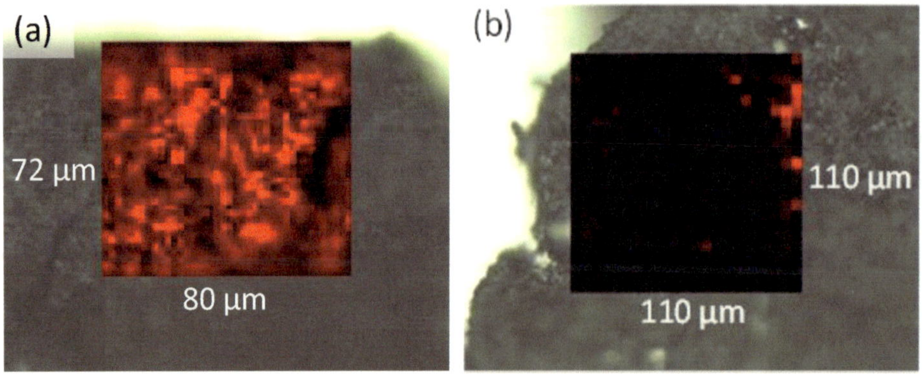

Figure 6.
Raman mappings of the LCC band (1790–1890 cm^{-1}) with 532 nm excitation in the areas (a) "A" and (b) "B". the brighter red corresponds to the stronger LCC band.

excitation, exhibiting overtones of the LCC band. Appearance of the strong 2nd and 3rd overtones, being labeled "2LCC" and "3LCC", respectively, indicates the resonance of LCCs with the incident photon energy.

Figure 6(a) and **(b)** show Raman mappings of the LCC bands with the 532 nm excitation in the areas "A" and "B", respectively. In the area "A", as shown in **Figure 6(a)**, the distribution of LCCs is relatively uniform, revealing the presence of LCCs over the area. In the area "B" (**Figure 6(b)**), on the other hand, the LCC peaks are rarely observed.

6. Raman scattering from the near-infrared 785 nm laser

As shown in **Figure 4**, the intensity of LCC band for the 785 nm excitation is much low compared to those for the shorter wavelength excitations. This may be owing to the weak resonance with this long wavelength (or low photon energy) laser. Electronic band gaps of LCCs depend on their lengths; the longer the length of LCCs is, the narrower the electronic energy gap of the LCC [12], suggesting that the population of long LCCs whose energy gaps match with the 785 nm (1.58 eV) laser in the present sample is much lower than those of shorter LCCs which resonate with the other shorter wavelength (633, 532, and 488 nm) lasers.

In Raman spectra from ordinary CNTs (i.e., no LCC inside) excited at 785 nm laser, weak peaks appear in a region from 1700 to 1900 cm^{-1}, which were assigned to an overtone of out-of-plane transversal optical mode (2oTO) and a combination mode of in-plane TO and longitudinal acoustic modes (iTO+LA) [24]. The vibrational frequencies of these modes show a dispersion (i.e., the frequencies change with the incident wavelength) and fall in a region of LCC modes at 785 nm excitation, as shown in **Figure 7**. The spectrum in **Figure 7(a)** is obtained from a LCC abundant region (A), and that in **Figure 7(b)** from a LCC-absent region (B). The peak at 1889 cm^{-1} in **Figure 7(a)** is considered to originate from the iTO+LA mode of carbon nanotubes, because the vibration frequency (1889 cm^{-1}) is too high for the LCCs. Concerning the peak at 1789 cm^{-1}, a large portion of its intensity come from the 2oTO mode of carbon nanotubes, but there may be some contribution from LCCs because the peak height (relative to that of iTO+LA) in **Figure 7(a)** is apparently higher than that in **Figure 7(b)**. Fine peaks and shoulders, which are absent in **Figure 7(b)**, are observed in **Figure 7(a)**. These additional peaks/shoulders are signals from LCCs.

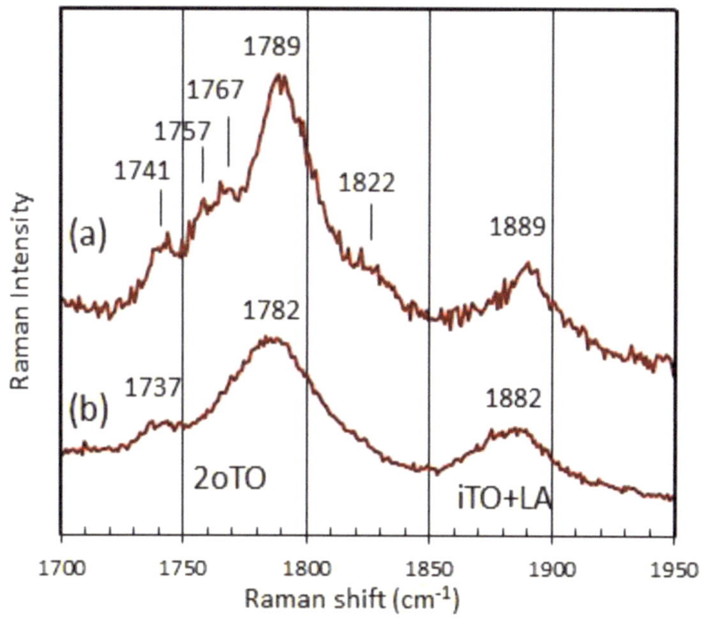

Figure 7.
Raman spectra in a LCC frequency range from (a) LCC-abundant and (b) LCC-absent regions with the 785 nm excitation.

7. Observable LCC vibration modes changing with the incident photon energies

Figure 8 shows a diagram displaying the observed Raman frequencies and intensities of LCC bands as a function of the incident photon energy. A tendency that the LCC frequency becomes lower with the decrease of the excitation photon energy is found; the observed LCC frequencies distribute from 1740 to 1820 cm^{-1}

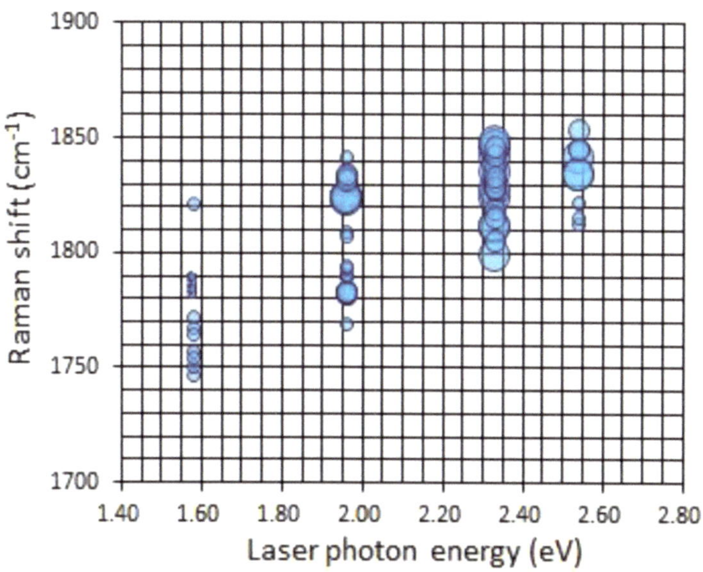

Figure 8.
LCC peak positions and intensity versus excitation energy. Size of circles represents roughly the intensity of LCC peaks; larger circles show higher intensity.

for the lowest photon energy 1.58 eV (785 nm), from 1770 to 1840 cm^{-1} for 1.96 eV (633 nm), from 1800 to 1850 cm^{-1} for 2.33 eV (532 nm), and concentrate to a band from 1810 and 1855 cm^{-1} for the highest photon energy 2.54 eV (488 nm). The change of LCC frequencies with the excitation photon energy is caused by the resonance of the energy gaps of LCCs with the incident photon. The vibrational frequency as well as the energy gap of LCC depends on its length [12, 25]; the longer the length is, the lower the vibrational frequency and the narrower the energy gap are. Lower-energy photons resonate with the longer LCCs and manifest lower-frequency vibrational modes. The incident photon dependence suggests that the energy gaps of LCCs in the present sample extend from 1.58 eV to at least 2.54 eV. The local environment inside a CNT is expected to modify the vibrational frequency as well as the energy gap due to the interaction between the LCC and CNT walls. At present, the effects of CNT hosts on the energy gap and the vibrational frequency of LCCs as a function of the LCC length are not thoroughly known, though the shifts in the band gap and the chain mode frequency depending on the diameter and the chirality of CNT hosts are reported [26].

8. Raman spectra of LCC/MWCNT composite with 785 nm excitation at different laser powers

Figure 9 shows Raman spectra obtained with the 785 nm excitation at different incidence powers from 0.25 to 5.2 mW. Broadening of the peaks and thus burring/disappearance of fine structures in the G and LCC bands are obviously discerned with the increase of incidence laser power. At low powers, satellite peaks (1520 and

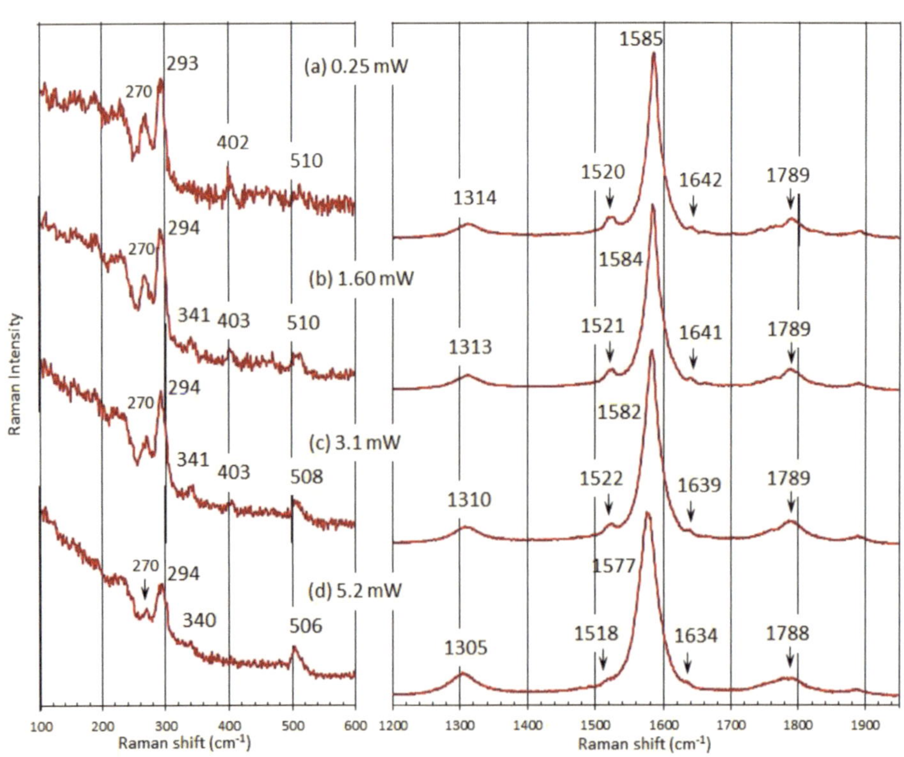

Figure 9.
Raman spectra in a low frequency RBM region and in a high frequency region including D, G and LCC bands (1200–1950 cm^{-1}) with 785 nm excitation at different incident laser powers.

1642 cm^{-1}) on both sides of the G-peak are clearly observed. These satellites are not identified yet. To the best of the authors' knowledge, no reports on the assignment of these satellites are found. Possibility of any combination modes of the G and the shear (a telescope and a rotary mode for MWCNT, 49 and 58 cm^{-1}, respectively [27]) should be accessed in the future study.

The peaks of the D and G bands shift to lower frequencies with the increase of laser power, as shown in **Figure 10(a)** and **(b)**, respectively. The shifts of these modes are caused by the temperature rise owing to the incidence laser. According to the previous report on the temperature dependence of the G-band frequencies of CNTs [28], the temperatures of the LCC/MWCNT sample shined by the 785 nm laser are estimated to be room temperature, 320 K, 400 K, and 610 K for **Figure 9(a)–(d)**, respectively. The temperature dependence of LCC and RBM frequencies is weak in this temperature range, as shown in **Figure 10(c)** and **(d)**.

In the low frequency region of **Figure 9**, radial breathing modes (RBMs) are observed even for MWCNTs. RBMs are commonly observed for SWCNTs (with diameter d less than ~ 1.5 nm) [29] and DWCNTs with a small inner CNT ($d <$ ~ 1 nm) [30], but the ordinary MWCNTs whose innermost tube is large, typically 4–10 nm, do not exhibit RBMs. The appearance of RBM in the present MWCNTs indicates the presence of small tubes ($d < 1$ nm) in the core of MWCNTs [23]. RBMs at 402 and 510 cm^{-1} observed in **Figure 9** suggest the presence of extremely small core-tubes with diameters 0.63 and 0.50 nm, respectively, according to the relation between the RBM frequency ν_R and the diameter, $\nu_R = 254/d$ [31]. The smallest tube

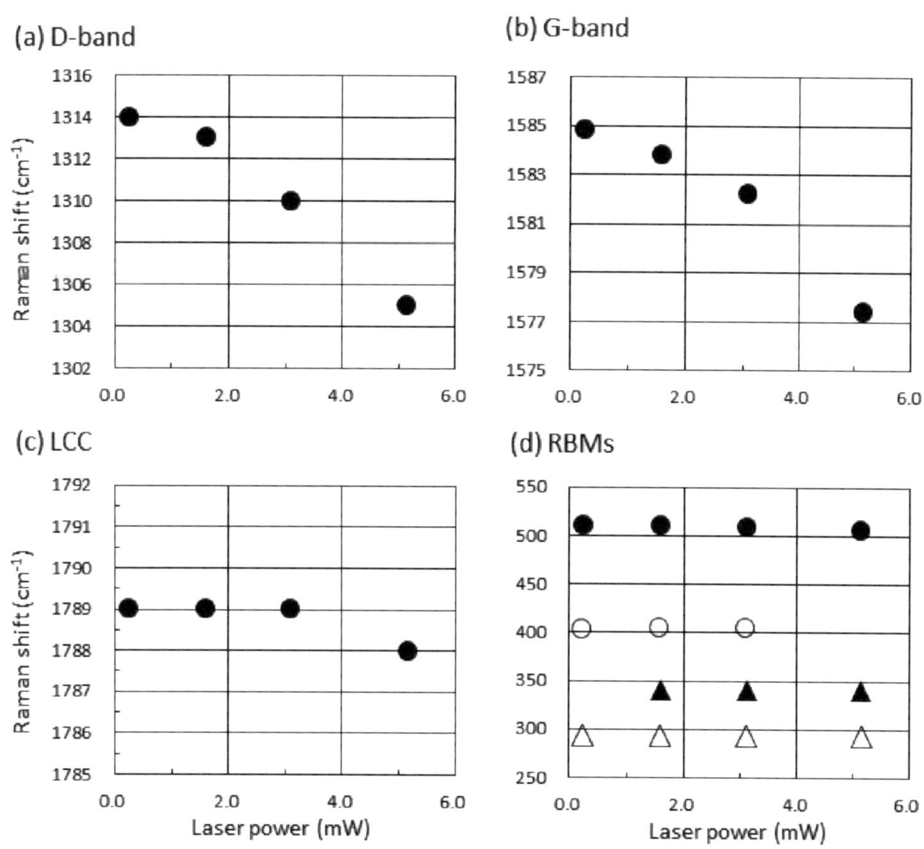

Figure 10.
Peak frequencies of (a) D, (b) G, (c) LCC, and (d) RBMs at 294, 341, 403, and 510 cm^{-1} as a function of the irradiating laser power.

giving the highest RBM of 510 cm^{-1} is tentatively assigned to the tube with a chiral index (4,3). The diameters estimated from Raman are consistent with the TEM observation, as shown in **Figure 3**.

9. Conclusions

Multiwall carbon nanotubes (MWCNTs) consisting of several to ten-something walls, which were produced by carbon arc-discharge in the atmospheric pressure with blowing inactive gases, contain long linear carbon chains (LCCs) whose length exceeds 20 nm. Signatures of LCCs in Raman scattering appear at a high frequency region from 1740 to 1855 cm^{-1}, depending on the wavelength of incident lasers. The longer the wavelengths of excitation lasers are, the lower the observed LCC frequencies are. Visible light lasers (633, 532, and 488 nm) show strong LCC-bands due to resonance Raman scattering. Even the 785 nm laser exhibits LCC signatures in a frequency range from 1740 to 1820 cm^{-1}, though their intensities are low. The incident photon-energy dependence signifies the energy gaps of LCCs extend from 1.58 eV to at least 2.54 eV. In a low frequency region, RBMs were also observed, indicating the presence of extremely small tubes in the core of MWCNTs. RBMs at 402 and 510 cm^{-1} suggest the diameters of innermost tubes are 0.63 and 0.50 nm, respectively. The diameters estimated from Raman are consistent with the TEM observation.

Acknowledgements

The LCC/MWCNT composite samples were provided by JFE Engineering Corporation.

Author details

Yahachi Saito[1*] and Koji Asaka[2]

1 Toyota Physical and Chemical Research Institute, Nagakute, Japan

2 Department of Applied Physics, Nagoya University, Nagoya, Japan

*Address all correspondence to: ysaito@toyotariken.jp

References

[1] A. El Goresy and G. Donnay, Science 161 (1968) 363-364.

[2] A. G. Whittaker and P. L. Kintner, Science 165 (1969) 589-591.

[3] A. G. Whittaker, Science 200 (1978) 763-764.

[4] P. P. K. Smith and P. R. Buseck, Science 216 (1982) 984-986.

[5] R. B. Heimann, J. Kleiman, and N. M. Salansky, Nature 306 (1983) 164-167.

[6] T. Wakabayashi, H. Tabata, T. Doi, H. Nagayama, K. Okuda, R. Umeda, I. Hisaki, M. Sonoda, Y. Tobe, T. Minematsu, K. Hashimoto, and S. Hayashi, Chem. Phys. Lett. 433 (2007) 296-300.

[7] S. Eisler, A. D. Slepkov, E. Elliott, T. Luu, R. McDonald, F. A. Hegmann, and R. R. Tykwinski, J. Am. Chem. Soc. 127 (2005) 2666-2676.

[8] W. A. Chalifoux and R. R. Tykwinski, Nature Chem. 2 (2010) 967-971.

[9] X. Zhao, Y. Ando, Y. Liu, M. Jinno, and T. Suzuki, Phys. Rev. Lett. 90 (2003) 187401 (4p).

[10] L. Shi, P. Rohringer, K. Suenaga, Y. Niimi, J. Kotakoski, J. C. Meyer, H. Peterlik, M. Wanko, S. Cahangirov, A. Rubio, Z. J. Lapin, L. Novotny, P. Ayala, and T. Pichler, Nature Mater. 15 (2016) 634-640.

[11] S. Toma, K. Asaka, M. Irita and Y. Saito, Surf. Interface Anal. 51 (2019) 131-135

[12] S. Yang and M. Kertesz, J. Phys. Chem. A 110 (2006) 9771-9774.

[13] M. Tommasini, D. Fazzi, A. Milani, M. Del Zoppo, C. Castiglioni, and G. Zerbi, J. Phys. Chem. A 111 (2007) 11645-11651.

[14] N. R. Agarwal, A. Lucotti, D. Fazzi, M. Tommasini, C. Castiglioni, W.A. Chalifoux, and R. R. Tykwinski, J. Raman Spectrosc. 44 (2013) 1398-1410.

[15] M. Jinno, Y. Ando, S. Bandow, J. Fan, M. Yudasaka, and S. Iijima, Chem. Phys. Lett. 418 (2006) 109-114.

[16] L. Shi, P. Rohringer, M. Wanko, A. Rubio, S. Waßerroth, S. Reich, S. Cambré, W. Wenseleers, P. Ayala, and T. Pichler, Phys. Rev. Mater. 1 (2017) 075601 (7p).

[17] Y. Zhang, J. Zhao, Y. Fang, Y. Liu, and X. Zhao, Nanoscale 10 (2018) 17824-17833.

[18] C.-S. Kang, K. Fujisawa, Y.-I. Ko, H. Muramatsu, T. Hayashi, M. Endo, H. J. Kim, D. Lim, J. H. Kim, Y. C. Jung, M. Terrones, M. S. Dresselhaus, and Y. A. Kim, Carbon 107 (2016) 217-224.

[19] Y. Nishi, Japan Patent No. JP3861857B2 (Dec. 27, 2006). Y. Nishi, H. Mukai, and D. Ozamoto, United States Patent No. US7625545B2 (Dec. 1, 2009).

[20] Y. Saito, R. Mizushima, S. Kondo and M. Maida, Jpn. J. Appl. Phys. 39 (2000) 4168-4173.

[21] Unpublished data

[22] J. Kürti, G. Kresse, and H. Kuzmany, Phys. Rev. B 58 (1998) R8869-R8872.

[23] X. Zhao, Y. Ando, L.-C. Qin, H. Kataura, Y. Maniwa, and R. Saito, Chem. Phys. Lett. 361 (2002) 169-174.

[24] V. W. Brar, G. G. Samsonidze, M. S. Dresselhaus, G. Dresselhaus, R. Saito, A. K. Swan, M. S. Ünlü, B. B. Goldberg, A. G. S. Filho, and A. Jorio, Phys. Rev. B 66 (2002) 155418 (10p).

[25] S. Yang, M. Kertesz, V. Zólyomi, and J. Kürti, J. Phys. Chem. A 111 (2007) 2434-2441.

[26] S. Heeg, L. Shi, L. V. Poulikakos, T Pichler, and L. Novotny, Nano Lett. 18 (2018) 5426-5431.

[27] P. C. Eklund, J. M. Holden, and R. A. Jishi, Carbon 33 (1995), 959-972.

[28] F. Huang, K. T. Yue, P. Tan, S.-L. Zhang, Z. Shi, X. Zhou, and Z. Gu, J. Appl. Phys. 84 (1998), 4022-4024.

[29] S. Bandow, S. Asaka, and Y. Saito, Phys. Rev. Lett. 80 (1998) 3779-3782.

[30] S. Bandow, G. Chen, G. U. Sumanasekera, R. Gupta, M. Yudasaka, S. Iijima, and P. C. Eklund, Phys. Rev. B 66 (2002) 075416 (8p).

[31] G. G. Samsonidze, R. Saito, N. Kobayashi, A. Grüneis, J. Jiang, A. Jorio, S. G. Chou, G. Dresselhaus, M. S. Dresselhaus, Appl. Phys. Lett. 85 (2004) 5703-5705.